Advances in Sustainability Science and Technology

The book series aims at bringing together valuable and novel scientific contributions that address the critical issues of renewable energy, sustainable building, sustainable manufacturing, and other sustainability science and technology topics that have an impact in this diverse and fast-changing research community in academia and industry.

The areas to be covered are

- Climate change and mitigation, atmospheric carbon reduction, global warming
- Sustainability science, sustainability technologies
- Sustainable building technologies
- Intelligent buildings
- Sustainable energy generation
- Combined heat and power and district heating systems
- Control and optimization of renewable energy systems
- Smart grids and micro grids, local energy markets
- Smart cities, smart buildings, smart districts, smart countryside
- Energy and environmental assessment in buildings and cities
- Sustainable design, innovation and services
- Sustainable manufacturing processes and technology
- Sustainable manufacturing systems and enterprises
- Decision support for sustainability
- Micro/nanomachining, microelectromechanical machines (MEMS)
- Sustainable transport, smart vehicles and smart roads
- Information technology and artificial intelligence applied to sustainability
- Big data and data analytics applied to sustainability
- Sustainable food production, sustainable horticulture and agriculture
- Sustainability of air, water and other natural resources
- Sustainability policy, shaping the future, the triple bottom line, the circular economy

High quality content is an essential feature for all book proposals accepted for the series. It is expected that editors of all accepted volumes will ensure that contributions are subjected to an appropriate level of reviewing process and adhere to KES quality principles.

The series will include monographs, edited volumes, and selected proceedings.

Christophe Gaie · Markus Mueck
Editors

Digital Sustainability: Legislations, Standards, Solutions and Related Case Studies

Springer

Editors
Christophe Gaie
CEMS-T
Paris, France

Markus Mueck
Intel Labs
Unterhaching, Germany

ISSN 2662-6829 ISSN 2662-6837 (electronic)
Advances in Sustainability Science and Technology
ISBN 978-981-92-0457-1 ISBN 978-981-92-0458-8 (eBook)
https://doi.org/10.1007/978-981-92-0458-8

This Springer imprint is published by the registered company Springer Nature Singapore Pte Ltd.
The registered company address is: 152 Beach Road, #21-01/04 Gateway East, Singapore 189721, Singapore

Y. Gaie

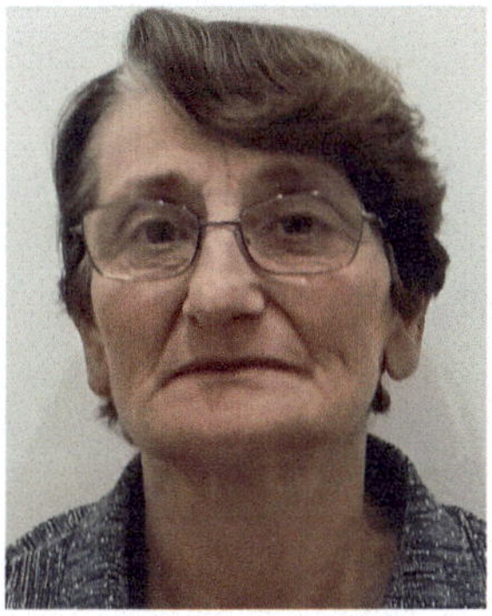

Mrs. Yvette Gaie is a former French civil servant who has developed painting and drawing skills. She is especially interested in illustrating nature in all its aspects, including animals, landscapes, and flowers.
The illustration depicts a bee, the emblem of ENSEIRB-MATMECA, the editor's engineering school, collecting free software to distribute worldwide. Free software is as essential to data processing as pollination is to humanity, lying at the very heart of software independence and development.

Foreword by Jan Ellsberger

In an era defined by rapid technological advancement and unprecedented global challenges, the pursuit of sustainability has become both a moral imperative and a strategic necessity. Digital transformation is reshaping our societies, economies, and institutions, offering immense opportunities, but also raising complex questions about our collective future.

It is within this context that *Digital Sustainability: Legislations, Standards, Solutions and Related Case Studies* makes a timely and invaluable contribution. The editors, Dr. Christophe Gaie and Dr. Markus Mueck, together with an impressive lineup of international experts, have compiled a comprehensive resource that bridges policy, technology, ethics, and practice.

This volume stands out for its holistic perspective. It not only examines the European Green Deal and the evolving regulatory landscape but also importantly explores the critical role of harmonized standards, the interplay between cybersecurity and sustainability, and the ethical dimensions of digital innovation. The inclusion of case studies, ranging from 6G mobile systems to ambient-powered IoT, demonstrates the real-world impact and complexity of sustainable digital transformation.

What is particularly impressive is the book's recognition that sustainability is multi-dimensional. Environmental leadership, economic resilience, social inclusion, and good governance are not competing priorities, but interdependent pillars of a sustainable digital future. The editors and authors do not shy away from the tensions and trade-offs inherent in this journey; instead, they offer frameworks and insights to help decision-makers navigate them with transparency and accountability.

On a personal note, I have the privilege of working closely with Dr. Markus Mueck in his capacity as Chairman of the ETSI Board. Our collaboration has given me direct insight into his dedication to advancing sustainable digital innovation and his commitment to fostering open, constructive dialogue across the standards community. This spirit of collaboration and excellence is evident throughout the pages of this book.

As Director-General of ETSI, I am acutely aware that the path to digital sustainability requires partnership across sectors, disciplines, and borders. This book exemplifies such a partnership, drawing on the expertise of researchers, industry leaders, policymakers, and practitioners from across Europe and beyond. It is a testament to the power of shared knowledge and collective action.

I am confident that this work will serve as an essential reference for all those engaged in shaping the digital future, whether in government, industry, academia, or civil society. It invites us not only to reflect, but to act: to design, implement, and govern digital technologies in ways that are responsible, inclusive, and sustainable.

I congratulate the editors and contributors for their vision and dedication. May this book inspire new ideas, foster meaningful partnerships, and accelerate our collective progress toward a more sustainable digital world!

Jan Ellsberger
Director-General
European Telecommunications
Standards Institute (ETSI)
Sophia Antipolis, France

Foreword by Bernard Barani

Sustainability has emerged as a key global policy issue. The 17 Sustainable Development Goals (SDGs) identified by the United Nations in 2015 provide a set of indicators to help policymakers frame their sustainability initiatives. They are part of the Agenda 2030 for Sustainable Development: under the motto "leave no one behind", the aim is to meet the current needs without compromising the ability of future generations to meet their own needs. In Europe, the Green Deal initiative initiated by the European Commission represents a strong policy drive to support the European economy and society in becoming more sustainable, whilst contributing at the same time to competitiveness and sovereignty.

Digital technologies have a key role to play in this context, and many initiatives have been started worldwide to address the two complementary facets of the issue: Sustainable ICT and ICT for sustainability. This is, however, a domain in its infancy compared to the decades of research and development underpinning competitive, affordable ICT solutions with ever-rising performances. It is indeed fairly extraordinary to have today's smartphone with a computing power about 1000 times faster than the CRAY-2, one of the most powerful machines of 1985. The lack of similar magnitudes of enhancements for sustainability issues has, in turn, prompted multiple research and development activities to advance this complex and multifaceted topic.

Against this short background, the book on *Digital Sustainability: Legislation, Standards, Solutions and Related Case Studies* represents a much-needed contribution to a debate that will undoubtedly expand further over the years to come. Whilst the book takes a broad perspective on Digital Sustainability, it rightly takes an in-depth view of communication networks, which probably represent the most comprehensive framework for addressing digital sustainability, as modern telecommunication networks play a critical background role in all aspects of everyday life. During the last decade, scientists and organisations worldwide have invested massively in designing and deploying 5G networks. In addition to enhanced mobile broadband, the target was to provide a suite of new services that would lay the foundations for the digital transformation of various vertical markets (e.g., smart industry, energy, automotive, transportation and logistics, health, media, gaming, etc.). Today, in the wake of these efforts, global initiatives are turning to the definition of 6G networks, whose commercial take-up is expected around 2030. 6G is particularly relevant to the theme of this book, as it is the very first attempt by the global community to take sustainability aspects into account from the onset at the research level, and to include as much as possible sustainability as part of the design drivers and principles. It reflects the rising involvement of the telecom and service providers ecosystem in sustainability, in the wake of important decisions impacting operational technologies and infrastructures, like the GSMA commitment to reach net-zero emissions by the middle of this century.

This reflection has been particularly developed under the European "Smart Networks and Services" Joint Undertaking initiative launched in 2021 by the European Commission under the Horizon Europe framework programme and in partnership with the 6G Smart Networks and Services Industry Association (6G-IA), gathering more than 400 key stakeholders of the European 6G ecosystem. Under this initiative, sustainability and the related SDGs have been classified under three categories of objectives and design drivers for networks:

- Economic sustainability covers issues such as profitability, investment-friendly technologies and infrastructure, resilience, productivity, and efficiency. This is a key aspect for stimulating the monetisation of the massive investments and for regaining control over key services for European businesses and society. It indirectly contributes to the public good by enabling the wide availability of connectivity and affordable public-interest services, such as safety and critical communication services.
- Environmental sustainability is directly related to the environmental impact throughout the full life cycle of network, terminal, and service implementation, covering construction, operations, and decommissioning. A particular aspect widely researched relates to energy consumption, which has a direct impact on network architectures and technologies.
- Societal sustainability is directly related to the societal value of networks and connectivity services. This aspect is critical, as it also relates to the overall acceptability of new technologies by the general public, an issue whose importance was probably underestimated during the early phases of 5G deployment. Digital inclusion, accessibility, privacy, safety, and well-being are carefully addressed under this topic.

Multiple technologies under investigation in the context of 6G networks directly impact these three objectives. For instance, innovative services like Integrated Sensing and Communications (ISAC) are providing new revenue opportunities from two perspectives, either by using sensing of the environment to optimise the use of network resources or to provide sensing data to third-party applications. Energy savings from third-party applications represent a significant potential contribution to environmental sustainability, and in that context, network opening with application-specific APIs, as developed under the important CAMARA initiative of ETSI, is key, beyond the business opportunities they offer. In the field of societal sustainability, it may be expected that the advent of TN/NTN integration will eventually solve the issue of accessibility for the 3.5 billion people in the world who do not have internet access, and under affordable conditions, thanks to the leveraging of the 3GPP ecosystem by the satcom community. Similarly, semantic communications offer interesting opportunities to convey specific aspects of the information to be transmitted and could be a powerful solution for future critical or safety communication services, such as those of PPDR systems.

Many of these issues are well reflected in the multiple SNS projects that have contributed to this book, including the SUSTAIN 6G sustainability "flagship" project launched by SNS. It is worth noting that European projects have pioneered the definition and design role of "Key Value Indicators" (KVI), a set of qualitative or quantitative (where possible) set of indicators that are designed to guide the development of network technology and systems under sustainability constraints. KVI's have been the subject of several publications and contributions to standardisation bodies such as the ITU and 3GPP. Still, it is clear that at this stage, the concept requires further development to reach the same maturity at the standardisation level as the classical Key Performance Indicators (KPI's) that ITU-R has traditionally used for verifying the compliance of proposed RITs and SRITs to IMT performance specifications. This is no doubt a domain that needs additional effort.

The multiple aspects of sustainability impose multiple design constraints on network technology and systems, and a common research thread across all emerging issues is the systematic quest for resource optimisation at every layer. Using the right resource at the right time without overprovisioning, tailoring resource usage as a function of traffic characteristic, type and quality-of-service requirements, spectrum usage, network environment, and type of access are typical parameters that have a clear impact on how future networks will support sustainability. These aspects call for flexible, versatile, and programmable networks, making best use of the myriad of data that may be captured by operational networks. In that context, Artificial Intelligence is called upon to play a key role in supporting both sustainable networks and networks for the sustainability of third-party applications. This has a significant impact on the architectural design of 6G. For instance, a data plane may be considered to support multiple issues related to the holistic, end-to-end operations of a network platform.

Beyond technological aspects, it is clear that regulations and standards are key to enabling the advent of sustainable solutions in the network domain. As networks are, by essence, global platforms, a comprehensive, end-to-end approach can only be achieved if corresponding standards are developed and applied globally. Key standardisation institutes like ETSI have already taken steps in this direction with the launch of the Technical Committee Environmental Engineering (TC EE). The TC EE addresses standards for reducing the eco-environmental impact of ICT equipment and is accountable for defining the environmental and infrastructural aspects of all telecommunication equipment and its surroundings. This is a domain where future work is needed, as networks are increasingly implemented over heterogeneous domains (hyperscalers, verticals, etc.), and a comprehensive approach to sustainability requires strong collaboration across a multiplicity of actors. Also, regulation has a critical role in reflecting the policy objectives of sustainable network implementation and operations, and in fostering the emergence of the corresponding technologies and standards.

In conclusion, digital sustainability is becoming a key driver of ICT policies, technology development, and standards development. Today, it is still an emerging domain where a large number of building blocks have still to be constructed, be it at the technological, standardisation, or regulatory level. This book contains a wealth of information on the state of the art and represents a major contribution to the way forward. I trust it will inspire the readers, and notably the younger generations, to further invest their talent, energy, and vision into this exciting and rewarding domain, which is key to serving the interests of the citizen and of future generations.

Bernard Barani
6G Industry Association Senior
Consultant, former European
Commission Official, and SNS JU
Board Member
Brussels, Belgium

Preface

Digital technologies are evolving at a breathtaking pace and are affecting all aspects of life. For example, Artificial Intelligence and related Machine Learning solutions are underway to fundamentally change the way that our society is functioning. Likewise, internet access and related mobile connectivity are becoming a commodity, with the tremendous underlying digital infrastructure becoming almost forgotten. Especially for the young generation, life without a smartphone is almost unimaginable. At the same time, this generation faces numerous challenges—including rising poverty in developed countries, global warming, and resource scarcity.

How can we create technologies that bring new hope?

The underlying challenges are indeed tremendous, especially in terms of sustainability. The present book takes a 360-degree perspective on this topic: Digital Sustainability is considered first from a policy angle, introducing specifically the European Green Deal—a generational task to *transform the EU into a fair and prosperous society*. Next, we will understand how Standards Development Organizations (SDOs), especially European Standardisation Organisations (ESOs), ETSI, CEN, and CENELEC, support the implementation of related regulations through the development of harmonised standards providing presumption of conformity with the harmonised legislation. We put a specific focus on the implementation of *The Ecodesign for Sustainable Products Regulation and Energy Labelling Regulation.*

Can we design coherent policy frameworks and technical standards capable of driving digital innovation toward sustainability?

Following the deep dive on policy related aspects of sustainability, we then discuss *Digital Architectures to foster sustainability,* we address the question on how to jointly consider Cybersecurity and Ethics objectives in the design of sustainable systems and we delve into a number of use cases: Sustainability objectives are discussed from the perspective of a Government Agency, from the perspective of the French Army and from an industrial perspective in the fields of Artificial Intelligence, Next-Generation Cellular Communication Systems (6G) and Ambient Internet-of-Things (IoT) technology. Finally, a discussion follows on how other regions in the world are tackling the sustainability challenge. We specifically outline the local actions of the United States, Brazil, India, and the UAE.

How can additional objectives and regional experiences build comprehensive solutions to achieve digital sustainability?

The present book has been created to serve as a valuable resource for researchers, industrial experts, policy makers, interested citizens, and other stakeholders to learn more about the topic of Digital Sustainability and to obtain a holistic understanding from distinct perspectives. Together, the questions raised throughout this book invite readers to reflect on how innovation, policy, ethics, technology, and regional experiences can converge to create comprehensive solutions for achieving digital sustainability.

Paris, France Markus Mueck
Unterhaching, Germany Christophe Gaie

Acknowledgments

The editors would like to express their sincere gratitude to all authors and contributors who have generously shared their expertise and insights to enhance the quality of this book. Their collective efforts have resulted in a comprehensive and informative resource that reflects the diverse perspectives of various European countries. We are particularly grateful for their willingness to collaborate and contribute to work that aims to promote a deeper understanding of European initiatives in the field of sustainability.

The editors also thank Prof. Robert Howlett, Prof. Lakhmi C. Jain, and Prof. Jean-Fabrice Lebraty, as well as the entire Springer editorial team, for their assistance in the editing process.

Thank you for choosing our book.

We hope you will find it a valuable resource!

Markus Mueck
Christophe Gaie

Introduction

In recent years, there has been tremendous advancement in the field of Digital Technologies, including broadband and high-throughput wireless systems, Artificial Intelligence solutions, and many more. Our society and way of life are currently experiencing substantial changes triggered by technology. Established players disappear, and new entrants grow at an unprecedented pace. Governments, Institutions, Corporations as well as citizens, need to rethink deeply rooted beliefs and culture to prepare for the future.

However, these technological advancements come at a price—especially including challenges related to sustainability. Policy makers across the world have realized that careful management of available resources, limitation of emissions of greenhouse gases, an intelligent approach to power generation, and many more challenges require attention at the highest level. This has led to a number of bold initiatives, especially in Europe, triggered by the European Commission's Green Deal—recognizing that it is a generational task to *transform the EU into a fair and prosperous society, with a modern, resource-efficient and competitive economy where there are no net emissions of greenhouse gases in 2050 and where economic growth is decoupled from resource use.*

While Europe has led the way with the Green Deal, other regions and organizations around the world are also promoting digital sustainability. The United Nations' Sustainable Development Goals (SDGs) provide a global framework for integrating technology with environmental, social, and economic objectives. In the United States, the Inflation Reduction Act incentivizes clean energy and sustainable innovation. Similarly, India has launched initiatives such as Digital India to foster inclusive development and improve administrative efficiency. These examples illustrate that digital sustainability is a global concern requiring context-specific approaches worldwide.

The present book provides a broad overview of initiatives taken by all stakeholders, including policymakers, industry, citizens, institutions, and many more, to address sustainability challenges in our digital age. A total of thirteen chapters address various perspectives, as is briefly outlined below.

Chapter 1 provides an *Introduction to Digital Sustainability*. Specifically the historic evolution of related discussions is analyzed, starting with early political discussions dating back to the 1960s; for example, in the influential report The Limits to Growth, published in 1972, the Club of Rome (a non-profit organization of intellectuals and business leaders concerned with global issues) highlighted the risks posed by the potential mismatch between exponential economic and population growth, and the finite supply of resources. Furthermore, recent research is considered which is addressing the question of how to assess sustainability. There are obvious quantitative metrics, such as power consumption, the level of greenhouse gas emissions, etc. However, not all aspects can be captured numerically, and thus recent research has introduced the concept of Key Value Indicators (KVIs), complementing the classical assessment based on Key Performance Indicators (KPIs).

Chapter 2 outlines a holistic summary of *The European Green Deal*, which was officially introduced on 11th December 2019 by the European Commission. Being a generational effort of immense proportions, the Green Deal needs to be understood as a framework encompassing a large number of initiatives and tools; those include a set of eight deeply transformative policies, a set of five initiatives for mainstreaming sustainability in all EU policies, a number of legislations, and many more. Under this framework, the EU Climate Pact allows for a hands-on involvement in the implementation of the European Green Deal. Recently, however, new challenges have become evident in Europe, such as industrial competitiveness, cost of living, purchasing power, etc. As a consequence, the President of the European Commission has acknowledged that Europe needs a "*new Clean Industrial Deal for competitive industries and quality jobs*". As it will be further detailed in the chapter, the Clean Industrial Deal complements, and in some cases refocuses, the Commission's 2019 Green Deal.

Chapter 3 discusses *The Ecodesign for Sustainable Products Regulation and Energy Labelling Regulation*—both are key components of the European Green Deal. Following the publication and implementation of several *Working Plans* related to *Ecodesign and Energy Labelling*, the *Ecodesign for Sustainable Products Regulation (ESPR)* was finally adopted and subsequently published in the EU's Official Journal in June 2024. We will explain how the ESPR became the materialization of the European Commission Circular Economy Action Plan, stating that "*in order to make products fit for a climate-neutral, resource-efficient and circular economy, reduce waste and ensure that the performance of front-runners in sustainability progressively becomes the norm, the Commission will propose a sustainable product policy legislative initiative*".

Chapter 4 explains the European approach relying on *Standardization Supporting Sustainability Objectives*. Indeed, both the *Ecodesign for Sustainable Products Regulation* and *Energy Labelling Regulation* are aligned with the New Legislative Framework (NLF). This alignment provides a unique opportunity for all stakeholders to contribute to the implementation process of the Regulation. While the legislation itself sets high-level principles, it also establishes the fact that the European Standardisation Organisations (ESOs) are tasked with writing the standards that implement the essential requirements of harmonised legislation. Once the harmonised standards are referenced, they are published in the Official Journal of the European Union and listed in a consolidated database. By following the standard, a presumption of conformity with the harmonised legislation exists. A Harmonised Standard is thus a key tool for manufacturers and other stakeholders to ensure compliance of related products with the essential requirements of the legislation; only if such compliance is demonstrated, access to the European Union Single Market is granted.

Chapter 5 provides an *Overview of Digital Architectures to Foster Sustainability*. We will dive into available architectural strategies and approaches in the context of digital systems to support sustainability objectives. These are indeed key requirements on multiple levels—government legislations impose more and more stringent sustainability requirements, manufacturers desire to position themselves as key innovator in the field of sustainability, and customers expect that manufacturers design products in accordance with green design principles. Example strategies include the shutdown of temporarily unused parts of a component or system, adapting the performance to the actual needs while minimizing expenses in terms of energy consumption, applying semantic and goal-oriented communication principles to minimize the over-the-air capacity requirements, etc.

Chapter 6 addresses the challenge of *Defining Synergies Between Cybersecurity and Sustainability*. In societal terms, cyberspace is not an independent entity but forms a part of the environment where humans function. At the same time, ICTs often provide new tools to achieve political and economic goals. One of the means used to achieve these goals is disinformation, through which the behaviour of individuals, societies, and, indirectly, the policies of states, can be influenced. Therefore, it is paramount that when using ICT for sustainability planning, no aspect of cybersecurity is overlooked. Both technical and human factors should be taken into account. It is also important to keep in mind that a significant proportion of cyber threats do not yet exist but will certainly emerge in the future. Therefore, cybersecurity is also about anticipating possible, even unlikely, scenarios.

Chapter 7 offers a deep dive into *Ethics and Sustainability*, both being closely intertwined. The underlying principle is that people should live in a manner that considers the needs of both present and future generations, as well as the natural world. It's a key factor of sustainable growth. This chapter examines the derived requirements in the context of digital systems, incorporating design guidelines, sourcing strategies, and considerations, as well as their convergence within the realm of digital systems, where technological development often deals with critical ethical and environmental questions. Highlighting that sustainable development extends beyond ecological impact, the chapter emphasizes and outlines key requirements for the moral development of digital systems. Topics addressed include ethical sourcing of materials, lifecycle impacts of digital technologies, human-centered design, regulatory frameworks, and governance initiatives.

Chapter 8 considers the topic of sustainability from a governmental perspective: *From Concept–Knowledge (C-K) and Theory of Inventive Problem Solving (TRIZ) to Net Technological Assessment (NTA): Structured Decision-Making for the European Green Deal*. We will discuss how governments can align digital innovation with the European Green Deal. Technical strategies—such as shutting down unused components, adapting performance to demand, or reducing communication overhead—illustrate sustainability at the system level, but they do not solve the challenge of prioritizing technologies under policy, budget, and ESG constraints. The chapter, therefore, introduces Net Technological Assessment (NTA) as its core contribution: a novel framework that integrates maturity, investment potential, and sustainability weighting to translate innovation options into accountable, Green Deal–aligned policy choices.

Chapter 9 takes another perspective on the topic of sustainability: *Emerging Tools in Security and Defense for Sustainable Systems: Stochastic Optimization, Quantum Decision Models, and Technological–Human Convergence.* The chapter builds on a collaborative effort, including the artificial intelligence and cognitive science lead at the French Defense Innovation Agency. It discusses how to address issues of sustainability in defense and security by linking approaches from applied mathematics, algorithmics, and cognitive science. The chapter offers a non-confidential synthesis of the activity.

Chapter 10 delves into *A Case Study: Sustainability in the Context of 6th Generation Mobile Communication Systems.* With cellular communications being ubiquitous and becoming a commodity, it is of key importance to consider sustainability, specifically the case of next-generation mobile systems (6G). Advocating a *sustainable by design* approach, we present the challenges of sustainability from the well-established perspective of three pillars: environmental, social, and economic sustainability. All three pillars have a fundamental role, spanning ecological effects, operational costs, business opportunities, and the controversial topic of social acceptance of new technologies.

Chapter 11 addresses *Artificial Intelligence: Sustainability Challenges, Opportunities, and the Case of 6G.* Being one of the major revolutions in human history, Artificial Intelligence comes with opportunities and threats. From the environmental impact of data centers to economic and social risks in labor markets, the sustainability of AI systems must progress hand in hand with their technical development. In this chapter, we provide an overview of the main challenges of sustainability in the context of AI. Considering all sustainability pillars related to environmental, social, and economic challenges, we introduce the main aspects of *Responsible AI*. Then, we focus on the fundamental question of where to host AI-related processing.

Chapter 12 discusses *A Case Study: Sustainability in the Context of Ambient-Powered IoT*. While previous chapters explore sustainability objectives at the level of large-scale Information and Communications Technology (ICT) systems, this chapter takes the perspective of device-level design and deployment. We focus on the concept of ambient-powered Internet of Things (IoT), where Energy Neutral Devices (ENDs) harvest energy from their environment, such as light, radio frequency, or thermal gradients, operate autonomously without wired power or frequent battery replacement, and communicate wirelessly with the infrastructure.

Chapter 13 takes a global perspective on sustainability—*Digital Pathways to Sustainability: Insights Beyond Europe.* We are considering regions in the world also engaging with similar ambitions compared to the European Union. In this chapter, we outline the local actions of the United States, Brazil, India, and the UAE to understand how each has developed distinct approaches. Each context reflects a different combination of governance structures, institutional capacity, and socio-political priorities. We discuss how Brazil uses digital tools to protect ecosystems and reduce inequalities, how India focuses on large-scale administrative efficiency and inclusion, how the UAE aligns digitalization with its net-zero 2050 agenda, and how the United States combines federal initiatives, state regulations, and industry-driven innovation in a complex and often politically divided landscape.

The present book leverages the expertise from multiple European Commission-funded research projects. It is only thanks to a close collaboration and an open dialogue across project boundaries that we can share the comprehensive material available in the subsequent chapters. The following research projects have partially supported the writing of the present book:

- *The Smart Networks and Services Joint Undertaking (SNS JU) project 6GARROW under the European Union's Horizon program, Grant Agreement Number 10119219, and under the Institute for Information and Communications Technology Promotion (IITP) grant funded by the Korean government (MSIT) (No. RS-2024-00435652);*
- *The Smart Networks and Services Joint Undertaking (SNS JU) project 6G-GOALS under the European Union's Horizon program, Grant Agreement Number 101139232;*
- *The Smart Networks and Services Joint Undertaking (SNS JU) project 6G-DISAC under the European Union's Horizon program, Grant Agreement Number 101139130;*
- *The Smart Networks and Services Joint Undertaking (SNS JU) project SUSTAIN-6G under the European Union's Horizon program, Grant Agreement Number 101191936;*
- *The Smart Networks and Services Joint Undertaking (SNS JU) project Ambient-6G under the European Union's Horizon program, Grant Agreement Number 101192113;*
- *The Smart Networks and Services Joint Undertaking (SNS JU) project 6G-XR under the European Union's Horizon program, Grant Agreement Number 101096838.*

The proposed research is intended to go beyond technological infrastructure and policy frameworks, recognizing that digital innovations are reshaping how people live, work, and interact. The authors aim to examine how technologies influence consumption habits, social behaviors, and access to opportunities, creating both possibilities and ethical dilemmas. This perspective helps to understand societal impacts and highlights that achieving digital sustainability requires not only energy efficiency and regulatory compliance, but also that each citizen pays attention to their own ethics and behavior.

Moreover, the editors underline that sustainability encompasses more than environmental considerations. It requires a holistic approach that involves economic, social, and governance dimensions. Economically, technologies must balance innovation with cost-effectiveness and equitable growth. Socially, digital systems should foster inclusion, accessibility, and fairness, avoiding the deepening of inequalities. Recognizing this multi-dimensional nature is essential to designing solutions that can be implemented in complex real-world contexts.

To this end, we are intimately convinced that defining trade-offs and managing tensions is a key concern. Technologies often present conflicting impacts: for example, artificial intelligence can optimize energy use but also require large-scale data centers, while next-generation mobile networks enhance connectivity but increase resource consumption. Understanding these trade-offs and providing transparent choices for each individual is essential for personal decision-making, as well as for developing government and institutional strategies that balance environmental, economic, and social objectives for the common good. By explicitly acknowledging these tensions, this book encourages a nuanced, multi-stakeholder approach to building sustainable digital futures.

Disclaimer

Views and opinions expressed are, however, those of the authors only and do not necessarily reflect those of all consortium parties of any of the above projects, nor those of their employers and of the European Union or the SNS JU (granting authority). Neither the European Union nor the granting authority can be held responsible for them.

Within the present book, extensive reference is given to applicable European Legislations and explanatory information available on European Commission websites. Key information is extracted and included in the relevant chapters for better readability. Please also note that this document is intended to share the authors' understanding of the current regulatory landscape of the European Union's digital legislation. It is not intended to constitute legal advice or lawful standing.

Markus Mueck
Christophe Gaie

Contents

1 **Introduction to Digital Sustainability** 1
Christophe Gaie

2 **The European Green Deal** 19
Markus Mueck, Stéphane Du Boispéan, and Kristof Claesen

3 **The Ecodesign for Sustainable Products Regulation and Energy Labelling Regulation** 41
Markus Mueck, Kristof Claesen, and Stéphane Du Boispéan

4 **Standardisation Supporting Sustainability Objectives** 71
Markus Mueck, Beniamino Gorini, Kristof Claesen, and Stéphane Du Boispéan

5 **Overview of Digital Architectures to Foster Sustainability** 121
Christophe Gaie and Laurent Denis

6 **Defining Synergies Between Cybersecurity and Sustainability** 139
Mirosław Karpiuk and Krzysztof Kaczmarek

7 **Ethics and Sustainability** 157
Dimitris C. Gkikas, Christophe Gaie, and Markus Mueck

8 **From Concept–Knowledge (C-K) and Theory of Inventive Problem Solving (TRIZ) to Net Technological Assessment (NTA): Structured Decision-Making for the European Green Deal** 181
Christophe Gaie and Jean Langlois-Berthelot

9 **Emerging Tools in Security and Defense for Sustainable Systems: Stochastic Optimization, Quantum Decision Models, and Technological–Human Convergence** 211
Jean Langlois-Berthelot and Christophe Gaie

10 **A Case Study: Sustainability in the Context of 6th Generation Mobile Communication Systems** 237
Mattia Merluzzi, Luigi Briguglio, Serge Bories, Gilles Callebaut, Emilio Calvanese Strinati, Lucas Pereira Carwile, Emma Complido, Chaima Ghribi, Bahare Masood Khorsandi, Marja Matinmikko-Blue, Davide Montagno Bozzone, Hosein Nikopour, Carmela Occhipinti, Chathura Sarathchandra, Lars Christoph Schmelz, and Emanuela Tangari

11 **Artificial Intelligence: Sustainability Challenges, Opportunities, and the Case of 6G** 265
Mattia Merluzzi, Luigi Briguglio, Serge Bories, David Brayford, Gilles Callebaut, Emilio Calvanese Strinati, Lucas Pereira Carwile, Emma Complido, Tim Dykes, Olivier Forceville, Chaima Ghribi, Utz-Uwe Haus, Bahare M. Khorsandi, Wei Mao, Marja Matinmikko-Blue, Davide Montagno Bozzone, Hosein Nikopour, Carmela Occhipinti, Chathura Sarathchandra, Lars Christoph Schmelz, Omid Semiari, Emanuela Tangari, Lili Wei, and Shu-Ping Yeh

12 **A Case Study: Sustainability in the Context of Ambient-Powered IoT** 297
Riku Jäntti, Mateen Ashraf, Gilles Callebaut, Jona Cappelle, Lukas D'angelo, Miltiadis Filippou, Kalle Koskinen, Guus Leenders, Onel López, Ulrich Muehlmann, Daniel Poehl, Priyesh Pappinisseri Puluckul, Lyssa Ramaut, Ritesh Kumar Singh, Bikramjit Singh, Jarne Van Mulders, Jenny Wiklund, and Boxuan Xie

13 **Digital Pathways to Sustainability: Insights Beyond Europe** 339
Fabrício Ramos Neves, Polyana Batista Silva, Vasantha Shanmugam, Vimala Dharmasivam, Lina Bariah, Merouane Debbah, and John M. Roman

Editors and Contributors

About the Editors

Dr. Christophe Gaie earned his Ph.D. in Telecommunications from Université Paris Sud—Saclay in 2010, following research collaboration between Motorola and Supélec. He also holds a double degree in Engineering and a Master of Research from INP Bordeaux—Université Bordeaux, obtained in 2006. Currently, he serves as the head of the Engineering and Digital Innovation Division at the Prime Minister's Office. Previously, he led the digital applications office at the Centre interministériel de services informatiques relatifs aux ressources humaines (CISIRH) and managed the interministerial exchange platform at the Direction Interministérielle du Numérique (DINUM). With over a decade of experience at the Direction Générale des Finances Publiques (DGFiP), he initially worked as an IT architect and later headed the Income Tax and Withholding Tax team. His main research interests include e-government, modernization of information and telecommunication systems, API development and usage, artificial intelligence, and fraud detection. He is also an associate researcher at the Magellan Laboratory and an officer in the Citizen Reserve of the French Army.

Dr. Markus Mueck received the Dipl.-Ing. and ing. dipl. degrees from the University of Stuttgart, Germany, and the Ecole Nationale Supérieure des Télécommunications (ENST), Paris, France, respectively in 1999. In 2006, he received a Doctorate degree from ENST in Communications. From 1999 to 2008, Dr. Mueck was a Senior Staff member and Technical Manager at Motorola Labs, Paris, France. In this role, he contributed actively to various standardization bodies, namely Digital Radio Mondiale, IEEE 802.11n, etc., and led the creation of the novel standardization group IEEE P1900.4 in the area of Cognitive Radio and Software Defined Radio (SDR). He also contributed to numerous European Research projects, namely as Technical Manager of IST-E^{2}R II (19 MEuros budget) and as overall technical leader for the definition of IST-E^3 (20 MEuros budget). In 2008, Dr. Mueck joined Infineon Technologies, Munich, Germany, where he was head of delegation for IEEE 802.16m and transitioned to Intel Mobile Communications in 2011 in the framework of an acquisition by Intel. Currently, Dr. Mueck is Director and Principal Engineer of Intel Laboratories. He acts as Chair of the European Telecommunications Standards Institute (ETSI) Board, he is Chair of the ETSI Operational Co-ordination Group Artificial Intelligence (OCG AI), coordinating the overall effort of the organization on AI, member of ETSI's Delegation to 3GPP PCG/OP, and Adj. Professor of the University of Technology, Sydney, Australia. Dr. Mueck was leading the effort on Artificial Intelligence and Machine Learning of the European-funded research Flagship Project Hexa-X, preparing the design of 6th Generation Mobile Networks (6G).

Contributors

Mateen Ashraf Center of Wireless Communication, University of Oulu, Oulu, Finland

Lina Bariah Abu Dhabi, United Arab Emirates

Polyana Batista Silva Guanambi, Brazil

Serge Bories University Grenoble Alpes and CEA-Leti, Grenoble, France

Davide Montagno Bozzone Hewlett Packard Italiana SRL, Milan, Italy

David Brayford Hewlett Packard Enterprise EMEA Research Lab Germany, Ratingen, Germany

Luigi Briguglio CyberSocial Lab. Srl Impresa Sociale, Rome, Italy

Gilles Callebaut Department of Electrical Engineering, KU Leuven, Ghent, Belgium

Jona Cappelle Department of Electrical Engineering, KU Leuven, Ghent, Belgium

Lucas Pereira Carwile CyberSocial Lab. Srl Impresa Sociale, Rome, Italy

Kristof Claesen Intel Corporation SA/NV, Kontich, Belgium;
Intel Corporation SA/NV, Brussels, Belgium

Emma Complido University Grenoble Alpes and CEA-Leti, Grenoble, France

Merouane Debbah Abu Dhabi, United Arab Emirates

Laurent Denis French Prime Minister Office, Paris, France

Vimala Dharmasivam Chennai, India

Stéphane Du Boispéan Intel Corporation SA/NV, Kontich, Belgium

Tim Dykes Hewlett Packard Enterprise EMEA Research Lab, Wokingham, UK

Lukas D'angelo Graz University of Technology, Graz, Austria

Miltiadis Filippou WINGS ICT Solutions S.A., Athens, Greece

Olivier Forceville Hewlett Packard France SAS, Puteaux, France

Christophe Gaie French Prime Minister Office, Laboratoire de Recherche Magellan, Université Jean Moulin, Lyon, France;
French Prime Minister Office, Paris, France

Chaima Ghribi Nokia, Massy, France

Dimitris C. Gkikas University of Patras, Patras, Greece

Beniamino Gorini Nokia Networks, Massy, France

Utz-Uwe Haus Hewlett Packard Enterprise EMEA Research Lab, Meyrin, Switzerland

Riku Jäntti Department of Information and Communications Engineering, Aalto University, Espoo, Finland

Krzysztof Kaczmarek Koszalin University of Technology, Koszalin, Poland

Mirosław Karpiuk University of Warmia and Mazury in Olsztyn, Olsztyn, Poland

Bahare M. Khorsandi Nokia Germany, Munich, Germany

Bahare Masood Khorsandi Nokia, Munich, Germany

Kalle Koskinen Department of Information and Communications Engineering, Aalto University, Espoo, Finland

Jean Langlois-Berthelot A.S.CENT Centre of Advanced Studies (UNIPA), Palermo, Italy

Guus Leenders Quicksand, Roeselare, Belgium

Onel López Center of Wireless Communication, University of Oulu, Oulu, Finland

Wei Mao Santa Clara, USA

Marja Matinmikko-Blue Centre for Wireless Communications (CWC), University of Oulu, Oulu, Finland

Mattia Merluzzi University Grenoble Alpes and CEA-Leti, Grenoble, France

Markus Mueck Intel Deutschland Gmbh, Neubiberg, Germany

Ulrich Muehlmann NXP Semiconductors Austria GmbH & Co KG, Gratkorn, Austria

Fabrício Ramos Neves Guanambi, Brazil

Hosein Nikopour Intel Labs, Santa Clara, CA, USA

Carmela Occhipinti CyberSocial Lab. Srl Impresa Sociale, Rome, Italy

Daniel Poehl NXP Semiconductors Austria GmbH & Co KG, Gratkorn, Austria

Priyesh Pappinisseri Puluckul IDLab, University of Antwerp – imec, Antwerp, Belgium

Lyssa Ramaut Department of Electrical Engineering, KU Leuven, Ghent, Belgium

John M. Roman Rochester, USA

Chathura Sarathchandra Interdigital, London, UK

Lars Christoph Schmelz Nokia Germany, Munich, Germany

Omid Semiari Intel Labs, Santa Clara, USA

Vasantha Shanmugam Chennai, India

Bikramjit Singh LMF Ericsson, Kirkkonummi, Finland

Ritesh Kumar Singh IDLab, University of Antwerp – imec, Antwerp, Belgium

Emilio Calvanese Strinati University Grenoble Alpes and CEA-Leti, Grenoble, France

Emanuela Tangari CyberSocial Lab. Srl Impresa Sociale, Rome, Italy

Jarne Van Mulders Department of Electrical Engineering, KU Leuven, Ghent, Belgium

Lili Wei Quebec, Canada

Jenny Wiklund Department of Information and Communications Engineering, Aalto University, Espoo, Finland;
Department of Bioproducts and Biosystems, Aalto University, Espoo, Finland

Boxuan Xie Department of Information and Communications Engineering, Aalto University, Espoo, Finland

Shu-Ping Yeh Intel Labs, Santa Clara, USA

Acronyms

3GPP	3rd Generation Partnership Project
3GPP SA	3GPP System Architecture
5G	5th Generation Communication Networks
5GC	5G Core
6G	6th Generation Communication Networks
6GIA	6G Smart Networks and Services Industry Association
ACCA	Association of Chartered Certified Accountants
ADC	Analog-to-Digital Converter
ADGM	Abu Dhabi Global Market
ADMM	Alternating Direction Method of Multipliers
AF	Application Function
AI	Artificial Intelligence
A-IoT	Ambient Internet of Things
AmBC	Ambient Backscatter Communication
AMF	Access and Mobility Management Function
AMP	Ambient Power
ANR	Agence Nationale de la Recherche
API	Application Programming Interface
AR	Augmented Reality
BASIC	Beginner's All-purpose Symbolic Instruction Code
BC	Backscatter Communication
BE	Best-Effort
BFSK	Binary Frequency-Shift Keying
BIOS	Basic Input/Output System
BS	Base Station
CB	Contextual Bandits
CE	Conformité Européene (European Conformity)
CEN	European Committee for Standardization
CENELEC	European Committee for Electrotechnical Standardization
CERN	European Organization for Nuclear Research
CERT-In	Indian Computer Emergency Response Team

CES	Constant Elasticity of Substitution
C-K	Concept–Knowledge
CMDP	Constrained Markov Decision Process
CMOS	Complementary Metal-Oxide-Semiconductor
CNN	Convolutional Neural Networks
CNRS	Centre national de la recherche scientifique
CO-END	Continuous Operation Energy Neutral Device
COP	United Nations Climate Change Conference
COTS	Commercial Off-The-Shelf
COVID	Coronavirus disease
CP	Control Plane
CPU	Central Processing Unit
C-RAN	Cloud/Centralized Radio Access Network
CRM	Critical Raw Materials
CvaR	Conditional Value-at-Risk
D2D	Device-to-Device
DAC	(i) Development Assistance Committee of the Organisation for Economic Co-operation and Development (OECD), (ii) Digital-to-Analog Converter
DB-END	Duty-Cycled Barebone Energy Neutral Device
DC	Direct Current
DEQ	Department of Environmental Quality, US
DevOps	A combination of Development and Operations that unites people, processes, and technology to shorten the software development lifecycle
DG	(European Commission) Directorate General
DIFC	Dubai International Financial Centre
DL	Deep Learning
DLI	Direct-Link Interference
DOE	Department of Energy, US
DOS	Disk Operating System
DP-END	Duty-Cycled Persistent Energy Neutral Device
DPP	Digital Product Passport
DQN	Deep Q-earning
DRO	Distributionally Robust Optimisation
DRX	Discontinuous Reception
DTX	Discontinuous Transmission
EAI	Enterprise Application Integration
EAIE	European Association for International Education
EBIA	Brazilian Artificial Intelligence Strategy
EC	European Commission
EDLC	Electric Double-Layer Capacitors
eDRX	Extended Discontinuous Reception
EDVAC	Electronic Discrete Variable Automatic Computer
EGDI	European Green Deal Index

EGDIP	European Green Deal Investment Plan
EH	Energy Harvesting
ELECTRE	ELimination and ChoiceExpressingREality
eMBB	Enhanced Mobile Broadband
EMC	Electromagnetic Compatibility
EMF	Electromagnetic Field
EN	European Norm
eNB	Evolved NodeB (base station in 4G LTE and 5G networks)
END	Energy Neutral Device
ENIG	Electroless Nickel Immersion Gold
ENVIAA	ENVironmental Impact Assessment of AI systems
EO	Executive Order
EOL	End of Life
EPA	Environmental Protection Agency, US
EPP	European People's Party
EPR	Extended Producer Responsibility
EPREL	European Product Registry for Energy Labelling
ERF	Energy Reuse
ErP	Energy-related Products
ES	ETSI Standard
ESG	Environmental, Social, and Governance
ESL	Electronic Shelf Label
ESO	European Standards Organisation
ESPR	Ecodesign for Sustainable Product Regulation
EtS	Energy to Solution
ETSI	European Telecommunications Standards Institute
EU	European Union
EUOJ	European Union Official Journal
EV	Electric Vehicle
FCC	Federal Communications Commission
FEAF	Federal Enterprise Architecture Framework
FET	Field-Effect Transistor
FLOPS	Floating Point Operations per Second
FPGA	Field-Programmable Gate Arrays
FRAM	Ferroelectric Random-Access Memory
GaN	Gallium Nitride
GBR	Guaranteed Bit Rate
GDP	Gross Domestic Product
GHG	GreenHouse Gas
GITAM	Green IT Adoption Model
gNB	Next-Generation Node B (base station for a 5G New Radio (NR) network)
GNN	Graph Neural Networks
GPIO	General-Purpose Input/Output
GPS	Global Positioning System

GPU	Graphical Processing Unit
GRI	Global Reporting Initiative
GRL	Graph Reinforcement Learning
GUI	Graphical User Interface
GWP	Global Warming Potential
HAPS	High-Altitude Platform Systems
HASL	Hot Air Solder Leveling
HB	House Bill
HEC	Hautes Etudes Commerciales (Higher Education in Commerce)
HEN	Harmonised European Norm
HLRS	High-Performance Computing Center
HPC	High-Performance Computing
HS	Harmonised Standard
HVAC	Heating, Ventilation, and Air Conditioning
I/O	Input/Output
I2C	Inter-Integrated Circuit
IaaS	Infrastructure as a Service
IC	Integrated Circuit
ICT	Information and Communications Technology
IEA	International Energy Agency
IEC	International Electrotechnical Commission
IEEE	Institute of Electrical and Electronics Engineers
IMF	International Monetary Fund
IMT	International Mobile Telecommunications
IoT	Internet of Things
IP	Internet Protocol
IRA	Inflation Reduction Act, US
IRS	Internal Revenue Service, US
IS	Information System
ISG	Industry Specification Group
ISM	Industrial, Scientific, and Medical
ISO	International Organization for Standardization
IT	Information Technology
ITU	International Telecommunication Union
ITU-R	International Telecommunication Union Radiocommunication Sector
ITU-T	International Telecommunication Union—Telecommunication Standardization Sector
JTC	Joint Technical Committee
JU	Joint Undertaking
$kgCO_2eq$	Kilograms of Carbon Dioxide Equivalent
KPI	Key Performance Indicator
KVI	Key Value Indicator
kWh	Kilowatt-hour
LB	Load Balancing

LCA	Lifecycle Assessment
LCOE	Levelized Cost of Energy (or Electricity)
LDO	Low-Dropout
LEO	Low Earth Orbit
LFP	Lithium Iron Phosphate
LGPD	General Law for the Protection of Personal Data
LiCBD	Light-controlled Backscatter Device
LLM	Large Language Model
LoRaWAN	Long Range Wide Area Network
LTE	Long Term Evolution
LTO	Lithium Titanium Oxide
LUMI	Large Unified Modern Infrastructure
LWA	Leaky-Wave Antenna
M&E	Monitoring and Evaluation
MBZUAI	Mohamed bin Zayed University of Artificial Intelligence
MCDH	Multi-Criteria Decision Analysis
MCU	Microcontroller Unit
MDC	Modular Data Center
MDG	Millennium Development Goals
MDP	Markov Decision Process
MEC	Multi-Access Edge Computing
MEERP	Methodology for the Ecodesign of Energy-Related Products
MICMAC	Matrice d'impacts croisés multiplication appliquée à un classement
MILP	Mixed-Integer Linear Program
MIMO	Multiple-Input Multiple Output
MIPS	Million Instructions Per Second
ML	Machine Learning
ML Ops	Machine Learning Operations
mMTC	Massive Machine Type Communications
MOCCAE	Ministry of Climate Change and Environment
MPP	Massively Parallel Processing
MPPT	Maximum Power Point Tracking
MTC	Machine Type Communications
MVL	Minimal Viable Legislation
MW	Megawatt
NB-IoT	Narrowband IoT
NCA	Nickel Cobalt Aluminum
NDLM	National Digital Literacy Mission
NEB	New European Bauhaus
NEF	Network Exposure Function
NeGP	National e-Governance Plan
NES	Network Energy Saving
NFC	Near-Field Communication
NFV	Network Function Virtualization

NFV MANO	Network Function Virtualization Management and Orchestration
NGEU	NextGenerationEU
NGNM	Next-Generation Mobile Networks Alliance
NGO	Non-Governmental Organization
NLF	New Legislative Framework
NMC	Lithium Nickel Manganese Cobalt Oxides
NPU	Neural Processing Unit
NR	New Radio
NSGA-II	Non-dominated Sorting Genetic Algorithm II
NTA	Net Technical Assessment
NTN	Non-Terrestrial Networks
NVM	Non-Volatile Memory
NWDAF	Network Data Analytics Function
OBBB	One Big Beautiful Bill
OECD	Organisation for Economic Co-operation and Development
OFDM	Orthogonal Frequency-Division Multiplexing
OJ	Official Journal
OJEU	Official Journal of the European Union
OPEX	Operational Expenditure
O-RAN	Open Radio Access Network
O-RAN SMO	Open Radio Access Network Service Management and Orchestration (SMO)
OS	Open Source
OSP	Organic Solderability Preservative
PaaS	Platform as a Service
PCB	Printed Circuit Board
PCE	Power Conversion Efficiency
PET	Polyethylene Terephthalate
PHY	Physical Layer
PMGDISHA	Pradhan Mantri Gramin Digital Saksharta Abhiyan
PMU	Power Management Unit
PPDR	Public Protection and Disaster Relief
PROMETHEE	Preference Ranking Organization Method for Enrichment Evaluation
PSM	Power Saving Mode
PUE	Power Usage Effectiveness
PUF	Physical Unclonable Functions
PV	Photovoltaic
PWM	Pulse-Width Modulated
QAOA	Quantum Approximate Optimization Algorithm
QoE	Quality of Experience
QoS	Quality-of-Service
R&D	Research and Development
RAM	Random-Access Memory
RAMAC	Random-Access Method of Accounting and Control

RAN	Radio Access Network
RCS	Rich Communication Services
REF	Renewable Energy Use
RegLab	Regulations Lab
RF	Radio Frequency
RFID	Radio-Frequency Identification
RF-WPT	Radio-Frequency Wireless Power Transfer
RIBA	Royal Institute of British Architects
RIS	Reconfigurable Intelligent Surface
RL	Reinforcement Learning
RMI	Responsible Minerals Initiative
RNN	Recurrent Neural Networks
ROI	Return on Investment
ROM	Read-Only Memory
RRI	Responsible Research and Innovation
RSRP	Reference Signal Received Power
rUC	Representative Use Case
SaaS	Software as a Service
SAI	Securing AI (ETSI Technical Committee)
SASB	Sustainability Accounting Standards Board
SAT	Social Acceptance of Technology
SBA	Service-based Architecture
SDG	Sustainable Development Goal
SDN	Software Defined Networking
SDO	Standards Development Organization
SINR	Signal-to-Interference-plus-Noise Ratio
SIR	Signal-to-Interference Ratio
SLA	Service Level Agreement
SM	Sleep Mode
SME	Small and Medium-sized Enterprises
SMF	Session Management Function
SMT	Surface-Mount Technology
SNS JU	Smart Networks and Services Joint Undertaking
SOA	Service-Oriented Architecture
SoA	State of the Art
SPI	Serial Peripheral Interface
SRAM	Static Random-Access Memory
STQC	Standardisation Testing and Quality Certification
SUNY	State University of New York
TC	Technical Committee
TCFD	Task Force on Climate-Related Financial Disclosures
TCO	Total Cost of Ownership
TDRA	Telecommunications and Digital Government Regulatory Authority, UAE
THC	Technological–Human Convergence

TISM	Total Interpretive Structural Model
TOGAF	The Open Group Architecture Framework
TPU	Tensor Processing Unit
TR	Technical Report
TRIZ	Theory of Inventive Problem Solving
TRL	Technology Readiness Level
TtS	Time to Solution
UAE	United Arab Emirates
UAV	Unmanned Aerial Vehicle
UE	User Equipment
UHF	Ultra High Frequency
UK	United Kingdom
UNCTAD	UN Trade and Development
UNESCO	United Nations Educational, Scientific, and Cultural Organization
UNFCCC	UN Framework Convention of Climate Change
UNIVAC	Universal Automatic Computer
UP	User Plane
UP-END	Ultra-Passive Energy Neutral Device
UPF	User Plane Function
URLLC	Ultra-Reliable Low Latency Communications
US	United States
Uu	UMTS air interface
V2X	Vehicle-to-Everything
ViT	Vision Transformers
VLC	Visible Light Communication
VNF	Virtualized Network Function
VPN	Virtual Private Network
VQE	Variational Quantum Eigensolver
VR	Virtual Reality
WFEO	World Federation of Engineering Organizations
WHO	World Health Organization (WHO)
Wi-Fi	Wireless Fidelity
WPT	Wireless Power Transfer
WRI	World Resources Institute
XR	Extended Reality
ZED	Zero Energy Devices

Chapter 1
Introduction to Digital Sustainability

Christophe Gaie

Abstract In this first chapter, a holistic overview of the topic sustainability is given, with a specific focus on sustainability in the context of Digital Systems. A summary of related policy initiatives is given, including the European Green Deal, legislation initiatives and others. Furthermore, a specific focus is placed on how to assess the level of sustainability; we will summarize quantitative metrics which are applied in the field but also recognize that a quantitative assessment is not possible in all cases and thus we rather use the concept of so-called Key Value Indicators, representing a more "intuitive" representation. The overall objective is to present a 360° overview on the state of the art related to sustainability, preparing the deep-dives following in the subsequent chapters.

Keywords Digitalization · European legislations · European Union single market · Sustainability

1 The Emergence of Sustainability

The concept of sustainability, while having its roots in earlier conservation efforts as depicted in [1], gained significant importance in the latter half of the twentieth century. This emergence was driven by growing anxieties among scientists and activists regarding pollution, the implications of uncontrolled economic growth, and the unsustainable exploitation of Earth's finite resources. In 1962, Carson [2] published *Silent Spring*, a book that highlights the dangers of pesticide use and marked a significant step forward in raising environmental awareness among the American public.

Then, in the influential report The Limits to Growth, published in 1972, the Club of Rome (a non-profit organization of intellectuals and business leaders concerned with global issues) highlighted the risks posed by the potential mismatch between

C. Gaie (✉)
French Prime Minister Office, Laboratoire de Recherche Magellan, Université Jean Moulin, Lyon, France

C. Gaie and M. Mueck (eds.), *Digital Sustainability: Legislations, Standards, Solutions and Related Case Studies*, Advances in Sustainability Science and Technology,
https://doi.org/10.1007/978-981-92-0458-8_1

exponential economic and population growth, and the finite supply of resources [3]. The report presented computer-generated models projecting several scenarios. Many of these scenarios predicted societal collapse in the twenty-first century if current trends continued. Scientists have compared the evolution of the world with the scenarios defined in the model, and they have noted that the "standard run" (business as usual) unfortunately remains the best-fitting model [4].

Afterward, the World Commission on Environment and Development's 1987 Brundtland Report played a major role by introducing and defining sustainability as a new concept [5]. Indeed, it defines our common future as "development that meets the needs of the present without compromising the ability of future generations to meet their own needs." This highlights the importance of balancing current needs with long-term environmental, social, and economic well-being. It also encourages responsible behavior that considers both present and future impacts in decision-making.

The Millennium Development Goals (MDGs) were set in 2000, following the Millennium Summit, and based on the OECD DAC international development goals, and consisted of eight areas of international development goals with a target date for achievement by 2015 [6]. The seventh goal was to address environmental sustainability, and included the targeted area of integrating sustainable development into national policies, halting the loss of biodiversity, improving access to clean drinking water and sanitation, and improving the lives of slum dwellers. While some progress was achieved in most areas, such as reducing the depletion of the ozone, increasing the extent of protected areas, reducing the number of people without access to drinking water and sanitation, we still face considerable challenges. Further, while primarily unforeseen at the turn of the millennium, the rapidly changing reality of digital technologies has emerged as significant drivers in sustainable development.

A few years later, Rockström et al. [7, 8] established the concept of planetary boundaries where they outline a "safe operating space" for humanity that will allow us to interact in the Earth's system without pushing the Earth system past critical thresholds. The planetary boundaries outline critical thresholds for several critical topics. These boundaries establish a critical threshold for processes within the Earth system, beyond which the risk of abrupt or irreversible environmental change occurs at an escalated level. The authors proposed nine planetary boundaries related to climate change, ocean acidification, etc. They described how transgressing these boundaries might produce adverse implications for human societies and ecosystems and, subsequently, bring about abrupt changes to Earth's climate system, biogeochemical cycles, and biodiversity.

Unfortunately, the research indicates that many of the boundaries have already been exceeded, highlighting the urgent need for action to transition to sustainable practices.

Kate Raworth extended the perspective of the planetary boundaries with Doughnut Economics. This visual model of sustainable development brings together the idea of planetary boundaries with the complementary notion of social boundaries [9]. The hole in the middle of the Doughnut denotes the percentage of people without access to the essentials of life (food, health care, education, equity, etc.), while the outer ring denotes the ecological ceilings (planetary boundaries) that must be acknowledged. Thus, a sustainable economy must exist within the Doughnut so that all needs are met within the ecological limits of the planet.

A very interesting approach to address resource limitations is building a circular economy. This concept was proposed and detailed in 2013 by the MacArthur [10] Foundation and aims to ensure sustainability through a regenerative approach to the economy. It requires rethinking the production model to eliminate waste and pollution, extend the lifespan of products and ensure efficient recycling, and continuously restore natural systems through specific attention to biodiversity, agriculture, and forestry. This regenerative system, as envisioned by Ellen MacArthur, keeps resources in use for as long as possible, extracting maximum value from them while in use, then recovering and regenerating products and materials at the end of each service life.

In 2015, the United Nations [11] implemented the 2030 Agenda for Sustainable Development. This new framework goes beyond the Millennium Development Goals and includes 17 Sustainable Development Goals (SDGs) that encompass a broader range of issues, including poverty, hunger, and health. An important new aspect of the SDGs is a focus on equity and inclusion, with a central goal of "leaving no one behind." The 2030 Agenda also explicitly integrates the three dimensions of sustainable development (economic, social, and environmental). This emphasizes the need for global and balanced approaches. Finally, the framework highlights the importance of partnership, as achieving these ambitious goals will require cooperation among governments, the private sector, civil society, and people.

The European Green Deal, launched in 2019, marks a significant turning point in the European Union's agenda for sustainable development [12]. This initiative translates strategies into action to guarantee climate neutrality by 2050 through the European Climate Law. The approach is designed to engage a wide range of interconnected sectors, including energy, transport, agriculture, and industry, with the objective of coherence and synergy. It also introduces fairness and inclusivity to facilitate the transition toward a sustainable economy. The agenda creates opportunities to drive ambitious policy initiatives, including the Circular Economy Action Plan, which aims to build a resource-efficient model, as well as the Sustainable and Smart Mobility Strategy dedicated to optimizing the transport sector. The success of these initiatives requires investment and financing, which can be supported by the Sustainable Europe Investment Plan that combines public and private funds for green projects and innovation.

Finally, current concerns focus on how digital technologies and climate innovations can enhance the adoption of sustainability [13, 14]. This digital transformation, often referred to as "Climate Tech," is reshaping industries, governments, and consumer behavior in ways that align with environmental goals. Current trends tackle, e.g., the use of artificial intelligence to optimize energy consumption or blockchain technology to ensure supply chain transparency. There is also significant interest in innovations aimed at optimizing energy production through smart grids and artificial intelligence, as well as improving the sustainability of agriculture by reducing inputs, for instance. All these innovations contribute to the establishment of a net-zero carbon economy, but their own ecological impact must be monitored, as they require the extraction of rare resources, the construction of powerful data centers, and the generation of significant heat, which negatively affects the climate. The political time line related to the implementation of sustainability objectives is illustrated in Fig. 1.

Definition of Sustainability in the Present Book Sustainability is a type of development that meets the needs of today without compromising the ability of future generations to meet their own needs. This objective should be achieved by balancing current needs with long-term environmental, social, and economic aspects.

This requires a shift toward regenerative practices, a circular economy, and global cooperation. These objectives should be reached by reducing the impact of digital technologies and respecting planetary boundaries.

2 The Development of Digitalization

The digitalization of society is a progressive evolution, moving from simply converting analog information into digital formats to transforming multiple processes. It began with information digitization, enabling the transition from analog to digital forms, such as archiving paper records for easier retrieval and making previously inaccessible information available online. This foundation then allowed for process digitalization, leveraging digital technologies to streamline and improve institutional operations, like payroll, procurement, tax declarations, and identity acquisition.

More recently, digital transformation has emerged, going even further. This approach represents a fundamental shift in an organization's strategy and value proposition. Digital transformation involves deep and coordinated changes across culture, workforce, and technology, often requiring significant prior work in both digitizing information and digitalizing processes. This holistic approach enables organizations not just to do things more efficiently, but to rethink and optimize their functioning, drastically simplifying operations and increasing impact. Examples include the replacement of traditional banks by fully online banks in the private sector, and in the public sector, the emergence of solutions like X-Road, which enables unified and secure data exchange and is the backbone of e-Estonia. Figure 2 illustrates this evolution and is inspired by [15]. Obviously, this is a simplified representation as the three concepts overlap.

The emergence of digitization comes from a long process that begins with fundamental work such as the development of the binary system by Leibniz, Boolean

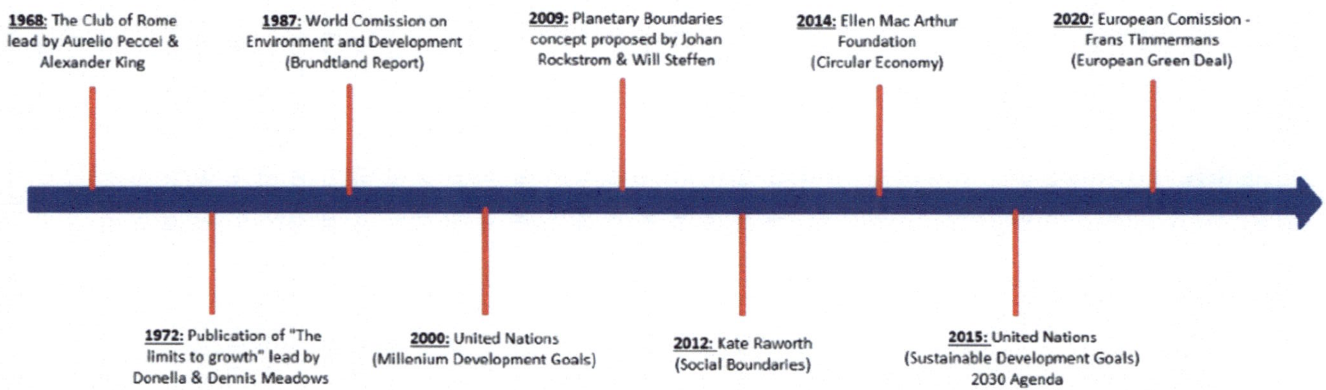

Fig. 1 Time line on political agenda related to sustainability

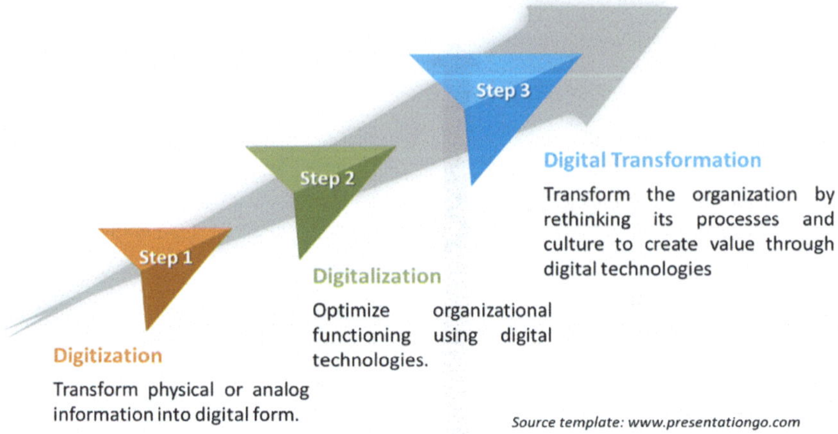

Fig. 2 Time line of digital transformation

algebra by Boole, and Shannon's work on information theory [16]. Their contributions notably laid the groundwork for digital computing and communication. A pivotal moment in the history of digitization is the construction of the Electronic Discrete Variable Automatic Computer (EDVAC), the first binary computer, by the University of Pennsylvania, delivered in 1949 [17]. Then the UNIVAC I (Universal Automatic Computer) was built in 1951. This first commercially produced electronic digital computer, utilized vacuum tubes and mercury delay-line memory, demonstrated the practical application of digital computing for business and government.

Later, digitalization shifted its emphasis to data storage and developed the ability to provide random access to memory, such as with the IBM RAMAC 305 developed in 1956. Data storage was then dramatically changed with the introduction of the IBM 350 Disk Storage Unit. This hard drive employed magnetic storage with spinning platters and allowed direct access to stored data. This was a significant advancement compared to earlier sequential access methods. The computation processing unit of the RAMAC 305 still utilized vacuum tubes (invented by Fleming in 1904), which limited its overall performance and reliability, but the hard drive was a major innovation. Despite this improvement, these early computers were characterized by their large physical size, significant power consumption, and slow access speeds compared to current technologies. The RAMAC 305 system, with its innovative hard drive, weighed more than a ton and marked a transition between pure vacuum tube computing and random-access data storage.

The next important step was the development of a second generation of computers that exploited the point-contact transistor (created by Bardeen and Brattain in 1947). These semiconductor devices could perform the switching functions of vacuum tubes with significantly smaller dimensions, lower power requirements, and higher reliability [18]. They facilitated the miniaturization of electronic circuits, reducing the physical footprint and cost of computing devices, and leading to advancements toward integrated circuits. The integration of multiple transistors on a single chip further augmented gains in computing power and efficiency. This enabled the rapid growth of digital technology observed in the following decades. The convergence of enhanced storage technologies and miniaturization laid the foundation for democratized computing.

The second generation of computers (1956–1963) marked a significant advancement with the introduction of transistors. These transistors, invented in 1947, replaced vacuum tubes, resulting in computers that were smaller, faster, and more reliable. This miniaturization, facilitated by the point-contact transistor, also reduced power consumption and cost. For example, the IBM 1401, a popular business computer of this period, utilized transistor-based circuitry, making it significantly more compact and efficient than its vacuum tube predecessors. Similarly, the Philco Transac S-2000, known for its use in scientific and military applications, demonstrated the increased speed and reliability achievable with transistors [19]. The use of transistors facilitated the development of integrated circuits, further amplifying computing power and efficiency. During this period, assembly language emerged, enhancing the capabilities of these machines. For instance, programmers began writing code closer to machine language, allowing for more efficient use of hardware resources compared to the previous generation's reliance on machine language. This convergence of improved storage and miniaturization laid the foundation for democratized computing.

Then, the third generation of computing (1964–1971) appeared and provided a significant advancement through the implementation of integrated circuits. These elements, fabricated on a single silicon substrate, miniaturized electronic circuits. They also enabled a substantial reduction in manufacturing costs of computing devices by integrating numerous transistors, resistors, and capacitors onto a single chip, leading to enhanced performance and reliability. For instance, the IBM System/360, a prominent example of this era, leveraged integrated circuit technology to offer a family of computers with varying performance levels, catering to diverse business needs. Furthermore, the development of computers like the Hewlett-Packard HP 2100 series, which utilized monolithic integrated circuits, demonstrated the increasing power and affordability of minicomputers [20].

These systems, employed in diverse applications such as data acquisition and process control, highlighted the impact of integrated circuits in expanding the reach of computing beyond traditional mainframe environments. Moreover, this generation witnessed the introduction of more user-friendly input and output devices, such as keyboards and video displays, replacing earlier, more cumbersome interfaces. These developments contributed to the broader accessibility of computing technology, facilitating its adoption in a wider range of scientific and commercial contexts.

Subsequently, the fourth generation of computers, beginning in 1971, was marked by the arrival of microprocessor chips. This generation of microprocessors led to the development of personal computers, graphical user interfaces (GUIs), and modern networking as we use them today. The year 1971 saw the release of the Intel 4004 and the Intel 8080, which was used in the Altair 8800, as hardware examples of the fourth generation of personal computers [21]. In this period, software advancements were also significant.

Operating systems like early versions of DOS, used on the IBM PC, provided user interfaces, while programming languages like BASIC, found on the Apple II, democratized software creation [22]. These advancements made tasks like word processing (WordStar) and spreadsheet calculations (VisiCalc) accessible to a wider audience. The foundation of the modern Internet, ARPANET, sent its first successful message on October 29, 1969, and began thereafter to connect research institutions building the future of networked computing.

During the 1980s, the fifth generation of computers emerged, offering significant advancements through the introduction of parallel processing and enhanced network capabilities. These developments were driven by the rise of multiprocessor systems, designed to execute multiple tasks concurrently and substantially increase computational power. Machines like the Sequent Balance 8000 showcased the viability of shared memory architectures. Complementing these hardware innovations, network technologies underwent rapid expansion, most notably with Tim Berners-Lee's proposal of the World Wide Web at CERN in 1989, followed by the development of the first web server and browser in 1990 [23]. These combined advancements facilitated the transition of computing from specialized environments to homes and offices, enabling massive information sharing and exchange. Parallel processing became a central focus, leading to innovations such as massively parallel processing (MPP) architectures, exemplified by the Thinking Machines' Connection Machine [24]. Furthermore, while occurring slightly after the end of the fifth generation time frame, IBM's Deep Blue defeating world chess champion Garry Kasparov in 1997 demonstrated the burgeoning capabilities of intelligent computing [25]. Finally, programming paradigms began to evolve from procedural languages like BASIC and C toward AI-focused languages such as Prolog, Lisp, and the growing adoption of Python.

Today, the implementation of sixth-generation computing is beginning, bringing the new potential of quantum mechanics. In particular, the creation of qubits allows for computations that would not be conceivable in classical binary systems. The foundation for this was established in 1994, when Peter Shor's algorithm showed the potential for quantum computers to factor large numbers exponentially faster than classical computers [26]. In the early 2000s, primitive examples of quantum computers emerged. For example, IBM began significant work in quantum computing in the 2010s, showcasing the IBM Q Experience and quantum devices to the public with limited capacity in 2016.

Google created the Sycamore processor, which achieved "quantum supremacy" by performing a calculation beyond the reach of classical supercomputers [27]. These developments offer potential for new applications in different domains such as cryptography and medicine. They also enable substantial advances in artificial intelligence. The sixth generation of communications also integrates nanotechnology, which is valuable for assisting in the miniaturization and efficient use of computing components. A major improvement of the sixth generation is the provision of higher-speed and lower-latency communication networks, including the launch of advanced wireless 6G networks and more dependable and secure Internet infrastructure [28].

A detailed in Table 1, there are digitalization tackles multiple dimensions of computer engineering such as computer processing, programming, and network. This synthesis describes the evolution from rudimentary technologies and data processing in the mid-twentieth century to a comprehensive system of interconnected technologies.

Definition of Digitalization in this Book Digitalization is the transformative process of converting information into a digital format and integrating it into IT systems.

This evolution, profoundly shaped by successive computer generations (from first-generation vacuum tubes to the emerging sixth-generation quantum era) extends beyond mere digitization (paper to digital). It encompasses the strategic implementation of digital technologies to streamline processes, enhance efficiency, create new value, and transform business and society.

Consequently, digitalization aims to provide increased efficiency, added value, and transformative change for all stakeholders.

3 The Rise of Digital Sustainability

Many researchers define sustainability as a nested combination of environment, society, and economy [1]. A refined approach considers digital as a fourth dimension, offering new perspectives on digital sustainability, as illustrated in Fig. 3. However, digital could also be seen as a transversal dimension that can affect each of the underlying domains. Proper utilization of digital technologies may reduce the environmental impact of transport, agriculture, and manufacturing, but could also worsen the situation due to increased pollution. Digital can also enhance democracy and societal inclusion, or increase exclusion through the digital divide as well as foster the economy by creating added value (e.g., websites, virtual visits, automation of administrative procedures). On the opposite, it may also destroy jobs and create unemployment if it not correctly regulated.

This section further delves into the emergence of digital sustainability and illustrates its benefits and drawbacks.

The origins of digital sustainability can be found in early critiques of the digital ecosystem, especially through the imagined vision of Marcus Dapp in 2004. He wrote a fictional keynote that details a scenario of what could happen in 2030 by applying the unsustainable practices of the late twentieth century and early twenty-first century [29]. His hypotheses are that corporations would aggressively seek

Table 1 Digitalization over team and related key features

Feature	1st generation (1940s–1950s)	2nd generation (1956–1963)	3rd generation (1964–1971)	4th generation (1970s–1980s)	5th generation (1980s–early 2000s)	6th generation (emerging)
Processing	Vacuum tubes	Transistors	Integrated circuits (ICs)	Microprocessors (single-core)	Parallel processing (multi-core, GPUs)	Quantum computing (qubits)
Programming	Machine language	Assembly language	High-level languages (FORTRAN, COBOL)	Procedural (C, assembly)	AI and logic-based (Prolog, python, ML, etc.)	Quantum algorithms (quantum circuits)
AI focus	None	None	Limited early concepts	Basic expert systems	Neural networks and deep learning	Quantum machine learning
Key innovations	ENIAC, UNIVAC I	Transistor-based computers (IBM 1401)	Minicomputers (IBM system/360), ICs	VLSI, personal computers (Altair 8800, IBM PC)	Internet, world wide web, MPP, deep learning	Quantum algorithms, advanced Qubit development
Network focus	None	None	Limited mainframe networking	ARPANET, LAN beginnings	Internet expansion, Wi-fi	Future
User interface	Punched cards, switches	Punched cards, magnetic tape	Keyboards, monitors	GUIs, mice	Advanced GUIs, natural language interfaces	Future
Memory	Vacuum tube storage, delay lines	Magnetic core memory	Semiconductor memory	RAM, ROM, floppy disks	Large-scale ram, solid-state drives	Future

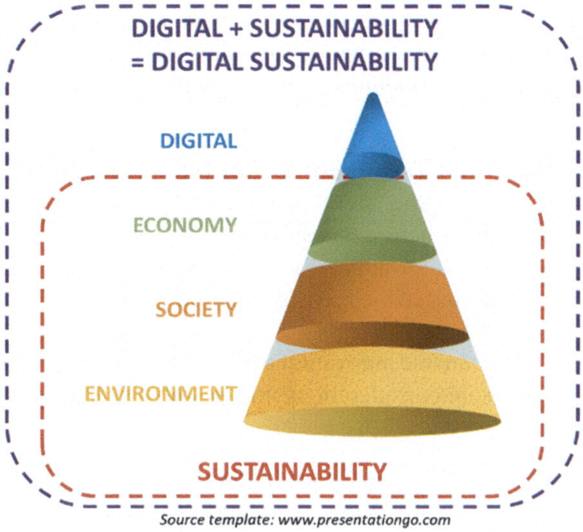

Fig. 3 Four dimensions of sustainability

access rights and market control over digital information, charging the collective large costs. He also critiqued practices on software patents and digital rights management that limit innovation and access to knowledge on equitable terms. Dapp's narrative supported the Commons Movement, characterized by Free Software and open access. Dapp's idea of knowledge as a public good contributed to conceptualizing digital sustainability.

Then, it was asserted by Bradley [30] that digital sustainability is a comprehensive approach including economic, social, and organizational dimensions. It does not only address the key concern of enabling digital preservation using the right technical context. Instead, it also seeks to promote an active lifecycle for digital items rather than relying solely on simple technical solutions, such as extending carrier durability or changing formats. This means looking at the development of both sustainable digital repositories and shared metadata schemas to facilitate the long-term usage of digital information. The author also underlines the economic realities related to digital preservation. To this end, he notes that sustainable access requires continuous and adaptive efforts, shared across institutions and stakeholders, to address the ever-changing technological and societal context.

The intersection of information technology and sustainability was explained by van Osch et al. [31], who indicate that the domain may be modeled through three distinct lenses with increasing scope: Green Information Technology (IT), Green IS (Information Systems), and sustainable innovation. First, Green IT focuses on minimizing the environmental impact of IT itself, reducing energy consumption and emissions throughout the lifecycle of IT hardware. Second, Green IS extends this view by examining how information systems can be employed to manage and mitigate an organization's overall environmental footprint, often relying on cost reduction and regulatory compliance. Finally, sustainable innovation adopts a broader and proactive approach that considers environmental and social challenges as opportunities for business innovation and competitive advantage. They all contribute to generating value across economic, environmental, and social dimensions.

Important Definitions Green Information Technology (IT) corresponds to minimizing the environmental footprint of information technology hardware and infrastructure (hardware focus).

Green Information Systems (IS) refers to the use of information systems to manage and reduce an organization's overall environmental impact (organizational focus).

Sustainable innovation is a proactive approach that integrates environmental and social considerations into business strategies (business focus).

These concepts have generated much research, such as the proposal of a framework that focuses on the practical adoption of Green IT by Molla. This framework describes how organizations can integrate environmentally sustainable IT practices [32]. Entitled GITAM, it details key factors like organizational readiness, perceived benefits, and regulatory pressures that drive the adoption of Green IT, especially to minimize the environmental footprint of IT hardware. In a larger scope, Chen et al. [33] explored, through an institutional lens, how to implement Green IT and Green IS. Notably, the authors highlight how IS can provide solutions to environmental challenges that go beyond reducing the negative impacts of IT hardware. They specifically relate Green IS and IT behavior to strategic focus areas such as pollution prevention, product stewardship, and sustainable development. Thus, they illustrate how IS can be optimized to manage and reduce the environmental impact of the entire organization.

Various researchers addressed the construction of these theoretical concepts. For example, Coroama et al. [34] implemented a field study to assess the feasibility of using advanced videoconferencing to substitute for greenhouse gas emissions from flying to conferences across continents. They created a resource-management conference held simultaneously in Switzerland and Japan, and compared the emissions from attendee travel and Information and Communication Technology (ICT) equipment under similar conditions. The results showed a reduction in greenhouse gas emissions of between 37 and 50%, while maintaining a significant increase in the overall number of participants. Key feedback is that attendees reported satisfaction with attending a conference at two locations, which indicates that this method can enhance both sustainability and user satisfaction compared to a traditional single-site conference format.

The understanding of ICT's importance for sustainability was detailed by Hilty and Aebischer [35]. The researchers described its potential to either contribute negatively or help mitigate environmental challenges. This emphasizes the necessity of interdisciplinary approaches to combine ICT with other domains and build sustainable innovations. The authors identified research areas where ICT can improve the efficient use of resources and highlighted the need to examine the societal impacts of digital technologies on environmental sustainability. Their research provides a broader perspective that complements the previously defined lenses of Green IT, Green IS, and sustainable innovation. By recognizing ICT as both a potential problem and a crucial solution, Hilty and Aebischer's work underscores the importance of a holistic approach that integrates technological innovation with environmental and social considerations.

Building upon this theoretical understanding, George et al. [36] further illustrated the practical application of digital sustainability through entrepreneurial innovation. They identified six managerial problems in sustainability (knowing, valuation, communication, coordination and trust, access and reach, institutions) and proposed a digital sustainability pathway for each to tackle them. This research is valuable as it highlights concrete and economically viable solutions to mitigate climate change and contribute to a more sustainable future. The proposed examples rely on entrepreneurial ventures that are well-adapted to provide innovative solutions to the rapid evolution of the planet.

More recently, Rosati et al. [37] proposed a broader approach that tackles the socio-economic and environmental impacts of digital technologies. They advocate for implementing sustainable practices throughout the digital lifecycle, from design and development to usage and disposal. They propose a clear definition of digital sustainability: "the design, development, configuration, deployment, and decommissioning of digital resources and artifacts toward improving the environment and economic welfare." Their book addresses multiple domains, such as Artificial Intelligence, Smart Cities, and the Circular Economy.

Definition of Digital Sustainability Digital sustainability refers to the responsible development and use of technological approaches to ensure environmental and economic resilience.

It aims to achieve sustainable development goals through the ethical use of digital tools to support social progress and fairness while minimizing environmental damage from a long-term perspective.

4 Assessing Sustainability

As discussed above, the concept of sustainability is broad and touches multiple dimensions including Digital, Economy, Society, and Environment as illustrated in Fig. 3. The question thus arises how to assess to which extent a certain product or service is addressing sustainability objectives. There are obvious numerical metrics which may be applied such as power consumption, usage of (physical) resources, and CO_2 emission levels. However, not all aspects of sustainability can be reflected by numerical metrics, such as societal aspects, inclusiveness, trustworthiness, and more.

To further address the question on how to assess sustainability, we consider the work of the European Flagship project Hexa-X [38] which studies the path toward sixth Generation (6G) Communication Systems. Although the scope of sustainability goes far beyond future communication systems, the concepts and ideas of Hexa-X are inspiring and may be applied more broadly to the question on how to perform a related assessment. Figure 4 illustrates the overall Hexa-X vision with 6G communication systems encapsulating the Digital World, the Physical World and the Human World [39]. Sustainability, together with Trustworthiness and Inclusion, is a key pillar of the proposed vision for future 6G Communication Systems.

With the broader vision illustrated in Fig. 4, Hexa-X recognizes that some aspects of sustainability—as well as Trustworthiness and Inclusion—may be characterized and measured based on Key Performance Indicators (KPIs), which are typically quantitative metrics. However, the KPI-based approach is not suitable to cover all aspects of the broad concepts of sustainability and thus Key Value Indicators (KVIs) are introduced to address this missing dimension. The basic principles are outlined in Fig. 5.

Key Value Indicators can be qualitative or—in some cases—quantitative measures that gauge the impact and achievement of value-driven goals, such as societal, economic, or environmental benefits, typically looking toward the future rather than just past performance like a traditional KPI. Unlike Key Performance Indicators

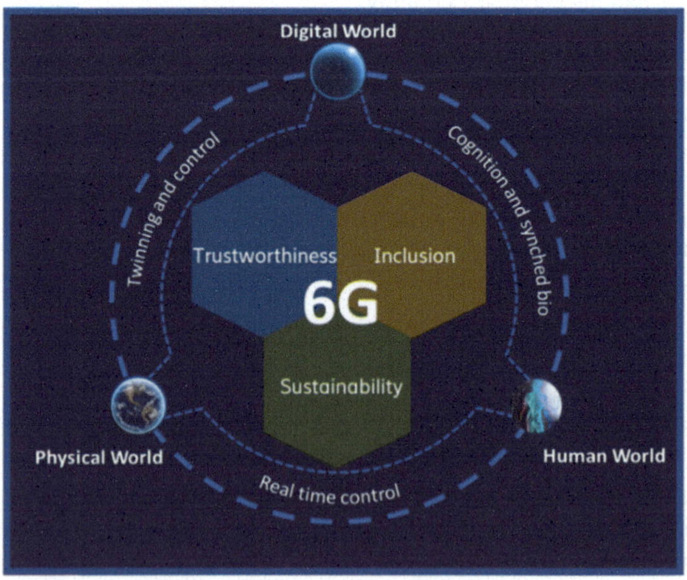

Fig. 4 Overall Hexa-X vision [Hexa-X-D1.1]

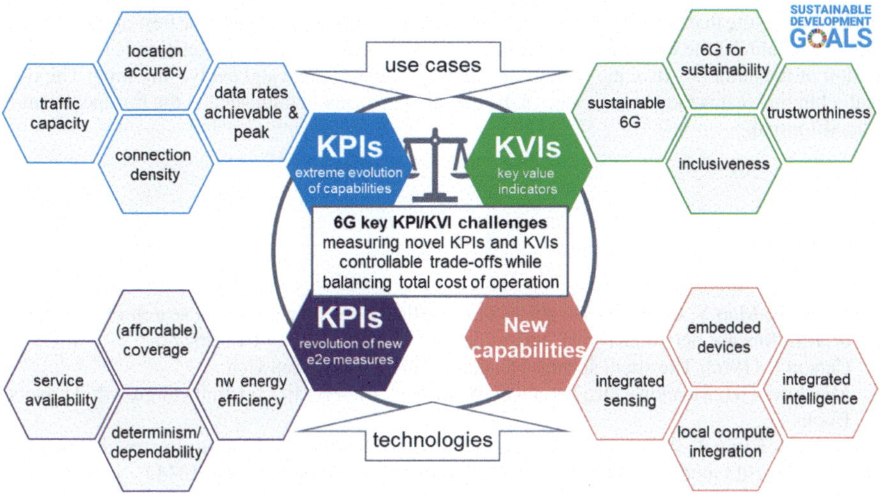

Fig. 5 Clustering of KPIs and KVIs [Hexa-X-D1.1]

(KPIs), which focus on what has happened or is happening, KVIs assess the value a business or project creates for various stakeholders and society as a whole.

As discussed above, the concept of Key Value Indicators is very suitable to address forward looking dimensions of the broader sustainability concepts. Throughout the present book, we will keep this approach in mind and apply assessment strategies based on KPIs and/or KVIs as appropriate.

5 Conclusions

The present chapter provides a comprehensive summary on the topic of sustainability which is a key pillar of the European Green Deal and the overall political agenda. The history of sustainability research is summarized and the related political debates are being discussed, we consider early discussions in the Club of Rome in 1968 over United Nations Millennium Development Goals established in 2000 and finally the establishment of the European Green Deal in 2020. We further more discuss the current holistic trend of digitalization across society at large and inherent sustainability challenges and requirements. Finally, the chapter discusses the assessment of sustainability and identifies the need to go beyond quantitative metrics only. Inspired by discussion of the European 6G Flagship Research project Hexa-X, it is proposed to apply Key Performance Indicators as well as Key Value Indicators for such purposes.

Acknowledgments The author would like to sincerely thank Dr. Markus Mueck for his invaluable support and insightful contributions throughout the development of this book chapter. His expertise, guidance, and constructive feedback greatly enhanced the quality and clarity of this chapter.

Disclaimer Note that the authors have assembled the information above to the best of their knowledge as available at the time when the present book was written. Any interpretation is the personal opinion of the authors. Also, the regulation framework can change and evolve any time. The only legally binding text is the one provided in the documents formally published in the European Union Official Journal.

References

1. Purvis B, Mao Y, Robinson D (2019) Three pillars of sustainability: in search of conceptual origins. Sustain Sci 14:681–695. https://doi.org/10.1007/s11625-018-0627-5
2. Carson R (1962) The silent spring. Fawcett Publications, Greenwich
3. Meadows DH, Meadows DL, Jørgen S, William R, Ill WB (1972) Limits to growth. Universe Books
4. Nebel A, Kling A, Willamowski R, Schell T (2024) Recalibration of limits to growth: an update of the World3 model. J Ind Ecol 28:87–99. https://doi.org/10.1111/jiec.13442
5. Burton I (1987) Report on reports: our common future: the world commission on environment and development. Environ Sci Policy Sustain Dev 29(5):25–29. https://doi.org/10.1080/00139157.1987.9928891
6. United Nations (2000) United Nations millennium declaration. (A/RES/55/2). https://www.un.org/en/development/desa/population/migration/generalassembly/docs/globalcompact/A_RES_55_2.pdf
7. Rockström J, Steffen W, Noone K, Persson A, Chapin FS, Lambin E, Lenton TM et al (2009) Planetary boundaries: exploring the safe operating space for humanity. Ecol Soc 14(2):316
8. Rockström J, Steffen W, Noone K et al (2009) A safe operating space for humanity. Nature 461:472–475. https://doi.org/10.1038/461472a
9. Raworth K (2012) A safe and just space for humanity: can we live within the doughnut? Oxfam
10. MacArthur E (2013) Towards the circular economy. J Ind Ecol 2(1):23–44
11. United Nations (2015) Transforming our world: the 2030 Agenda for sustainable development. Resolution adopted by the General Assembly on 25 September 2015. A/RES/70/1
12. Communication from The Commission to The European Parliament, The European Council, The Council (2019) The European economic and social committee and the committee of the regions: the European green deal. COM(2019) 640 final. https://eur-lex.europa.eu/legal-content/EN/TXT/?uri=celex:52019DC0640. Accessed on 15 March 2025
13. Matos S, Viardot E, Sovacool BK, Geels FW, Xiong Y (2022) Innovation and climate change: a review and introduction to the special issue. Technovation 117:102612. https://doi.org/10.1016/j.technovation.2022.102612
14. Pan SL, Carter L, Tim Y, Sandeep MS (2022) Digital sustainability, climate change, and information systems solutions: opportunities for future research. Int J Inf Manag 63:102444. https://doi.org/10.1016/j.ijinfomgt.2021.102444
15. Reinitz B (2020) Consider the three Ds when talking about digital transformation. https://er.educause.edu/blogs/2020/6/consider-the-three-ds-when-talking-about-digital-transformation. Accessed 30 March 2022
16. Nahin PJ (2013) The logician and the engineer: how George Boole and Claude Shannon created the information age. Princeton University Press
17. Press G (2015) A very short history of digitization. Forbes https://www.forbes.com/sites/gilpress/2015/12/27/a-very-short-history-of-digitization/
18. Knell M (2021) The digital revolution and digitalized network society. Rev Evol Polit Econ 2:9–25. https://doi.org/10.1007/s43253-021-00037-4
19. Garner R (2013) STARS: early popular computers, 1950-1970 [scanning out past]. Proc IEEE 101(9):2134–2140. https://doi.org/10.1109/JPROC.2013.2270899

20. O'Regan G (2016) Minicomputers and later mainframes. In: Introduction to the history of computing, undergraduate topics in computer science. Springer, Cham. https://doi.org/10.1007/978-3-319-33138-6_9
21. Crow TS (2004) Evolution of the graphical processing unit. Master's thesis, University of Nevada, Reno, NV
22. Tanenbaum AS, Bos H (2015) Modern operating systems. Pearson Education, Inc.
23. Kleinrock L (2008) History of the internet and its flexible future. IEEE Wirel Commun 15(1):8–18. https://doi.org/10.1109/MWC.2008.4454699
24. Hillis WD (1985) The connection machine. MIT press
25. Fazi MB (2019) Can a machine think (anything new)? Automation beyond simulation. AI Soc 34:813–824. https://doi.org/10.1007/s00146-018-0821-0
26. Shor PW (1994) Algorithms for quantum computation: discrete logarithms and factoring. In: Proceedings 35th annual symposium on foundations of computer science, pp 124–134. https://doi.org/10.1109/sfcs.1994.365700
27. Arute F, Arya K, Babbush R et al (2019) Quantum supremacy using a programmable superconducting processor. Nature 574:505–510. https://doi.org/10.1038/s41586-019-1666-5
28. Salh A, Audah L, Shah NSM, Alhammadi A, Abdullah Q, Kim YH, Al-Gailani SA, Hamzah SA, Esmail BAF, Almohammedi AA (2021) A survey on deep learning for ultra-reliable and low-latency communications challenges on 6G wireless systems. IEEE Access 9:55098–55131. https://doi.org/10.1109/ACCESS.2021.3069707
29. Marcus Dapp (2004) ETH Zurich—a pioneer in digital sustainability! https://web.archive.org/web/20150721213621/ http://www.essays2030.ethz.ch/onlineversion/065-072.pdf. Accessed 1 May 2025.
30. Bradley K (2007) Defining digital sustainability. Libr Trends 56(1):148–163. https://doi.org/10.1353/lib.2007.0044
31. van Osch W, Avital M (2010) From green IT to sustainable innovation. In: AMCIS 2010 proceedings, p 490. https://aisel.aisnet.org/amcis2010/490
32. Molla A (2008) GITAM: a model for the adoption of green IT. In: ACIS 2008 proceedings, p 64. https://aisel.aisnet.org/acis2008/64
33. Chen AJ, Watson RT, Boudreau M-C, Karahanna E (2011) An institutional perspective on the adoption of green IS and IT. Austr J Inform Syst 17(1):572. https://doi.org/10.3127/ajis.v17i1.572
34. Coroama VC, Hilty LM, Birtel M (2012) Effects of internet-based multiple-site conferences on greenhouse gas emissions. Telematics Inform 29(4):362–374. https://doi.org/10.1016/j.tele.2011.11.006
35. Hilty LM, Aebischer B (2015) ICT for sustainability: an emerging research field. In: Hilty L, Aebischer B (eds) ICT innovations for sustainability. Springer, Cham, pp 3–36. https://doi.org/10.1007/978-3-319-09228-7_1
36. George G, Merrill RK, Schillebeeckx SJD (2021) Digital sustainability and entrepreneurship: how digital innovations are helping tackle climate change and sustainable development. Entrep Theory Pract 45(5):999–1027. https://doi.org/10.1177/1042258719899425
37. Rosati P, Lynn T, Kreps D, Conboy K (2024) Digital sustainability: key definitions and concepts. In: Lynn T, Rosati P, Kreps D, Conboy K (eds) Digital sustainability. Palgrave studies in digital business and enabling technologies. Palgrave Macmillan, Cham. https://doi.org/10.1007/978-3-031-61749-2_1
38. Hexa-X project web-pages. https://hexa-x.eu/
39. Hexa-X Deliverable D1.1: 6G Vision, use cases and key societal values. https://hexa-x.eu/wp-content/uploads/2021/02/Hexa-X_D1.1.pdf

Chapter 2
The European Green Deal

Markus Mueck, Stéphane Du Boispéan, and Kristof Claesen

Abstract The European Green Deal represents a generational initiative and comprehensive growth strategy with the objective to transform the European Union (EU) into a *fair and prosperous society, with a modern, resource-efficient and competitive economy where there are no net emissions of greenhouse gases in 2050 and where economic growth is decoupled from resource use.* Under the slogan *Fit for 55*, the European Commission has *adopted a set of proposals to make the EU's climate, energy, transport and taxation policies fit for reducing net greenhouse gas emissions by at least 55% by 2030, compared to 1990 levels.* The European Green Deal thus does not relate to a single political initiative but rather represents a framework under which a multitude of actions and legislations are being developed and addressed. Since the official communication by the European Commission on 11 December 2019, substantial progress has been achieved on a number of initiatives which will be discussed throughout the present chapter. Still, the effort is ongoing and is constantly adapted to the political environment, e.g. to the current trend on simplifying the regulatory landscape.

Keywords European green deal · European climate pact · European policy

M. Mueck (✉)
Intel Deutschland Gmbh, Neubiberg, Germany
e-mail: markus.muck@gmail.com

S. Du Boispéan · K. Claesen
Intel Corporation SA/NV, Kontich, Belgium
e-mail: Kristof.claesen@intel.com

C. Gaie and M. Mueck (eds.), *Digital Sustainability: Legislations, Standards, Solutions and Related Case Studies*, Advances in Sustainability Science and Technology,
https://doi.org/10.1007/978-981-92-0458-8_2

1 Introduction

The Communication of the Commission of 11 December 2019 officially introduced the *European Green Deal* [1]. Considering current climate challenges, it is understood to be a generational task to achieve the stated objectives, including the transformation of the European Union fostering a *modern, resource-efficient and competitive economy eliminating net emissions of greenhouse gases by 2050 and achieving a decoupling of growth from resource use.* In this communication, the European Commission furthermore announces the launch of a European Climate Pact, a corresponding communication was issued on 9 December 2020 [2, 3]. The objective of the Pact is to engage with concerned stakeholders and civil society with the objective to have a commitment related to climate action and a renewed behaviour based on sustainability principles.

The European Green Deal not only provides sector-specific regional policies to encourage environmentally friendly practices, but also fundamentally transforms society beyond environmental ambitions and affects production and consumption habits, labour market structures, skills needed, and where capital is spent. The European Green Deal gives priority to environmental issues in how things are created and how businesses and individuals interact with the world. Such changes will directly affect every day choices made by individuals and companies as the products individuals buy should have less of an impact on their environment, or every business action should minimize the impact of those actions on their environment. In developing an overall strategy which includes climate change, energy, mobility, industry, agriculture and financial systems, the overall result of the Green Deal is an integrated approach for establishing ecological sustainability in economic development, social sustainability and territorial development throughout Europe.

Politically, the European Green Deal was the result of complex political discussions and was not necessarily self-evident. In summer 2019, Ursula von der Leyen became President of the European Commission after weeks of discussions between the Heads of State and Governments of the European Union on the one hand, and the political groups in the European Parliament on the other hand. While being a conservative herself (from the centre-right European People's party), Ursula von der Leyen relied on the votes of three other political groups in order to secure her election as President: the centrist group of Renew, the centre-left Socialist & Democrats, and the Greens. This explains why the Green Deal was the first priority of the von der Leyen Commission between 2019 and 2024. It is to be understood as a political compromise between the different groups that constitute de facto the governing coalition in the European Parliament. This will explain why a Commission led by a centre-right figure focussed so much on the Green Deal.

Furthermore, the European Green deal needs to be seen in the context of related global initiatives. As a global leader, the European Union will interact with other regions and advocate for related objectives and actions. A specific focus is laid on the Paris Agreement [1, 4] as an indispensable multilateral contractual framework to address climate change. The European Union's objective to establish bilateral agreements, specifically focusing on the G7 and G20 as priority stakeholders. Further opportunities for influencing global debates include the 2020 Beijing and Leipzig EU-China summits, and the 2020 African Union and the EU summit.

The European Green Deal has very challenging long-term objectives. However, the authors maintain that the legitimacy and achievement of this regulation depend largely upon actual implementation. Different obstacles need to be managed. They require to pass from strategic initiatives into concrete actions. This explains why public and private stakeholders need to build coherent solutions across multiple policy areas, levels of government, and economy sectors. To this end, frameworks must be created that are able to coordinate (regulatory, financial, innovation, community engagement) measures consistently throughout the entire Green Deal. Therefore, the rest of this chapter will outline the way in which the regulation is now being implemented, with an emphasis upon identifying the key areas of action and the operational perspectives associated with each of them.

In this context, the Communication of the Commission [1] introduces the specific need for

1. Introducing deeply transformative policies; a total set of (eight) of such policies are identified and
2. Introducing targeted initiatives such that sustainability becomes mainstream in all EU policies; a set of (five) such initiatives are identified.

The concerned (eight) policies are summarized in Fig. 1. The related initiatives are summarized in Fig. 2.

Since the initial presentation of the European Green Deal on 11 December 2019, a substantial number of initiatives have been taken over time and more will follow.

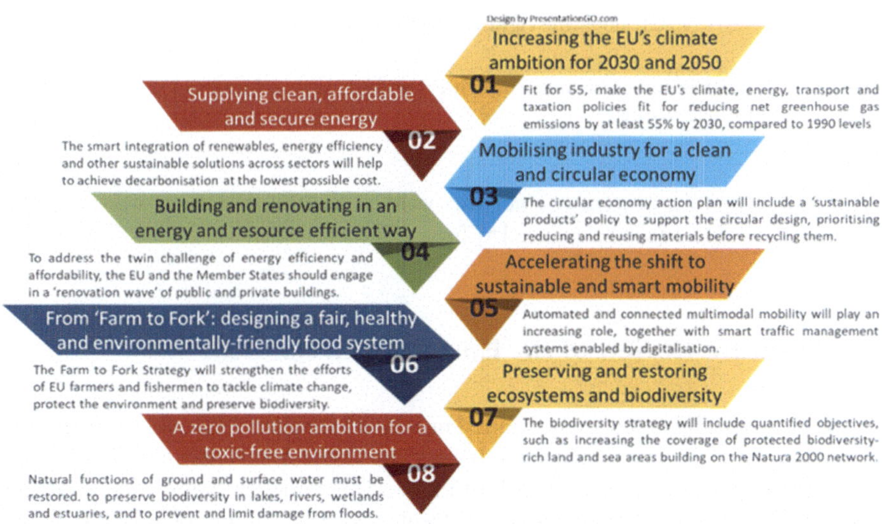

Fig. 1 Designing a set of (eight) deeply transformative policies [1]

Fig. 2 (Five) initiatives to ensure that sustainability becomes mainstreaming all EU policies [1]

Table 1 Initiatives for implementing the European Green Deal (selection)

Date	Initiative
18 August 2024	Entry into force of the **Nature Restoration Law** [6]
30 July 2024	Entry into force of the new **Directive on repair of goods** [7]
28 June 2024	The **net-zero industry act** enters into force [8]
17 June 2024	The Council adopts the **Nature restoration law** [9]
13 June 2024	New **regulation** is adopted ensuring **better and sustainable connectivity in Europe** [10]
30 May 2024	The Council adopts the **right-to-repair directive** [11]
28 May 2024	**Partnership on sustainable critical and strategic minerals** [12] signed by the EU and Australia

At the time when the present chapter was written, a series of actions were taken as illustrated in Table 1 (see reference [5] for a full list).

2 Literature Review of the European Green Deal

This section highlights the ideas detailed in important papers of the literature review concerning the European Green Deal. Interesting research was performed by Fetting [13] and concluded to a European Sustainable Development Network report. The report details how the European Green Deal provides a comprehensive approach based on environmental, economic, and social aspects. The report further explains that achieving successful long-term transformation requires both internal cohesion within institutions and effective coordination with other parties. Moreover, public and private actors must cooperate to translate EGD goals into practical action. Finally, the report underlines the importance of evaluating progress, which reinforces efforts towards achieving climate protection.

Other research by Siddi [14] highlights that the European Green Deal integrates and covers multiple sectors. As a matter of fact, it notably addresses the dimensions of climate, energy, industry, agriculture, and finance. The authors underline that the EGD does not treat these sectors independently but rather connects and aligns them with sustainability goals. It also underlines that climate neutrality serves as the central guiding principle. It is a central objective that defines transition strategies such as replacing fossil energy systems with renewable energy; promoting new mobility solutions or replacing traditional industrial or agricultural practices.

In order for the successful implementation of the European Green Deal to occur, many researchers concur that there needs to be sufficient governance structures throughout Europe (including EU institutions, Member States and the private sector) that can help transform the strategic aims contained in the Green Deal into tangible results at the local level. A strong emphasis should also be placed on providing a cohesive regulatory framework across all the various sectors where the European Green Deal is being planned; efficient financing through all available resources; as well as adequate participation by all relevant stakeholders [15]. Additionally, the literature highlights that because of the current fragmentation of research and analytical methods within Europe, there is currently a lack of large-scale system-level assessments of the performance of the implementation of the European Green Deal. Therefore, many authors are advocating for the creation of integrated, interdisciplinary frameworks backed by robust monitoring tools and standardized indicators so that comparisons can be made of the progress of the various regions and sectors towards the achievement of their goals.

According to scholars, the successful operationalization of the European Green Deal will necessitate the development of a strong monitoring framework, along with an adaptive management approach able to adequately capture advancements made in the various policy areas associated with the Deal. The study conducted by Olczyk and Kuc-Czarnecka [16] acknowledges these limitations regarding the current Green Deal assessment capability and presents a solution through the creation of the European Green Deal Index. This composite index is intended to allow for consistency in tracking the performance of each Member State against the other through the use of an integrated set of thematic components of the Green Deal. Thus, it will provide cross-country comparisons and longitudinal assessments. The authors also highlight that the development of these types of harmonized monitoring instruments is critical to developing adaptive governance, which will aid policymakers in tracking implementation dynamics, and modifying strategies throughout the EU.

The literature identifies, on a sectoral level, a multitude of not only opportunities but also barriers to the development and implementation of the European Green Deal across several pillars of European Union policy. For instance, the national transition pathways of the energy sector are influenced by existing infrastructure, concerns regarding energy security and financial capability, which, in turn, create a risk of uneven implementation progress among Member States. According to Tomaszewski [17], national energy policies may not align with Green Deal objectives due to structural dependence on fossil fuels, the need for high investment and social acceptance issues; therefore, to support the national transition pathways of the energy sector, there is a need for specific investment strategies and innovation support.

Additionally, for both agricultural and land use, an important factor in achieving the sustainability objectives is the need to balance environmental effectiveness with economic viability and social cohesion. According to Boix-Fayos and De Vente [18], current policy options have failed to successfully integrate environmental effectiveness with farmers' livelihoods and rural development needs. Hence, the authors recommend coherent and integrated policy options that simultaneously consider climate mitigation, ecosystem restoration, agricultural competitiveness and socio-economic resilience; therefore, policy analysis supports the need for context-specific policy design and cross-sectoral integration to address the barriers (technical, financial and socio-economic) to implementing the Green Deal.

3 The European Climate Pact

Under the umbrella of the European Green Deal, the European Climate Pact was initiated on 9 December 2020 [2, 19]. The objective is to foster a stronger link with and involvement of related stakeholders, involving regions, communities on the local level, industry stakeholders and schools. A set of pledges are expected to be designed by those stakeholders leading to a behavioural change across the entire chain, from the individual citizen to multinational organizations.

Participation to initiatives under the European Climate Pact are possible as illustrated in Fig. 3 [19].

Fig. 3 Participation to initiatives under the EU climate pact

Fig. 4 EU climate pact values [2]

Taking part in the EU Climate Pact requires registering a pledge to respect the Pact's values, which are illustrated in Fig. 4.

The EU Climate Pact thus allows for a hands-on involvement in the implementation of the European Green Deal. The role of a Pact Ambassador should be specifically highlighted. The role is open to anyone and applications are possible at the EU Climate Pact website. As an Ambassador under the EU Climate Pact, it is expected to actively lead in climate and environmental protection actions and to act as a linkage between civil society, stakeholders and the European Commission.

4 Implementing the European Green Deal

At the time when the present chapter was written, the European Commissions list nine key work areas [5] supporting the implementation of the European Green Deal. These work areas form part of a framework that allows for the effective execution of the European Green Deal (EGD). Together, they translate the EGD's overarching objectives into concrete and actionable strategies through a set of interrelated activities. The objective of combining regulatory, financial, technological and societal measures is to address the full spectrum of the EGD's climate transition process.

Moreover, these initiatives tend to create an informal platform where citizens, researchers, investors and policymakers naturally collaborate to solve concrete problems. These interactions 'from the ground' generate a virtuous circle that fosters innovation and supports the development and deployment of new technologies, thereby contributing to the achievement of the EGD's long-term sustainable growth objectives.

The nine work areas are described below.

4.1 Climate Action and the Green Deal

Achieving climate neutrality by 2050 is a stated key objective of the European Green Deal. The European Climate Law [20] is making the corresponding objectives legally binding including a net greenhouse gas emissions objective of—55% or more by 2030 compared to 1990 levels—the so-called *fit for 55* initiative.

To achieve the corresponding decarbonization objectives, an integrated approach is needed involving all sectors spanning across energy, industry, farming, transport and more. Furthermore, the European Union efforts need to be integrated into a global response. For this purpose, the European Union builds on the Framework Convention of Climate Change (UNFCCC) and its **Paris Agreement** to collaborate with partners across the globe on climate related actions. Complementing these initiatives, the European Union is implementing necessary steps for climate adaptation addressing the challenges of the unavoidable impacts of climate change.

The European Commission has published a Communication on 2040 climate targets in February 2024 [20, 21]. The objective is to achieve a 90% reduction of net greenhouse gas emissions by 2024 compared to the levels of 1990. At the same time, the European Commission is joining the European Climate Pact with the pledge to make its operations climate neutral by 2030.

The European Commission website related to Climate action and the Green Deal [22] introduces a set of seven actions which are currently taken to achieve the stated objectives. Those are summarized in Fig. 5.

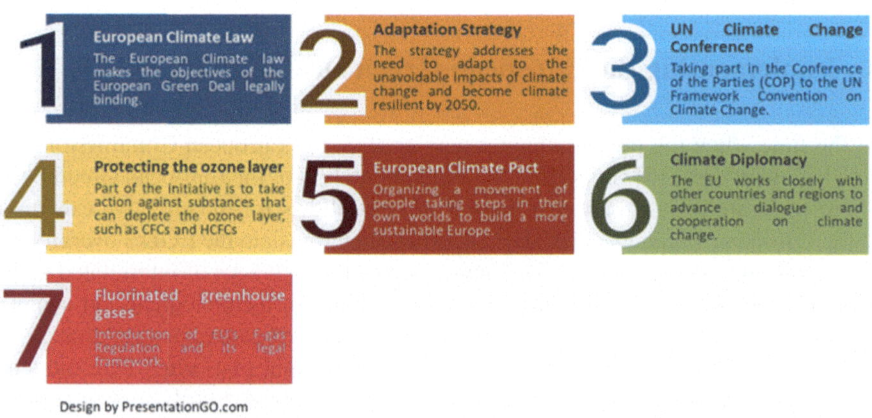

Fig. 5 Initiatives supporting the climate action and the Green Deal [22]

4.2 Energy and the Green Deal

Use of energy and its production account for over 75% of the greenhouse gas emissions of the European Union. To reach the European Union 2030 climate objective and the longer-term objective on achieving carbon neutrality by 2050, it is thus essential to decarbonize the European Union's energy system.

The European Green Deal introduces the following three key principles for a clean energy transition with the objective to reduce greenhouse gas emissions and to enhance the overall quality of life of European Union citizens:

- Ensuring the European Union energy supply to be affordable and secure;
- Implementing a European Union energy market that is fully integrated, interconnected and digitized;
- Prioritizing energy efficiency, with a focus on usage of renewable sources for the power sector and energy performance of buildings.

The European Commission has taken the following steps towards the stated objectives:

- Energy systems are becoming interconnected and renewable energy sources are prioritized for integrated grids;
- Modern infrastructure and technology innovation are being promoted;
- Increased importance of eco-design of concerned products and energy efficiency principles;
- Fostering cross-sectors smart integration and decarbonization of the gas sector;
- Supporting European Union countries to tackle energy poverty and empowering consumers correspondingly;
- Global promotion of European Union technologies and energy standards;
- Further developing Europe's offshore wind energy.

Upper steps will help the European Union to achieve its net greenhouse gas emissions objective, i.e. a reduction of 55% or more by 2030 compared to the levels of 1990 levels.

The European Commission website related to Energy and the Green Deal [23] introduces a set of six actions which are currently taken to achieve the stated objectives. Those are summarized in Fig. 6.

4.3 Protecting the Environment and Oceans with the Green Deal

The European Commission emphasizes the importance of preserving and protecting Europe's seas, oceans and environment which are a source of natural and economic wealth [24].

The European Green Deal introduces the following related priorities:

Fig. 6 Initiatives supporting the action on energy and the Green Deal [23]

- The ecosystems and biodiversity of the European Union require protection;
- Pollution of water, air and soil needs to be reduced;
- Further steps for transitioning towards a circular economy;
- Management of waste needs to be improved;
- Fisheries and blue economy sectors of the European Union need to reach sustainability objectives.

Upper steps will help the European Union to achieve its net greenhouse gas emissions objective, i.e. a reduction of 55% or more by 2030 compared to the levels of 1990 levels.

The European Commission website related to protecting the environment and oceans with the Green Deal [24] introduces a set of eleven actions which are currently taken to achieve the stated objectives. Those are summarized in Fig. 7.

4.4 Agriculture and Rural Development

As stated by the European Commission website [25], the European Union builds on 9.1 million farms in the European Union (as of 2020) providing highest levels of food quality and healthiness. Following actions supporting generational renewal in this sector, 377 thousand young farmers will engage in agricultural activities; The European Union has reached €228.6 billion in exports in 2023 with specific contributions through cereal preparations, dairy products and wine.

In the context of the European Green Deal, the European Commission takes action to foster sustainability, competitiveness and resilience of European Union agriculture, creating lively rural areas.

The European Union main objectives related to agriculture policy are summarized below:

- In support of farmers, actions will be taken to achieve higher revenue while producing safe food of sufficient quantity and respecting the rules of the European Union, specifically related to traceability, sustainability, animal welfare, etc.

Fig. 7 Initiatives supporting the action on protecting the environment and oceans with the Green Deal [24]

- Farm businesses require stabilization of their income while acknowledging often limited predictability of production conditions; corresponding support systems need to be introduced;
- The next generation needs to be attracted to engage in farming;
- The European Union requires a competitive, innovative, sustainable farming sector; a corresponding appealing investment framework needs to be established;
- Foster viable rural communities, with diverse economies;
- Jobs throughout the food chain need to be created and maintained.

The European Commission website related to agriculture and rural development [25] introduces a set of nine actions which are currently taken to achieve the stated objectives. Those are summarized in Fig. 8.

4.5 Transport and the Green Deal

The transport system [26] is a critical element in the European business environment and supporting global supply chains. Over 10 million people are employed in this sector in the European Union and it is responsible for approximately 5% of the European Union's GDP. On the other hand, negative effects of the transport systems also need to be acknowledged, including as the emission of greenhouse gas, noise, congestion, accidents on roads and other. Transport emission total approx. 25% of the European Union's overall greenhouse gas emissions with the trend being rather

Fig. 8 Initiatives supporting the action on agriculture and rural development [25]

on the rise. To meet the European Union's objective to become climate neutral by 2050 requires ambitious changes across the transport system to achieve at least a 90% reduction in transport related greenhouse gas emissions by 2050.

With the so-called *fit for 55* and other initiatives, the European Commission has adopted a set of proposals to make the EU's transport related policies fit for reducing net greenhouse gas emissions.

The objectives are being addressed through a strategy focusing on sustainability and resilience of smart mobility solutions.

The plans on Sustainable Mobility target the following [26]:

- A minimum of 30 million cars and 80 thousand lorries meeting zero-emissions requirements by 2030;
- A minimum 100 cities and communities within the European Union meeting climate neutrality requirements by 2030;
- Scheduled collective travel under 500 km to meet carbon neutrality requirements by 2030 within the European Union;
- Large aircraft meeting zero-emission requirements to become ready for market by 2035;
- Sustainability based urban mobility plans to be implemented by large and medium-sized cities by 2030;
- Doubling traffic on high-speed rail by 2030;
- Doubling rail freight traffic by 2030;
- A 25% increase of transport by short sea shipping and inland waterways by 2030.

The plans on Smart Mobility target the following [26]:

- Enabling seamless multimodal passenger transport through integrated electronic ticketing by 2030;
- Paperless freight transport;
- Large-scale deployment of automated mobility by 2030.

Fig. 9 Initiatives supporting the action on transport and the Green Deal [27]

The plans on Resilient Mobility target the following [26]:

- Transport infrastructure related investments across the European Union Member States;
- Operation of a multimodal Trans-European Transport Network based on principles of sustainability and smart transport solutions enabled by high-speed connectivity by 2050;
- Mobility provided for all passengers and across all based on principles of affordability and accessibility;
- Improvement of conditions for transport workers;
- Achieving a close to zero death toll for all modes of transport in the European Union by 2050.

The European Commission website related to transport and the Green Deal [27] introduces a set of two actions which are currently taken to achieve the stated objectives. Those are summarized in Fig. 9.

4.6 Industry and the Green Deal

With the European Green Deal representing the European Union's new strategy for growth [28]—while targeting a fairer and more prosperous society based on a modern economy relying on principles of resource efficiency and competitiveness and avoiding greenhouse gas emissions by 2050—it is essential to take advantage of the potential in global markets in terms of sustainability of products and services and technologies of low levels of emissions. All aspects of industry are concerned by this vision and involving the entirety of industrial value chains, especially including energy intensive sectors.

This new vision on an European Industrial Strategy will enable green and digital transitions and contribute to the European Union's global competitiveness. It is understood that the reduction of the carbon footprint requires support to further develop and deploy clean and affordable technology solutions and to adapt related business models. Building on the lessons learned from the COVID-19 pandemic, the industry of the European Union will be in a position to lead accelerated green and digital transitions.

Fig. 10 Initiatives supporting the action on industry and the Green Deal [28]

The European Commission is actively taking steps to achieve its net greenhouse gas emissions objective, i.e. a reduction of 55% or more by 2030 compared to the levels of 1990 levels.

In February 2024, the European Union Industrial Carbon Management Strategy [29] was adopted by the European Commission. It employs sustainability principles to support technologies to capture, store and re-use carbon, thus contributing significantly to the target of climate neutrality by 2050. In this context, carbon capture technologies need to be scaled up and the required infrastructure needs to be created for a Single Market for CO_2 in Europe.

The European Commission website related to Industry and the Green Deal [28] introduces a set of six actions which are currently taken to achieve the stated objectives. Those are summarized in Fig. 10.

4.7 Research and Innovation for the European Green Deal

Under the European Green Deal objective to become the first climate-neutral continent by 2050, it is a generational opportunity to modernize the economy and society of the European Union and to reorient those towards a fairer and more sustainable future.

In this context, research and innovation will play a central role [30], specifically.

- To provide support enabling appropriate navigation through and acceleration of the necessary transitions;
- To support the deployment, demonstration and de-risking of appropriate solutions;
- To facilitate personal engagement of citizens in social innovation.

The European Union has established the current Horizon Europe framework to provide substantial funding for research and innovation activities. Funding and Tenders Portal offer the possibility for participation. Under the Green Deal call, the European Commission has made one billion Euros available to support related research and activities. Among all submitted project proposals, 73 projects were selected for funding. They are expected to contribute to the European Union's response to the climate crisis and help protect the unique ecosystems and biodiversity of Europe.

Horizon Europe is the European Union's research and innovation programme established over the time period 2021–2027. At the time when the present chapter was written, initial discussions were starting to shape the follow-up programme. Horizon Europe will be instrumental to support climate neutrality objectives and to implement a transition based on principles of ecological and economic inclusiveness. The mechanisms provided by Horizon Europe will be combined with other Programmes of the European Union thus leveraging public and private investment. As a result, new technologies, solutions and disruptive innovation according to principles of sustainability will be developed and spread across Europe and world-wide. Projects related to climate change are thus of specific importance in Horizon Europe, representing 35% of the available budget.

New research and innovation partnerships are being introduced under Horizon Europe with the objective to support the implementation of the European Green Deal. A specific focus is laid on the fields of society, economy and environment. A close working relationship has been set up by the European Union with industry and countries to support partnerships multiple areas including transport, low carbon steel, clean hydrogen, circular bio-based sectors, the built environment and biodiversity.

The European Union missions in Horizon Europe aim to capture citizens' imagination and inspire confidence in the transformations and challenges ahead. Four out of the five mission areas in Horizon Europe support the European Green deal by

- Restoration of the ocean and waters of the European Union;
- Transitioning towards smart cities operating according to principles of climate neutrality;
- Establishing a European soil deal;
- Adaptation and transformation of the society according to needs of climate change.

In the context of the European Green Deal, it is essential to allow for new and out-of-the-box ideas to support a sustainable future. The Innovation Principle [31] is a promising approach to be used for serving this purpose, covering the entire three phases of the policy making cycle: (i) agenda-setting, (ii) legislation and (iii) implementation.

4.8 Finance and the Green Deal

Over the next decade, the European Commission's intention is to provide a minimum of one trillion Euros of sustainable investment to support the goals set out in the European Green Deal. Those investments include 30% of the European Union's multiannual budget (2021–2028) and the European Union's NextGenerationEU (NGEU) instrument for the recovery from the COVID-19 pandemic.

European Union countries are obliged to allocate a minimum of 37% of the financing they receive under the 672.5 billion Euros Recovery and Resilience Facility such that investments and reforms do not create significant harm to related climate objectives of the European Union. Green bonds will be issued by the European Commission to raise 30% of the funds under NextGenerationEU (NGEU).

European Union countries, regions, local governments and cities are supported by the Cohesian Policy of the European Union to implement large European Green Deal related investments. It is required that they devote a minimum of 30% of the funds received from the European Regional Development Fund correspondingly. Additionally, 37% of the Cohesion Fund will be allocated to support the objective to achieve climate neutrality by 2050.

The Green Deal Investment Plan (EGDIP) has been introduced by the European Commission as part of the Green Deal—also called Sustainable Europe Investment Plan (SEIP). The Just Transition Mechanism is part of this initiative focusing on a fair and just transition to a green economy. Significant investments will be triggered over the period 2021–2027 supporting citizens of regions most impacted by the transition.

Further contributions to this initiative will be provided by the InvestEU programme, including long-term funding by leveraging substantial private and public funds. The InvestEU fund is mandated by the InvestEU Regulation to target a minimum of 30% of investment related to climate objectives. A so-called Just Transition Scheme is implemented with the objective to generate additional investment supporting Just Transition territories, thus complementing the Just Transition Fund and the loan facility of the public sector. The Taxonomy Regulation provides additional finance measures for classifying green investments. Private sector investments in green and sustainable projects will be accelerated by those measures, thus contributing to the European Green Deal.

Upper steps will help the European Union to achieve its net greenhouse gas emissions objective, i.e. a reduction of 55% or more by 2030 compared to the levels of 1990 levels.

The European Commission website related to Finance and the Green Deal [32] introduces a set of six actions which are currently taken to achieve the stated objectives. Those are summarized in Fig. 11.

Fig. 11 Finance and the Green Deal [32]

4.9 New European Bauhaus

The New European Bauhaus (NEB) policy and funding initiative targets green transitions in built environments and beyond, ensuring that they are enjoyable, attractive and convenient for all. It is acknowledged that even the smallest communities deserve living spaces improving the well-being and sense of belonging of citizens. Corresponding solutions will be implemented based on principles of sustainability, inclusiveness and beauty while respecting diversity of places, traditions and cultures.

The New European Bauhaus (NEB) builds on the following:

- To engage citizens at a grassroots level, focusing on neighbourhoods;
- To provide tools and guidance;
- To offer tailor-made solutions for different communities;
- To incorporate diverse views into the process of design and implementation;
- To prioritize citizens and social inclusion but also the economy to accelerate competitiveness and EU strategic autonomy.

5 Criticism

The focus of the European Commission on environment policy was from the beginning not always a consensual approach, including within the own party of President von der Leyen (European People's Party, EPP). Over the years of the 2019–2024 mandate and as other issues starting to arise (cost of living, purchasing power, industrial competitiveness), the EPP slowly shifting from a support to criticism of the Green Deal [33]. The symbol of this change of attitude is the 2023 vote on the legislation dealing with restoration of nature, when a part of the EPP in the European Parliament voted against the legislation, together with other political groups from the right and against their traditional coalition partner from the centre and centre left.

This was followed by a shift of rhetoric during the 2024 campaign for the European elections and since then, with some prominent policymakers, including President von der Leyen herself, switching from a 'Green Deal' to a policy using the wording 'Clean industry'. This seems to indicate that the priority is now more in investing in some industrial technologies that are deemed essential for the Green transition. In other words, the priority seems to be more on the industrial investment and slightly less on regulation. The results of the European elections in 2024, with a light shift of the European Parliament more to the right of the EPP, confirms also this tendency. At the same time, the centre and centre left coalition partners of the EPP, which renewed the 2019 coalition, insisted on keeping the Green Deal acquis in place [34].

At the same time, the question of industrial competitiveness started to be raised more frequently in the EU political discussions, with some industry groups complaining about the lack of supportive framework to invest in the EU, and others insisting on keeping the Green Deal acquis. It seems that the 2024–2029 mandate of the von der Leyen Commission will be influenced by these tensions: keeping the

Green Deal acquis, strengthening industrial competitiveness and investing in new industries that support the Green Deal objectives.

As a consequence, there are three categories of concerns regarding the implementation of the European Green Deal. First, many actors within Europe have expressed alarm about the costs they will have to bear in order to meet the new requirements established by the regulatory guidelines. For example, small and medium-sized businesses, in addition to being particularly vulnerable to market fluctuations, are concerned that they may simply not be able to absorb all these new expenses. Additionally, the current situation of global conflict adds further complications for these businesses. Indeed, the imposition of new regulations may lead to countermeasures, such as increases in tariff rates on exports.

From a social perspective, the acceptability of Green Deal measures has emerged as a growing challenge. Tensions have become visible in sectors such as agriculture, transport and housing, where the effects of the transition are perceived as uneven. These dynamics have contributed to a form of 'green backlash', underlining the importance of accompanying climate policies with effective just transition mechanisms and social compensation measures.

Finally, institutional criticisms point to the complexity of the regulatory landscape and to the difficulties associated with implementation at national and regional levels. The risk of uneven application across Member States raises concerns about a fragmented internal market and a potential 'multi-speed' Green Deal, reinforcing the need for coordination, simplification and stronger implementation capacity.

6 The Green Deal Revisited—The New 'Clean Industrial Deal'

On 18 July 18, 2024, Commission President Ursula von der Leyen presented her political guidelines [35] for her second term at the helm of the European Commission, entitled 'Europe's Choice'. In the guidelines, von der Leyen argues that the EU '*must and will stay the course on the goals set out in the European Green Deal*'. However, she also acknowledges that Europe needs a '*new Clean Industrial Deal for competitive industries and quality jobs*'. The Clean Industrial Deal complements, and in some cases refocuses, the Commission's 2019 Green Deal.

Indeed, while von der Leyen states that the EU will remain committed to achieving the Green Deal's ambitions, it is clear that the 'Clean Industrial Deal' brings in a renewed focus on competitiveness, reflective of the changing political climate in Europe.

The Clean Industrial Deal [36] was presented on February 26, 2025, and is the European Commission's answer to '*rising geopolitical tensions, slow economic growth and technological competition*' that Europe is facing. The goal of the Clean Industrial Deal is therefore to bring together '*climate action and competitiveness under one overarching growth strategy*'. The ambition of the Commission is to mobilize over one hundred billion Euros to support EU manufacturing and create 500 thousand new jobs.

The Commission outlines that the Deal provides measures to support all stages of the production process, with a focus on:

- industries that are energy-intensive, e.g. steel, metals and chemicals, which require urgent support to decarbonize, to operate according to clean energy principles and to address high costs, unfair competition on a global level and complex regulations;
- the clean-tech sector representing the core of future competitiveness and being necessary for the transformation of industry, and for achieving circularity and decarbonization.

The main elements of the Clean Industrial Deal can be summarized as follows:

- A specific Affordable Energy Action Plan to address the need for affordable energy;
- An industrial Decarbonization Accelerator Act and a review of the Public Procurement Framework will strengthen demand for clean products;
- Financing the clean transition, by mobilizing one hundred billion Euros in support of clean manufacturing in the European Union;
- Several steps will be taken to support circularity and access to materials, including the establishment of a system for companies to aggregate demand for Critical Raw materials, of an EU centre for critical raw materials to jointly purchase raw materials and a Circular Economy Act with the objective to accelerate the transition to a circular economy;
- Impact on a global scale will be achieved through Clean Trade and Investment Partnerships, the introduction of trade defence and other instruments and by strengthening the carbon border adjustment mechanism;
- A union of skills and reinforcing education and training programmes will be established to provide skills and quality jobs.

The European Commission will roll out the necessary initiatives in the years ahead to deliver on the ambitious Clean Industrial Deal. On 2 July 2025, the Commission presented their first overview of the progress on delivering on the Clean Industrial Deal [37]. In the 'First Package of Clean Industrial Deal Delivery', the Commission states that '*progress has been made in all six core business drivers .... The measures strike a balance ..., so that Member States can provide targeted financial support to decarbonisation efforts and for building sovereign value chains ...*'.

Looking ahead, '*the Commission remains steadfast in its commitment to deliver on the Clean Industrial Deal and on making decarbonisation a driver of Europe's competitiveness and prosperity, ... It will deliver the rest of Clean Industrial Deal actions .... Each package will be ... reinforcing collective efforts to enhance Europe's industrial competitiveness and progress towards the EU's decarbonisation goals*'.

The full impact of the 'industrial twist' to the Green Deal provided by the Clean Industrial Deal remains to be seen as the Commission continues to implement the Clean Industrial Deal initiatives, while navigating the tension between its strong environmental commitments and its renewed focus on competitiveness.

7 Conclusions

As a generational initiative and comprehensive growth strategy, the European Green Deal is serving as a guiding compass for European Policy making and has led to the implementation of key initiatives such as deeply transformative policies as illustrated in Fig. 1 of the present chapter. These are complemented by initiatives for mainstreaming sustainability in all EU policies as illustrated in Fig. 2. To foster a stronger link with and involvement of citizens, communities, etc., the European Climate Pact was initiated on 9 December 2020. The Green Deal is not only a set of sector-specific regulations, but a broad societal transformation, influencing production and consumption patterns, labour markets, skills development and investment priorities across the EU.

The actual implementation of the Green Deal is organized through (currently nine) key work areas. All of these initiatives are designed to support the objective to transform the European Union (EU) into a *fair and prosperous society, … where there are no net emissions of greenhouse gases in 2050 and where economic growth is decoupled from resource use.* Still, criticism emerges related to a too strong reliance on regulation measures. It is indeed observed that the level of action has recently shifted towards strengthening industrial competitiveness and investing in new industries. Some further challenges include ensuring social acceptance and guaranteeing the operational applicability of climate regulations.

The applicability of the Green Deal and its industrial successor will depend on the clarity of the legislative output as well as the effort put into financial investment. The ability of the EU to maintain public support and global leadership will also be fundamental. Thus, the success of the Climate Pact will be measured by its impact on long-term climate targets as well as on the concrete benefits for citizens. For instance, it should enable access to affordable energy, quality jobs, and sustainable supply chains. Finally, the Green Deal can only succeed through practical adaptation, the promotion of green innovations, and the active engagement of society as a whole to turn ambitious goals into concrete results.

Even Small Steps Forward Are Preferable to Immobility.

Acknowledgements The authors would like to sincerely thank Dr. Christophe Gaie for his invaluable support and insightful contributions throughout the development of this book chapter. His expertise, guidance and constructive feedback greatly enhanced the quality and clarity of this chapter.

Disclaimer Note that the authors have assembled the information above to the best of their knowledge as available at the time when the present book was written. Any interpretation is the personal opinion of the authors. Also, the regulation framework can change and evolve any time. The only legally binding text is the one provided in the documents formally published in the European Union Official Journal.

References

1. Communication from the Commission (2019) The European green deal. COM, p 640. https://eur-lex.europa.eu/legal-content/EN/TXT/HTML/?uri=CELEX:52019DC0640
2. Communication from the Commission to the European Parliament, the Council, the European Economic and Social Committee and the Committee of the Regions European Climate Pact (2020) European commission. https://eur-lex.europa.eu/legal-content/EN/TXT/?uri=COM%3A2020%3A788%3AFIN
3. The European Climate Pact, European Commission. https://climate-pact.europa.eu/index_en
4. The Paris Agreement, United Nations. https://www.un.org/en/climatechange/paris-agreement
5. The European Green Deal Striving to be the first climate-neutral continent, European Commission. https://commission.europa.eu/strategy-and-policy/priorities-2019-2024/european-green-deal_en
6. Degraded ecosystems to be restored across Europe as Nature Restoration Law enters into force, European Commission. https://environment.ec.europa.eu/news/nature-restoration-law-enters-force-2024-08-15_en
7. Directive on repair of goods, European Commission. https://commission.europa.eu/law/law-topic/consumer-protection-law/directive-repair-goods_en
8. Regulation (EU) 2024/1735 of the European Parliament and of the Council of 13 June 2024 on establishing a framework of measures for strengthening Europe's net-zero technology manufacturing ecosystem and amending Regulation (EU) 2018/1724 (Text with EEA relevance). https://eur-lex.europa.eu/legal-content/EN/TXT/?uri=CELEX%3A32024R1735
9. Nature restoration law: Council gives final green light, European Commission. https://www.consilium.europa.eu/en/press/press-releases/2024/06/17/nature-restoration-law-council-gives-final-green-light/
10. Trans-European transport network (TEN-T): Council gives final green light to new regulation ensuring better and sustainable connectivity in Europe, European Commission. https://www.consilium.europa.eu/en/press/press-releases/2024/06/13/trans-european-transport-network-ten-t-council-gives-final-green-light-to-new-regulation-ensuring-better-and-sustainable-connectivity-in-europe/
11. Circular economy: Council gives final approval to right-to-repair directive, European Commission. https://www.consilium.europa.eu/en/press/press-releases/2024/05/30/circular-economy-council-gives-final-approval-to-right-to-repair-directive/
12. EU and Australia sign partnership on sustainable critical and strategic minerals, European Commission. https://ec.europa.eu/commission/presscorner/detail/en/ip_24_2904
13. European Sustainable Development Network (ESDN) (2020) The European green deal: an overview of current policy and implementation status. ESDN Report No. 2. https://www.esdn.eu/fileadmin/ESDN_Reports/ESDN_Report_2_2020.pdf
14. Siddi M (2021) The European green deal: asseasing its current state and future implementation. UPI REPORT 114:14
15. Ozcelik N, Rey-García M, Mato-Santiso V (2025) Enablers and barriers to European green deal implementation: a systematic review and framework proposal. Sustain Product Consump 61:164–180. https://doi.org/10.1016/j.spc.2025.10.018
16. Olczyk M, Kuc-Czarnecka M (2025) European green deal index: a new composite tool for monitoring European Union's green deal strategy. J Clean Prod 495:145077. https://doi.org/10.1016/j.jclepro.2025.145077
17. Tomaszewski K (2020) The polish road to the new European green deal: challenges and threats to the national energy policy. Polityka Energetyczna Energy Policy J 23(2):5–18
18. Boix-Fayos C, De Vente J (2023) Challenges and potential pathways towards sustainable agriculture within the European green deal. Agric Syst 207:103634. https://doi.org/10.1016/j.agsy.2023.103634
19. The European Climate Pact, European Commission. https://climate-pact.europa.eu/about/about-pact_en

20. European Climate Law, European Commission. https://climate.ec.europa.eu/eu-action/european-climate-law_en
21. 2040 climate target, European Commission. https://climate.ec.europa.eu/eu-action/climate-strategies-targets/2040-climate-target_en
22. Climate action and the Green Deal, European Commission. https://commission.europa.eu/strategy-and-policy/priorities-2019-2024/european-green-deal/climate-action-and-green-deal_en
23. Energy and the Green Deal, European Commission. https://commission.europa.eu/strategy-and-policy/priorities-2019-2024/european-green-deal/energy-and-green-deal_en
24. Protecting the environment and oceans with the Green Deal, European Commission. https://commission.europa.eu/strategy-and-policy/priorities-2019-2024/european-green-deal/protecting-environment-and-oceans-green-deal_en
25. Agriculture and rural development, European Commission. https://commission.europa.eu/topics/agriculture-and-rural-development_en
26. The Transport and Mobility Sector, European Commission. https://ec.europa.eu/commission/presscorner/detail/en/fs_20_2350
27. Transport and the Green Deal, European Commission. https://commission.europa.eu/strategy-and-policy/priorities-2019-2024/european-green-deal/transport-and-green-deal_en
28. Industry and the Green Deal, European Commission. https://commission.europa.eu/strategy-and-policy/priorities-2019-2024/european-green-deal/industry-and-green-deal_en
29. Questions and Answers on the EU Industrial Carbon Management Strategy, European Commission. https://ec.europa.eu/commission/presscorner/detail/en/qanda_24_586
30. Research and innovation for the European Green Deal, European Commission. https://research-and-innovation.ec.europa.eu/strategy/strategy-research-and-innovation/environment-and-climate/european-green-deal_en
31. Ensuring EU legislation supports innovation, European Commission. https://research-and-innovation.ec.europa.eu/law-and-regulations/ensuring-eu-legislation-supports-innovation_en
32. Finance and the Green Deal, European Commission. https://commission.europa.eu/strategy-and-policy/priorities-2019-2024/european-green-deal/finance-and-green-deal_en
33. The year the EPP pushed the EU from 'green' to 'clean', Euractiv, December 2024. https://www.euractiv.com/section/eet/news/the-year-the-epp-pushed-the-eu-from-green-to-clean/
34. Our European Green Deal, Socialists and Democrats Group. https://www.socialistsanddemocrats.eu/our-european-green-deal
35. Europe's Choice, Political Guidelines for the Next European Commission 2024–2029. https://commission.europa.eu/document/download/e6cd4328-673c-4e7a-8683-f63ffb2cf648_en?filename=Political%20Guidelines%202024-2029_EN.pdf
36. The clean industrial deal: a joint roadmap for competitiveness and decarbonisation. https://eur-lex.europa.eu/legal-content/EN/TXT/?uri=CELEX%3A52025DC0085
37. Delivering on the Clean Industrial Deal I. https://commission.europa.eu/document/download/ae2ea9ea-d037-4920-bbf6-a4183b747e34_en?filename=COM_2025_378_1_EN_ACT_part1_v5.pdf

Chapter 3
The Ecodesign for Sustainable Products Regulation and Energy Labelling Regulation

Markus Mueck, Kristof Claesen, and Stéphane Du Boispéan

Abstract The European Green Deal is a major framework under which a number of legislations have been initiated and maintained. While it is challenging to provide a comprehensive analysis of all components of the Green Deal, we choose to focus specifically on the *Ecodesign for Sustainable Products Regulation* and *Energy Labelling Regulation*. Both regulations are key for stakeholders to understand sustainability-related requirements for accessing the European Union Single Market. Specifically, the Ecodesign Regulation introduces a number of product categories (so-called *Lots*) which were originally introduced for clarifications on a working level but in the mean-time are broadly applied across industry. It is explained how the list of such product categories evolves over time according to so-called *Ecodesign Working Plans*. For identified product categories, specific legislations are being introduced to address market access requirements to be fulfilled by manufacturers.

Keywords Regulation · Ecodesign · Circular economy · Market access · *European* Union single market

1 Introduction

Among the many legislations relating to the European Green Deal, we choose to specifically focus on the *Ecodesign for Sustainable Products Regulation* [1] (repealing the prior Directive 2009/125/EC [2]) and the *Energy Labelling Regulation* [3]. These are essential legislations for manufacturers with compliance being required to place products on the European Union Single Market.

M. Mueck (✉)
Intel Deutschland GmbH, Neubiberg, Germany
e-mail: markus.muck@gmail.com

K. Claesen · S. Du Boispéan
Intel Corporation SA/NV, Kontich, Belgium
e-mail: Kristof.claesen@intel.com

C. Gaie and M. Mueck (eds.), *Digital Sustainability: Legislations, Standards, Solutions and Related Case Studies*, Advances in Sustainability Science and Technology,
https://doi.org/10.1007/978-981-92-0458-8_3

The history of "ecodesign" can be traced back to the late 1980's (see for instance Hübner [4]) when not only the environmental impact of production processes but also of products themselves became the subject of regulatory scrutiny. In the EU, for instance, laws were introduced to improve energy efficiency of specific products, such as boilers (92/42/EEC) or refrigerators and freezers (96/57/EC) as of the early 1990's. In this context, the EU had introduced a "Community Ecolabel" in 1992. This voluntary mark for products that meet high environmental standards still exists today as the "EU Ecolabel".

However, the clearest step towards ecodesign policy as we know it today might be the 2001 Commission's Green Paper on Integrated Product Policy [5]. The intention of the Integrated Product Policy was to reduce products' environmental impact across their life cycle, focusing primarily on "ecodesign of products and the creation of information and incentives for an efficient take-up and use of greener products". The Green Paper was followed in 2003 by a Commission Communication entitled "Integrated Product Policy—Building on Environmental Life-Cycle Thinking" [6], which further enshrined the regulatory focus on a product's entire life cycle.

The Integrated Product Policy eventually led to the creation of the 2005 Directive establishing a framework for ecodesign requirements for energy-using products (2005/32/EC). The ambition of the ecodesign framework was twofold: on the one hand the Commission wanted to be able to set ecodesign requirements to reduce product environmental footprints through product-specific implementing measures while, on the other hand, the Directive also aimed to avoid market fragmentation due to potentially diverging national requirements and therefore to ensure the free movement of compliant products within the EU internal market.

In 2009, the Commission updated the 2005 Directive (a "recast") to further strengthen, clarify, and develop the EU's ecodesign legal framework. In the following years, this "Ecodesign Directive (2009/125/EC)" became the basis for a range of product specific ecodesign requirements developed and adopted by the European Commission. As explained by the Commission, the Ecodesign Directive is key for the "*strategy on Integrated Product Policy" and will lead to "new opportunities for manufacturers, consumers and society* ..." [7].

Conformity with the product-specific ecodesign requirements is a prerequisite for the CE marking and as such for products being allowed to be placed on the EU market. As the 2009 Directive states, "*before a product … is placed on the market …, a CE marking shall be affixed and an EC declaration of conformity issued whereby the manufacturer … ensures … that the product complies with all relevant provisions* ....." Furthermore, all products (i.e. units) placed on the EU market need to comply once the regulation becomes effective, which means that older, previously designed, non-compliant products either need to be redesigned or they can no longer be sold after the rules become effective.

Alongside the 2009 "Ecodesign Directive", which allowed the Commission to set product specific requirements, the EU also adopted a Directive specifically on energy labelling (2010/30/EU). This Energy Labelling Directive enabled the Commission to define energy labelling requirements for specific product groups (i.e. the well-known A-G energy ratings displayed in stores and included with products across

Europe) and can be seen as a complement to the "Ecodesign Directive". This Energy Labelling Directive has seen several amendments and updates, in particular through the Energy Labelling Framework Regulation of 2017 (2017/1369).

As of 2024, the Ecodesign for Sustainable Products Regulation (2024/1781) replaces the 2009 Ecodesign Directive. It represents the latest update to the EU's ecodesign framework. The Ecodesign for Sustainable Products Regulation (ESPR) in combination with the Energy Labelling Framework Regulation will serve as baseline for upcoming ecodesign initiatives, with a focus on product sustainability and energy efficiency which are entering the European Union market.

On a high level, the *Ecodesign for Sustainable Products Regulation* [1] is creating a framework for "*Ecodesign requirements …, … improving the environmental sustainability of products in order to make sustainable products the norm…*" which will lead to a reduction of the "*overall carbon footprint and environmental footprint of products …*" [1]. In this context, the term *ecodesign* is defined as follows [1]:

> *"Ecodesign" means the integration of environmental sustainability considerations into the characteristics of a product and the processes taking place throughout the product's value chain.*

Complementing the Ecodesign Regulation, the *Energy Labelling Framework Regulation* [3] is applying to "*energy-related products …*" providing for "*the labelling of those products*" with a specific focus on energy … *regarding energy efficiency, the consumption of energy and of other resources … enabling customers to choose more efficient products in order to reduce their energy consumption*". In this context, the term energy-related product is defined as follows [2]:

> *"energy-related product" or "product" means a good or system with an impact on energy consumption …, including parts with an impact on energy consumption … intended to be incorporated into products.*

As illustrated in Fig. 1, energy labels are intended to help consumers choose energy efficient products and as a result drive manufacturers towards innovations that will lead to more energy efficient technologies. According to the EU, the label is "recognised by 93% of consumers and 75% use it to get informed when buying labelled products" [8]. The label provides an energy efficiency rating (e.g. A–G) as well as supplementary information, for instance a water ingress protection rating or reliability indicators. The label typically also links to the European Product Database for Energy Labelling (EPREL), where more detailed information can be found. The label, which provides ratings and comparisons for consumers, complements the ecodesign requirements which focus on minimum requirements for products entering the EU market.

Both the Ecodesign for Sustainable Product Regulation (ESPR) and the Energy Labelling Framework Regulation provide the overarching framework for the Commission to adopt product-specific requirements and, in case of the ESPR, also horizontal requirement relevant for a number of different product groups (see below). As the product-specific regulations often include detailed technical requirements, these regulations rely on standardisation processes to provide clarity and consistency on, for instance, detailed measurement methodologies or technical definitions.

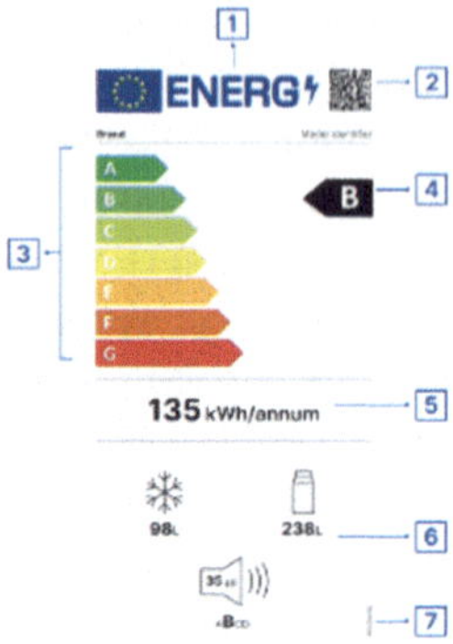

1. **The language-neutral logo.** The bolt replacing the last letter in the word ENERG replaces a desinence that varies across the various EU languages (e.g. Energy, Energie, Energia, Energía, Énergie, Energija, Enerģija, Energi, etc.)
2. QR code with link to EPREL
3. Energy efficiency classes (with a scale from A to G)
4. Energy efficiency class of this product model
5. Energy consumption
6. Indication of additional non-energy parameters (noise emissions, water consumption, capacity, repairability, reliability class, etc.)
7. Reference to the regulation

Fig. 1 Understanding the energy label [8]

Indeed, both the *Ecodesign for Sustainable Products Regulation* and *Energy Labelling Regulation* are aligned with the New Legislative Framework (NLF) [9, 10]. This alignment provides a unique opportunity for all stakeholders to contribute to the implementation process of the Regulation as illustrated in Fig. 2. While the legislation itself set high-level principles, it also established the fact that the European Standardisation Organisations (ESOs) are tasked with writing the standards that implement the essential requirements of harmonised legislations. Once the harmonised standards are referenced, they are published in the Official Journal (OJ) of the European Union and listed in a related database [11]. Further information and a more detailed introduction to the NLF are available in [10].

By following the standard, a presumption of conformity with the harmonised legislations exists. Currently, the European Commission is contemplating to further strengthen the use of Common Specifications; this trend indeed triggers concern especially among industrial and other stakeholders, because it is unclear to which extent stakeholders may be able to contribute to the detailed market access requirements (as it is currently possible by contributing to the development of Harmonised Standards within European Standardization Organizations) under this alternative regime.

In the following sections, we propose a state-of-the-art of the literature concerning ecodesign. Then, an overview will be provided of the ecodesign framework before the adoption of the ESPR, followed by a deeper overview of the innovations as they are introduced by the *Ecodesign for Sustainable Products Regulation* [1].

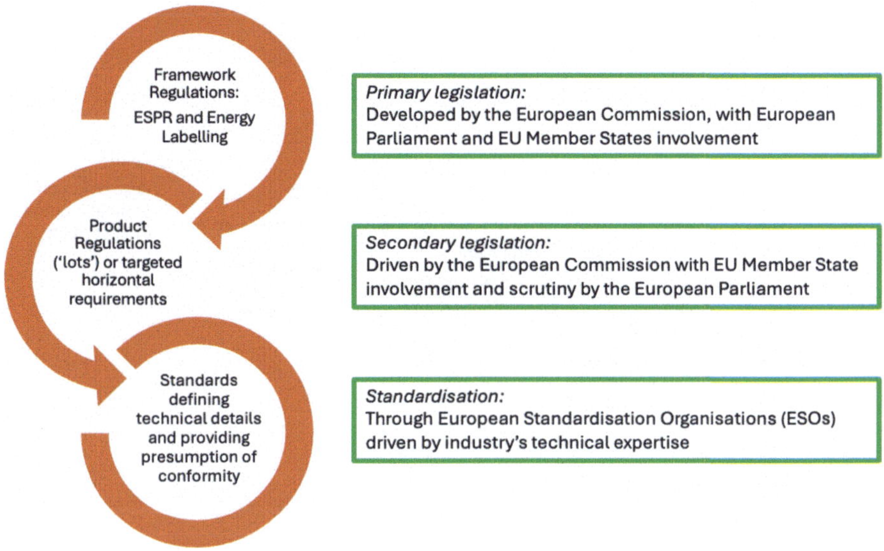

Fig. 2 High-level illustration on the layers of the ecodesign regulatory framework and relationships between primary legislations, secondary legislations, and standardisation

2 State-of-the-Art of Ecodesign

The history of ecodesign can be traced back to the early 1970s, as researchers and policymakers pointed out the effects of resource scarcity and pollution on the planet [12–14]. From this moment, researchers began to seek alternatives to traditional production methods in order to protect environmental limits in industrially developing countries. It was at this time that early strides were made towards creating an "environmentally sensitive" design method based upon minimizing the amount of materials and energy consumed in the design and manufacturing of products (the "greening" of design).

Then, in the early to mid-1980s, the Brundtland Report was published, which developed the concept of Sustainable Development and advised both designers and manufacturers of the need to take into account long-term environmental impacts, responsibility, and intergenerational equity. However, the tools available to implement the principles of sustainable development were still new and limited.

Since the late 1980s and throughout the 1990s, ecodesign has continued to evolve through the development of systematic approaches (environmentally conscious manufacturing, industrial ecology, and ideas relating to "R Strategies"[1]). This emphasized an increased interest in pollution prevention and cleaner production methods.

[1] R strategy refers to reduce, reuse, and recycle.

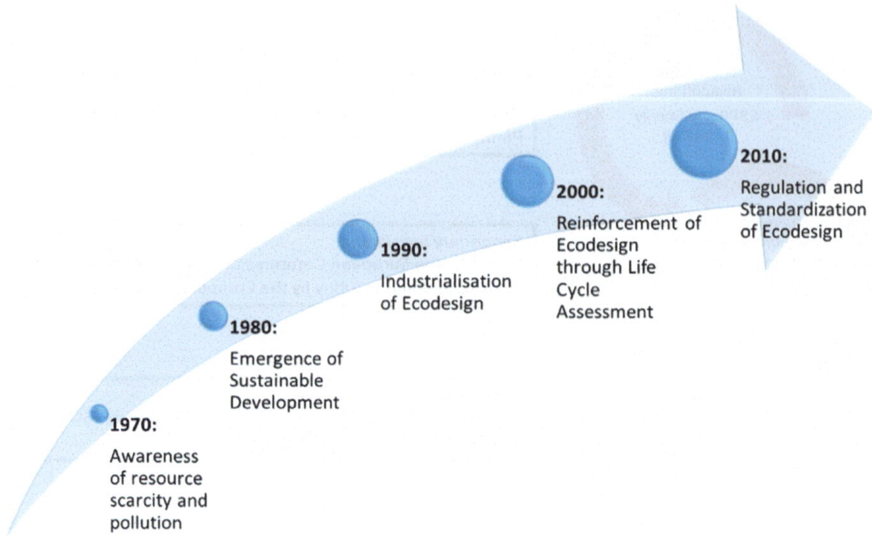

Fig. 3 Brief timeline of ecodesign

In the 2000s, life-cycle thinking towards the design of products and processes became central to ecodesign, with an increasing number of organisations adopting life cycle assessment (LCA) as well as cradle to grave[2] perspectives and philosophies in the design process. As a result, ecodesign no longer was limited to the manufacturing phase of product development but also included the use and end-of-life phases. Finally, regulations and standardization were clearly established in Europe during the 2010s.

The timeline of ecodesign is illustrated in Fig. 3.

Early discussions of ecodesign in electronics were attributed to Holt [15]. The author pointed out the necessity of incorporating ecological considerations into electronic product design from the early stages of development. This contribution stressed that even the vague decisions made in the early design of electronic products significantly influence material use, energy efficiency, and end-of-life impacts, constituting one of the first domain-specific applications of ecodesign principles.

Roy [16] also provided one of the earliest structured analyses of ecodesign as an evolving discipline. The publication underlined the necessity of integrating environmental considerations at an early stage of product design. To this end, it recommended implementing a continuous improvement approach: experimentation, then consolidation, and finally maturity. He notably gave examples of ecodesigned products, such as bicycles, cars, and houses, illustrating that ecodesign innovations must move from the experimental stage to mature products.

[2] Cradle-to-grave is a life-cycle concept that describes assessing a product's environmental impacts across all stages of its life, from creation to disposal.

Extending the life-cycle concept, Tu and Hsu [17] were among the first to develop ecodesign strategies explicitly grounded in product life-cycle design. Their work proposed a structured approach to integrating environmental objectives into corporate research and development activities through the use of life cycle assessment (LCA) models. By applying tools such as material and energy flow analysis, toxic emissions assessment, and the ecodesign strategy wheel, they demonstrated how environmental targets could be systematically embedded into product development. The study emphasized design for disassembly and end-of-life take-back as key elements of sustainable product development, illustrating the effectiveness of life cycle-based ecodesign through a case study on the green design of a notebook computer.

Aoe [18] advanced ecodesign research by integrating it with the concept of eco-efficiency, particularly in electrical and electronic products. This study introduced eco-efficiency indicators that were applied to concrete case studies. These indicators evaluated environmental performance relative to product functions. The results demonstrated that the majority of products achieved improved functionality while reducing environmental impacts. This indicates that environmental measures may also bring economic benefits. The final part of the study provides further justification for the use of ecodesign as a business strategy and discusses potential reductions in material use, energy consumption, and total life-cycle cost of ecodesigned products.

From a designer-oriented perspective, Lofthouse [19] examined the practical requirements for implementing ecodesign tools in industrial design practice. In this paper, the author highlighted the need to align tools with the designer's perspective (workflows and cognitive processes). It emphasizes the importance of considering usability and decision support rather than simple analytical complexity.

Building on these insights, she proposed a holistic framework for ecodesign tools tailored to industrial designers. The framework integrates education, product-specific information, and inspiration. It communicates in a way that fits designers' ways of working and provides a starting point for ecodesign. The paper also describes a web-based prototype, "Information/Inspiration", which concretely implements the framework. By using this tool, designers better understand ecodesign principles, explore ideas, and become more aware of environmental issues. As a consequence, they are more likely to imagine sustainable products.

Then, an overview of ecodesign research was proposed in 2006 by Karlsson and Luttropp [20]. The publication synthesizes developments from various sectors, including industry and academia. The authors describe how ecodesign has evolved into a multidisciplinary field. They also outline the dependence on many regulations concerning ecodesign, as well as a growing adoption by the industrial sector. The research states that ecodesign integrates human sustainability priorities with business considerations. It aims to reduce environmental impacts while inspiring innovative product solutions. The authors emphasize that achieving sustainable product development is an ongoing process that requires goal setting, early specification, and careful organization of product development. The success of this approach requires to reinforce the link between resources and functionalities. In addition, achieving sustainability requires to integrate environmental considerations into everyday life and engage all the society in a deep transformation.

An observational study was also carried out by Vallet et al. [21]. It examines the modernization of activities through the use of ecodesign tools and analyses how experienced designers use these tools in real-world situations. This work provides empirical insights into ecodesign implementation. It also demonstrates that ecodesigning largely relies on classic design processes. Indeed, key ecodesigning activities (environmental assessment and strategy definition) are primarily driven by the designers' own abilities and experience and do not depend on dedicated tools. Additionally, ecodesign tools prove particularly useful for setting goals, exploring design solutions, and supporting final evaluations. Overall, the study highlights that successful ecodesign implementation depends not only on tools. It also depends on designers' experience, a supportive organizational framework, and the ability to adapt to different project contexts.

A comprehensive review of ecodesign tools, methods, and literature was conducted by Pigosso et al. [22]. It is based on a systematized literature review and a bibliometric analysis, summarizing more than 500 publications spanning 30 years. Through this analysis, the authors found that there are numerous ways to approach ecodesign. These can be grouped into nine major thematic areas of research and three significant historical periods. Additionally, the research shows a clear evolution in terms of tools. Indeed, researchers have progressed from using qualitative and independent tools to integrated tools based on the life-cycle approach. They now employ a more strategic way of using ecodesign tools. Furthermore, the authors underline that ecodesign now goes beyond product design to include the entire business process through systems thinking. The analysis also points out that manufacturers still have difficulty effectively using ecodesign tools. Overall, the analysis shows that ecodesign has evolved as an organized, developmental innovation process and is not merely a collection of independent techniques.

Afterwards, Rossi et al. [23] performed an extensive review of available ecodesign methodologies and tools in order to identify and understand barriers related to the successful implementation of ecodesign practices within industry. Their findings reveal numerous obstacles associated with the use of ecodesign. First, they observed resistance from organizations that do not have enough experienced staff with ecodesign expertise. Second, ecodesign tools may not have sufficient usability for users. Third, ecodesign tools and their requirements may be inconsistent with organizational business objectives. Fourth, ecodesign tools are poorly integrated into existing product development processes. The authors indicated that several may be improved to overcome these issues. First, companies should align ecodesign with their own strategies and goals. Second, companies should incorporate ecodesign into established product development processes. Third, organizations should invest in training to increase ecodesign awareness and skills. Finally, the authors concluded that there is a need for a clearly defined structure that connects ecodesign and industry practices.

Bundgaard et al. [24] explicitly addressed the regulatory dimension of ecodesign by analysing the evolution of the EU Ecodesign Directive (2009/125/EC) from a primary focus on energy efficiency towards broader resource efficiency. Their study demonstrated that resource efficiency requirements are overshadowed by energy-related criteria. The authors also supported this finding with case studies on vacuums and imaging equipment that show the significance of considering resource efficiency early on in product development. Finally, Bundgaard et al. pointed out the lack of tools available to measure ecodesign impact as an area for improvement. This study is significant among the literature by showing how policy frameworks impact ecodesign priorities and thus industrial behaviour.

Then, virtual prototyping was detailed by Landi et al. [25]. They explained its role in facilitating ecodesign and energy labelling at the early stages of product development. Their work proposed a methodology that virtualizes regulatory energy tests, enabling designers to assess energy performance before physical prototyping. By applying numerical simulations to household appliances, the approach allowed accurate estimation of energy consumption and energy class with minimal error. The study proves that virtual analysis improves the design tasks and ensures compliance with regulations. Lastly, the results indicate that this approach can enhance time to market, reduce testing costs, and optimize resource consumption.

Afterwards, Kim et al. [26] synthesized a decade's worth of scientific and practical examples to summarize the most current overview of ecodesign. Their research provides a deep understanding of how ecodesign has been used and applied sustainably over time. This shows that ecodesign has shifted over time from purely looking at environmental impacts to considering social and economic impacts—equally or more so than environmental impacts. For example, they identified trends associated with systems thinking, data and analytics, and digital technologies, in addition to challenges in resource efficiency, lifespan, and understanding user behaviour as challenges in ecodesign. They emphasized that due to the complexity of these issues, it is imperative to define system boundaries prior to addressing complex design problems. Therefore, they conclude that ecodesign can enable sustainable innovation (i.e. development and application) in complex technical and social systems but that there must be greater industry-wide uptake and social sustainability considerations included when leveraging ecodesign.

A comprehensive meta-level synthesis of ecodesign research was provided by Schäfer and Löwer [27] based on a review of 106 existing review articles. They found that ecodesign has matured from a focus on end-of-pipe environmental approaches to a systemic and life cycle-oriented design paradigm. Their review showed the increasing number and geographical concentration of ecodesign research in Europe; identified continuing barriers and success factors outside of design teams, including management, policymakers, customers, and educators; highlighted the need for further validation and real-world testing of a wide variety of methods and tools; and emphasized that there are still substantial gaps in both the implementation of ecodesign projects and the incorporation of social sustainability into ecodesign projects. Thus, this work presents an integrated view of the field and identifies several key challenges for future ecodesign research and practice.

Finally, Bhamra and Hernandez [28] provided a review of three decades of design for sustainability. Their analysis moves from a historical perspective of sustainability as a distinct concept to its contemporary use as a holistic approach to sustainable design. They observed that ecodesign remains a fundamental and relevant concept in both design and industry. Indeed, connections between design research and global sustainability objectives are still increasing to align with regulations and policies. Additionally, the authors mentioned how the principles behind the circular economy and collaborative efforts across multiple industry sectors have influenced how businesses approach their ecodesign practices. Lastly, they mentioned the impact that new digital technologies, specifically AI and IoT, are having on emerging approaches to sustainable design practices.

References	Interest	Limits
Janahi et al. [12]	It provides a systematic overview of eco-innovative methods and processes, linking ecodesign to organization and strategy in relation to technology and performance in terms of sustainable development	The models focus on manufacturing/production strategy rather than design-level tools It limits their application to early-stage product designer processes
Sartal et al. [13]	The paper reviews the evolution of sustainable manufacturing and examines the opportunities offered by Industry 4.0 technologies to support sustainable development	The broad conceptual scope has limited measurable validation of the benefits for sustainable development across multiple industries
Alderman [14]	The article highlights how understanding the relationship between product regulations and technical standards can provide a competitive advantage. Indeed, it enables earlier compliance and market readiness	However, the discussion is largely practical and descriptive, with limited analytical depth
Holt [15]	The publication provides one of the earliest examples of an application of ecodesign principles for electronic products. It also highlights the importance of early design decisions on environmental impact	The content presents a limited scope and methodological depth It mainly focuses on awareness rather than systematic tools or frameworks
Roy [16]	The paper provides a foundational conceptualization of ecodesign as an evolving design philosophy responding to environmental pressures	The proposal is largely theoretical It lacks empirical validation and concrete implementation guidance
Tu and Hsu [17]	The publication integrates ecodesign with life-cycle design and inverse manufacturing, emphasizing R&D integration	The content is focused on manufacturing contexts. It only offers limited discussion of organizational and market constraints
Aoe [18]	The paper makes the link between ecodesign and eco-efficiency. It reinforces the economic interest towards environmental improvements	The proposal is centred on electrical and electronic products. It also tackles a very specific dimension of eco-design

(continued)

(continued)

References	Interest	Limits
Lofthouse [19]	The paper addresses the needs of designers, emphasizes usability and enables integration of ecodesign tools into design practice	The work does not provide quantitative assessment and environmental performance metrics
Karlsson and Luttropp [20]	The paper offers a comprehensive overview of ecodesign research, policy, and practice, highlighting field maturation	The research does not provide detailed methodological analysis
Vallet et al. [21]	The publication provides empirical insights into how experts actually use ecodesign tools in practice	The proposal focuses on expert users, which may not represent typical industrial designers
Pigosso et al. [22]	The paper classifies three decades of ecodesign tools and methods and identities major trends	The work is mainly descriptive. It offers limited critical assessment of tool effectiveness in practice
Rossi et al. [23]	The publication identifies difficulties to industrial implementation and proposes operational strategies	The research is limited sector-specific validation
Bundgaard et al. [24]	The paper expands ecodesign discussion to regulatory evolution, particularly resource efficiency within EU policy	The work has a strong EU focus. The applicability in other contexts may be limited
Landi et al. [25]	The publication demonstrates the role of virtual prototyping in supporting ecodesign and energy labelling compliance	The work highlights the importance of digital tools. It only lightly addresses organizational and human factors
Kim et al. [26]	The paper identifies new research perspectives. It addresses especially systems thinking and digitalization	The study is largely conceptual and exploratory. It offers limited empirical evidence
Schäfer and Löwer [27]	The publication synthesizes multiple review papers, offering a high-level consolidation of ecodesign knowledge	The work details a meta-level analysis. It does not provide solutions to implement
Bhamra and Hernandez [28]	The research indicates how sustainability may evolve in terms of innovation, policy, and practice	The broad analysis reduces the specificity on ecodesign tools and methods

3 Ecodesign Product Regulations Before the ESPR

The 2024 Ecodesign for Sustainable Products Regulation (ESPR) replaces (repeals) the prior Ecodesign Directive (2009/125/EC). New product regulations will be adopted under the newly established ESPR framework, while existing product-specific ecodesign requirements have been adopted under the previous Ecodesign Directives. The status in 2023 of available product group-related legislations—referred to as "lots"—is summarized in Tables 1, 2, 3, and 4.

Based on the Ecodesign Directives, the Commission can introduce product-specific requirements for energy-related products when (a) the product represents a relevant volume of sales and trade, (b) the product has a high environmental impact, and (c) there is high potential for improvement in terms of environmental impact while avoiding excessive costs (see for instance Art 15 of Directive 2005/32/EC [30]).

In order to decide more specifically which products should be subject to ecodesign requirements, a "Methodology Study Eco-design of Energy-using Products" [31] was completed in 2005, coinciding with the first Ecodesign Directive of 2005. The aim of the study was to further develop "*... a methodology allowing to evaluate whether ... energy-using products fulfil certain criteria ...*". This methodology has been updated regularly (2011, 2013, and 2024) and is now known as the "Methodology for the Ecodesign of Energy-Related Products" or "MEErP", to be used for implementing measures in the context of the Ecodesign of EuP Directive 2005/32/EC.

A useful indicator of European Commission's expected ecodesign regulatory activity has been the various "working plans" the European Commission published periodically since the establishment of the ecodesign legal framework. These working plans are a legal requirement, as first stipulated in Article 16 of the 2005 Ecodesign Directive, the "*Commission shall... establish a working plan...*" which shall set out "*an indicative list of product groups ... considered as priorities*" for the adoption of product specific measures. Besides new product groups that are introduced, the working plans also include necessary reviews of already adopted ecodesign product regulations.

At the time when the present chapter was written, the following working plans were available:

- 2009–2011 Ecodesign and Energy Labelling Working Plan [32];
- 2012–2014 Ecodesign and Energy Labelling Working Plan [33];
- 2016–2019 Ecodesign and Energy Labelling Working Plan [34];
- 2022–2024 Ecodesign and Energy Labelling Working Plan [35];
- 2025–2030 ESPR and Energy Labelling Working Plan [36].

The latest working plan was adopted on 16 April 2025 as discussed by the European Commission in a related press release [37] and will be covered in the next section on the ESPR.

Table 1 Summary of existing EU measures related to ecodesign, energy labelling, and tyre labelling (Sept. 2023) [29]

Product group	Relevant acts/ legislations	Type of measure	Link to available acts/legislations
Horizontal: framework legislation	Directive 2009/ 125/EC	Ecodesign rules	http://eur-lex.europa.eu/legal-content/ EN/TXT/?qid=1516983157666&uri= CELEX:32009L0125
	Regulation (EU) 2017/1369	Energy labelling rules	http://eur-lex.europa.eu/legal-content/ EN/TXT/?uri=celex:32017R1369
Horizontal: guidelines on self-regulation/ VAs	Recommendation (EU) 2016/2125	Voluntary	http://eur-lex.europa.eu/legal-content/ EN/TXT/?uri=uriserv:OJ.L_.2016.329. 01.0109.01.ENG&toc=OJ:L:2016: 329:TOC
Horizontal: standby/off mode consumption	Regulation (EC) 1275/2008 (applies till 9 May 2025)	Ecodesign rules	http://eur-lex.europa.eu/legal-content/ EN/TXT/?qid=1517329941706&uri= CELEX:32008R1275
	Regulation (EU) 2023/826 (applies from 9 May 2025)	Ecodesign rules	https://eur-lex.europa.eu/legal-content/ EN/TXT/?uri=CELEX%3A32023R0 826&qid=1681803866374
Welding equipment	Regulation (EU) 2019/1784	Ecodesign rules	https://eur-lex.europa.eu/legal-content/ EN/TXT/?uri=uriserv:OJ.L_.2019.272. 01.0121.01.ENG&toc=OJ:L:2019: 272:TOC
Power transformers	Regulation (EU) 548/2014	Ecodesign rules	http://eur-lex.europa.eu/legal-content/ EN/TXT/?qid=151115201.0001. 01.ENG
Electric motors + variable speed drives (VSDs)	Regulation (EU) 2019/1781	Ecodesign rules	https://eur-lex.europa.eu/legal-content/ EN/TXT/?uri=uriserv:OJ.L_.2019.272. 01.0074.01.ENG&toc=OJ:L:2019: 272:TOC
Water pumps	Regulation (EU) 547/2012	Ecodesign rules	http://eur-lex.europa.eu/LexUriServ/ LexUriServ.do?uri=CELEX:32012R 0547:EN:NOT
Circulators	Regulation (EC) 641/2009	Ecodesign rules	http://eur-lex.europa.eu/legal-content/ EN/TXT/?qid=1444920251241&uri= CELEX:32009R0641
Industrial fans	Regulation (EU) 327/2011	Ecodesign rules	http://eur-lex.europa.eu/legal-content/ EN/TXT/?qid=1521113646166&uri= CELEX:32011R0327
Professional refrigeration equipment	Regulation (EU) 2015/1095	Ecodesign rules	http://eur-lex.europa.eu/legal-content/ EN/TXT/?uri=uriserv:OJ.L_.2015.177. 01.0019.01.ENG
	Regulation (EU) 2015/1094	Energy labelling rules	https://eur-lex.europa.eu/legal-content/ EN/TXT/?uri=CELEX%3A32015R1 094&qid=1695308263313

(continued)

Table 1 (continued)

Product group	Relevant acts/ legislations	Type of measure	Link to available acts/legislations
Air heating/ cooling products	Regulation (EU) 2016/2281	Ecodesign rules	http://eur-lex.europa.eu/legal-content/EN/TXT/?uri=uriserv:OJ.L_.2016.346.01.0001.01.ENG&toc=OJ:L:2016:346:TOC
External power supplies	Regulation (EU) 2019/1782	Ecodesign rules	https://eur-lex.europa.eu/legal-content/EN/TXT/?uri=uriserv:OJ.L_.2019.272.01.0095.01.ENG&toc=OJ:L:2019:272:TOC

3.1 Ecodesign and Energy Labelling Working Plan 2009–2011

The first Ecodesign Working Plan 2009–2011 [32] was published 4 years after the adoption of the 2005 Ecodesign Directive. However, the 2005 Directive already indicated a number of product groups that should be considered for product-specific measures as a "transitional" measure and in anticipation of the first working plan.

Products mentioned directly in the 2005 Ecodesign Directive include (1) heating and water-heating equipment, (2) electric motor systems, (3) lighting (domestic and tertiary sectors), (4) domestic appliances, (5) office equipment (domestic and tertiary sectors), (6) consumer electronics, (7) heating or ventilating or air conditioning (HVAC) domestic systems, and (8) electronic and electrical products in standby modes.

During the preparatory work for the first working plan, 57 product groups which fell within the scope of the 2005 Ecodesign Directive—but were not already introduced as part of the transitional period—were identified. A further shortlist was created leading to the following product groups introduced in the first working plan:

Air-conditioning and ventilation systems	Equipment for network, data processing, and data storing
Equipment for electric and fossil-fuelled heating	Equipment for refrigerating and freezing
Equipment for food preparing	Equipment for sound and imaging
Industrial and laboratory furnaces and ovens	Transformers
Machine tools	Equipment for water-using

Table 2 Summary of existing EU measures related to ecodesign, energy labelling, and tyre labelling (2023) [29]

Product group	Relevant acts/ legislations	Type of measure	Link to available acts/legislations
Computers	Regulation (EU) 617/2013	Ecodesign rules	http://eur-lex.europa.eu/legal-content/EN/TXT/?qid=1521115326014&uri=CELEX:32013R0617
Servers and data storage products	Regulation (EU) 2019/424	Ecodesign rules	https://eur-lex.europa.eu/legal-content/EN/TXT/?uri=uriserv:OJ.L_.2019.074.01.0046.01.ENG&toc=OJ:L:2019:074:TOC
Simple set-top boxes—repealed	Regulation (EU) 107/2009	Ecodesign rules	http://eur-lex.europa.eu/legal-content/EN/TXT/?qid=1521115794453&uri=CELEX:32009R0107
Vacuum cleaners	Regulation (EU) 666/2013	Ecodesign rules	http://eur-lex.europa.eu/legal-content/EN/TXT/?qid=1521115222853&uri=CELEX:32013R0666
Mobile phones and tablets	Regulation (EU) 2023/1670	Ecodesign rules	https://eur-lex.europa.eu/eli/reg/2023/1670/oj
	Regulation (EU) 2023/1669	Energy labelling rules	https://eur-lex.europa.eu/legal-content/EN/TXT/?uri=CELEX%3A32023R1669
TVs/electronic displays	Regulation (EU) 2019/2021	Ecodesign rules	https://eur-lex.europa.eu/legal-content/EN/TXT/?uri=uriserv:OJ.L_.2019.315.01.0241.01.ENG&toc=OJ:L:2019:315:TOC
	Regulation (EU) 2019/2013	Energy labelling rules	https://eur-lex.europa.eu/legal-content/EN/TXT/?uri=uriserv:OJ.L_.2019.315.01.0209.01.ENG&toc=OJ:L:2019:315:TOC
Light sources and control gears	Regulation (EU) 2019/2020	Ecodesign rules	https://eur-lex.europa.eu/legal-content/EN/TXT/?uri=uriserv:OJ.L_.2019.315.01.0209.01.ENG&toc=OJ:L:2019:315:TOC
	Regulation (EU) 2019/2015	Energy labelling rules	http://eur-lex.europa.eu/legal-content/EN/TXT/?qid=1575537561243&uri=CELEX:32019R2015
Dishwashers	Regulation (EU) 2019/2022	Ecodesign rules	https://eur-lex.europa.eu/legal-content/EN/TXT/?uri=uriserv:OJ.L_.2019.315.01.0267.01.ENG&toc=OJ:L:2019:315:TOC
	Regulation (EU) 2019/2017	Energy labelling rules	https://eur-lex.europa.eu/legal-content/EN/TXT/?uri=uriserv:OJ.L_.2019.315.01.0267.01.ENG&toc=OJ:L:2019:315:TOC
Washing machines + washer-dryers	Regulation (EU) 2019/2023	Ecodesign rules	https://eur-lex.europa.eu/legal-content/EN/TXT/?qid=1575536811417&uri=CELEX:32019R2014
	Regulation (EU) 2019/2014	Energy labelling rules	https://eur-lex.europa.eu/legal-content/EN/TXT/?qid=1575536811417&uri=CELEX:32019R2014

(continued)

Table 2 (continued)

Product group	Relevant acts/ legislations	Type of measure	Link to available acts/legislations
Tumble driers	Regulation (EU) 932/2012	Ecodesign rules	http://eur-lex.europa.eu/LexUriServ/LexUriServ.do?uri=CELEX:32012R0932:EN:NOT
	Regulation (EU) 392/2012	Energy labelling rules	http://eur-lex.europa.eu/legal-content/EN/TXT/?qid=1521114567232&uri=CELEX:32012R0392
Domestic cooking appliances	Regulation (EU) 66/2014	Ecodesign rules	http://eur-lex.europa.eu/legal-content/EN/TXT/?qid=1521114342246&uri=CELEX:32014R0066
	Regulation (EU) 65/2014	Energy labelling rules	http://eur-lex.europa.eu/LexUriServ/LexUriServ.do?uri=OJ:L:2014:029:0001:0032:EN:PDF

3.2 Ecodesign and Energy Labelling Working Plan 2012–2014

The Working Plan 2012–2014 is based on the updated Ecodesign Directive of 2009, and it builds on the results achieved "*... since mid-2005 for the ... product groups listed in Article 16(2) of the ... 2005 Ecodesign Directive on Energy-using Products (EuPs) and on the first working plan*" which was issued for the period 2009–2011.

Before the Working Plan 2012–2014, a total of 18 indicative product groups had been identified (8 in the 2005 Ecodesign Directive and 10 in the Working Plan 2009–2011). As a result, the Commission had launched thirty-seven preparatory studies and had already adopted seventeen implementing measures for specific types of products (twelve ecodesign regulations and five energy labelling regulations). Furthermore, thirty-nine standardisation mandates had been launched for these product groups.

The breadth of work needed to develop and implement Ecodesign requirements now also becomes increasingly clear as the Commission highlights in the working plan that "*the increasing number of measures ... requires ... attention. Each implementing measure ... implies the development of transitional calculation methods and measurement standards, implementation guidelines, ...*".

As stated in the Working Plan, "*between 2012 and 2014, the Commission will focus on the ... implementation of the Regulations adopted under the 2005 Ecodesign Directive and the first Working Plan. However, the Commission intends to start the preparatory work for ... implementing measures for product groups identified ....*" Within the Working Plan 2012–2014, the Commission identifies the following new product groups:

Window products	Smart appliances/meters
Steam boilers	Appliances for wine storage

(continued)

(continued)

Power cables	Water-related products
Equipment for enterprises' servers, data storage, and ancillary	

The annex of the working plan provides an overview of all the various product groups and assigns each product group with a "lot" number. The responsibility for the "lots" is divided between various Directorates General (DG's) of the Commission, in particular DG Energy and what was known in 2012 as DG Enterprise. The numbers for the lots are assigned for each DG separately, resulting in overlapping and at times confusing lot numbering (e.g. televisions is assigned lot 5 for DG ENER while machine tools are also assigned lot 5 for DG ENTR in the annex of the working plan).

3.3 *Ecodesign and Energy Labelling Working Plan 2016–2019*

In 2016, it has been over 10 years since the 2005 Ecodesign Directive was introduced. In the 2016–2019 Working Plan, by 2020 the Commission estimates to "*deliver energy savings of around 175 Mtoe per year …. For consumers, this translates into 490 Euros savings per household per year …. Moreover, this policy is estimated to deliver approximately 55 billion Euros per year extra revenue for industry, wholesale and retail sectors, …which could translate into up to 800 000 direct additional jobs …. As such, it also contributes to energy security by reducing the import of energy … by the equivalent of 1.3 billion barrels of oil each year and by reducing CO_2 emissions by 320 million tonnes annually*".

The adoption of the Ecodesign Working Plan 2016–2019 coincides with the adoption of the EU's first circular economy action plan (December 2015). As such, ecodesign is increasingly seen against the wider background of improving the circular economy, or as the Commission states in the Working Plan: It "*contributes to the Commission's new initiative on the Circular Economy, which promotes a transition towards a more circular economy in the EU … covering the whole lifecycle of products and materials. There is an increasing need, …, to improve resource efficiency in the EU. Product design is a key aspect …, as it can have significant impacts across the product life cycle e.g. in making a product more durable, easier to repair, reuse or recycle. The Ecodesign directive already covers all significant environmental impacts … but the focus so far has been on energy efficiency improvements.*

> *…, Ecodesign should make a much more significant contribution …, for example by … tackling material efficiency issues such as durability and recyclability*"

At the time of introduction of the 2016–2019 Working Plan, the twenty-eight Ecodesign Regulations, sixteen Energy Labelling Delegated Regulations, and three recognised Voluntary Agreements were already introduced by the Commission.

Table 3 Summary of existing EU measures related to ecodesign, energy labelling, and tyre labelling (2023) [29]

Product group	Relevant acts/ legislations	Type of measure	Link to available acts/legislations
Household fridges and freezers	Regulation (EU) 2019/2019	Ecodesign rules	https://eur-lex.europa.eu/legal-content/EN/TXT/?uri=uriserv:OJ.L_.2019.315.01.0187.01.ENG&toc=OJ:L:2019:315:TOC
	Regulation (EU) 2019/2016	Energy labelling rules	https://eur-lex.europa.eu/legal-content/EN/TXT/?qid=1575537791838&uri=CELEX:32019R2016
Refrigerating appliances with sales function	Regulation (EU) 2019/2024	Ecodesign rules	https://eur-lex.europa.eu/legal-content/EN/TXT/?qid=1593616801922&uri=CELEX:32019R2018
	Regulation (EU) 2019/2018	Energy labelling rules	http://eur-lex.europa.eu/legal-content/EN/TXT/?uri=uriserv:OJ.L_.2014.337.01.0008.01.ENG
Ventilation units	Regulation (EU) 1253/2014	Ecodesign rules	http://eur-lex.europa.eu/legal-content/EN/TXT/?uri=uriserv:OJ.L_.2014.337.01.0008.01.ENG
	Regulation (EU) 1254/2014	Energy labelling rules	http://eur-lex.europa.eu/legal-content/EN/TXT/?uri=uriserv:OJ.L_.2014.337.01.0027.01.ENG
Space and combination heaters	Regulation (EU) 813/2013	Ecodesign rules	http://eur-lex.europa.eu/legal-content/EN/TXT/?qid=1521111746792&uri=CELEX:32013R0813
	Council Directive 92/42/EEC	Ecodesign rules	https://eur-lex.europa.eu/legal-content/EN/TXT/?uri=CELEX:31992L0042
	Regulation (EU) 811/2013	Energy labelling rules	http://eur-lex.europa.eu/legal-content/EN/TXT/?qid=1477486490380&uri=CELEX:32013R0811
Water heaters/ storage tanks	Regulation (EU) 814/2013	Ecodesign rules	http://eur-lex.europa.eu/legal-content/EN/TXT/?qid=1521112004143&uri=CELEX:32013R0814
	Regulation (EU) 812/2013	Energy labelling rules	http://eur-lex.europa.eu/LexUriServ/LexUriServ.do?uri=OJ:L:2013:239:0083:0135:EN:PDF
Local space heaters	Regulation (EU) 2015/1188	Ecodesign rules	http://eur-lex.europa.eu/legal-content/EN/TXT/?uri=uriserv:OJ.L_.2015.193.01.0076.01.ENG
	Regulation (EU) 2015/1186	Energy labelling rules	http://eur-lex.europa.eu/legal-content/EN/TXT/?uri=uriserv:OJ.L_.2015.193.01.0020.01.ENG

Table 4 Overview of existing EU ecodesign, energy labelling, and tyre labelling measures (2023) [29]

Product group	Relevant acts/ legislations	Type of measure	Link to available acts/legislations
Solid fuel space heaters	Regulation (EU) 2015/1185	Ecodesign rules	http://eur-lex.europa.eu/legal-content/EN/TXT/?uri=uriserv:OJ.L_.2015.193.01.0001.01.ENG
	Regulation (EU) 2015/1186	Energy labelling rules	http://eur-lex.europa.eu/legal-content/EN/TXT/?uri=uriserv:OJ.L_.2015.193.01.0020.01.ENG
Solid fuel boilers	Regulation (EU) 2015/1189	Ecodesign rules	http://eur-lex.europa.eu/legal-content/EN/TXT/?uri=uriserv:OJ.L_.2015.193.01.0100.01.ENG
	Regulation (EU) 2015/1187	Energy labelling rules	http://eur-lex.europa.eu/legal-content/EN/TXT/?uri=uriserv:OJ.L_.2015.193.01.0043.01.ENG
Air conditioners	Regulation (EU) 206/2012	Ecodesign rules	http://eur-lex.europa.eu/legal-content/EN/TXT/?qid=1521112322269&uri=CELEX:32012R0206
	Regulation (EU) 626/2011	Energy labelling rules	http://eur-lex.europa.eu/legal-content/EN/TXT/?qid=1521112473531&uri=CELEX:32011R0626
Tyres	Regulation (EU) 2020/740	Energy labelling rules	https://eur-lex.europa.eu/legal-content/EN/TXT/?uri=uriserv:OJ.L_.2020.177.01.0001.01.ENG&toc=OJ:L:2020:177:FULL
Imaging equipment	COM (2013) 23	Voluntary	https://eur-lex.europa.eu/legal-content/EN/TXT/?qid=1593617256261&uri=CELEX:52015SC0089
Game consoles	COM(2015) 178	Voluntary	http://eur-lex.europa.eu/legal-content/EN/TXT/?qid=1432808656922&uri=CELEX:52015DC0178

Furthermore, around forty standardisation mandates were launched for these product groups.

New product groups included in the Working Plan 1016–1019 are as follows:

Building automation and control systems	Solar panels and inverters
Electric kettles	Refrigerated containers
Hand dryers	High-pressure cleaners
Lifts	

Alongside the products mentioned above, the Commission also introduces a specific workstream on ICT products "*that will … take into account their circular economy potential…*". The reason for a separate workstream relates to the specificities of such products, as "*for ICT products it has proven very difficult to make a reliable estimate of their energy savings potential, ... Moreover, ..., questions have arisen as to the suitability of the Ecodesign/energy labelling process (which takes on average around 4 years) for ... minimum energy and resource efficiency criteria*".

The ICT products that the Commission aims to investigate in first instance are as follows:

Gateways (home network equipment)	Base stations
Mobile/smart phones	

In the 2016–2019 Working Plan, the Commission also recognises that already adopted ecodesign measures have review clauses, many of which will be due during the timeframe of the working plan, adding to the Commission's workload.

Finally, the Commission also aims to improve market surveillance and international cooperation, as there is the estimation "*that 10–25% of products ... do not comply with Ecodesign and energy labelling requirements. As a result, around 10% of ... energy savings are being lost. ... this corresponds to about seventeen Mtoe of primary energy per year*".

3.4 Ecodesign and Energy Labelling Working Plan 2022–2024

The transition from looking primarily at energy efficiency of products towards a more holistic view of a product's environmental footprint really comes to the fore into the early 2020s when the Commission publishes its Working Plan 2022–2024. In fact, when the working plan was tabled the Commission has already adopted its proposal for an Ecodesign for Sustainable Products Regulation (ESPR). The Working Plan 2022–2024 is intended to bridge the gap until the entry into force of the ESPR, which at that time still needed to go through the EU's decision-making processes.

The working plan itself also recognizes the transition towards a more holistic, circular view of a product's environmental footprint: "*While energy efficiency has ... been the ... focus of Ecodesign work ..., other aspects have increasingly been ... integrated over time, ..., several measures adopted in 2019 include new circular economy elements with requirements on reparability, recyclability, ease of end-of-life disassembly and reuse*".

The new working plan explicitly states that is it looking to push the boundaries further towards a holistic circularity approach as "*within ... the existing EU Ecodesign rules and in synergy with the focus on energy efficiency, this working plan strengthens the focus on the circularity aspects of Ecodesign, ... As such, ..., new product-specific requirements on material efficiency aspects can and will be explored*".

When the Working Plan 2022–2024 was published, "*about 30 'ErP' product groups are regulated through some 50 measures. They apply to billions of products ..., and have direct implications for suppliers, retailers, businesses and consumers*

…. About half of the EU total final energy use is consumed in products … subject to this legislation". The impact of these regulations, according to the Commission, has been significant: "*…the cumulative effect of EU rules … in 2020 reduced EU primary energy demand by 7% or 1037 TWh/year (ca. 170 Mt CO_2 equivalent GHG reduction)…. Energy savings from the measures … will increase substantially in coming years, …, with an average exceeding 1500 TWh/year over the period 2021–2030. … The EU rules in 2020 delivered energy consumer benefits to the tune of EUR 60 billion/year (ca. 0.4% of the EU GDP)—EUR 210/year per household*".

As the product regulations already adopted need to be reviewed to ensure the measures remain fit for purpose due to developing technologies, the amount of work related to these previously adopted regulations becomes significant for the Commission, as the Working Plan states: "*… the aggregate effort required for such critical "maintenance" efforts becomes significant as the scope of the legislation increases …, and that this work needs to be better factored in …. Otherwise, it will entail delays that have important consequences in terms of foregone benefits.*"

Another challenge identified by the working plan relates to the role of technical standardisation, because of "*… the diversity of technologies and user behaviour, the timely development of harmonized standards… is an inherent and often difficult, resource intensive challenge.*"

In light of the vast amount of product regulations scheduled for review during the working plan's timeframe (38 reviews before the end of 2024), the number of new products introduced remains relatively limited:

Low-temperature emitters	Universal external power supplies
Professional laundry appliances	Electric vehicle chargers
Professional dishwashers	

Besides the revisions and the introduction of new product groups, the working plan also initiates the transition towards a more prominent inclusions of material efficiency requirements, such as measures on durability, recyclability, ability to repair, reuse or upgrade, and recycled content. In particular, the Commission intends to initiate studies to investigate product specific requirements on recycled content; on durability, firmware, and software; and on scarce, environmentally relevant, and critical raw materials.

Finally, the Commission also intends to further update the recently established European Product Registry for Energy Labelling, or EPREL database. Since January 2019, suppliers including manufacturers, importers, and authorised representatives had the legal obligation for registering any products according to energy labelling rules in EPREL before market placement. While EPREL was in operation at that time, important functionalities still needed to be addressed.

Example Ecodesign requirements for smartphones, mobile phones other than smartphones, cordless phones, and slate tablets.

In June 2023, Commission Regulation 2023/1670 [38] and Commission Delegated Regulation 2003/1669 [39] were published, respectively, laying down ecodesign requirements and specifying energy labelling requirements for smartphones and tablets. These Regulations were adopted under the 2009 Ecodesign Framework but already anticipate the expanded scope of the ESPR.

Smartphones, feature phones, cordless phones, and slate tablets placed on the EU market will be required to meet the ecodesign requirements from 20 June 2025 onwards. Energy labelling requirements will apply accordingly for smartphones and slate tablets.

The Ecodesign Regulation specifies—among others—requirements related to:

- Disassembly, ..., the type of workshop environment and skill level needed to disassemble the device
- Availability of spare parts, ... as well as the delivery times
- Type of maintenance information ...
- Reliability requirements, ...
- Requirements related to ... updates to the Operating System
- Information requirements related to certain critical raw materials ...

The Energy Labelling Regulation requires the energy label to supplied with products and visible at the point of sale. The Regulation also specifies the energy classes (A-G) and the rating schemes related to durability and repairability of the products.

4 Ecodesign for Sustainable Products Regulation

In December 2019, the European Commission published the European Green Deal, its flagship strategy aimed at making Europe the first climate-neutral continent by 2050. The transition towards a more circular economy was part of the Green Deal and was further elaborated on in the Commission's new Circular Economy Action Plan published in 2020. Within the Circular Economy Action Plan, the Commission announced that it "*will propose a sustainable product policy legislative initiative*" with the objective to "*make products fit for a climate-neutral, resource-efficient and circular economy, reduce waste and ensure that the performance ... in sustainability progressively becomes the norm, the Commission*" [40]. This legislative initiative eventually became the proposal by the Commission for an Ecodesign for Sustainable Products Regulation, published in March 2022, illustrated in Fig. 4.

Following negotiations over a two years period, the Ecodesign for Sustainable Products Regulation (ESPR) was finally adopted and subsequently published in the EU's Official Journal in June 2024, entering into force on 18 July 2024 and replacing

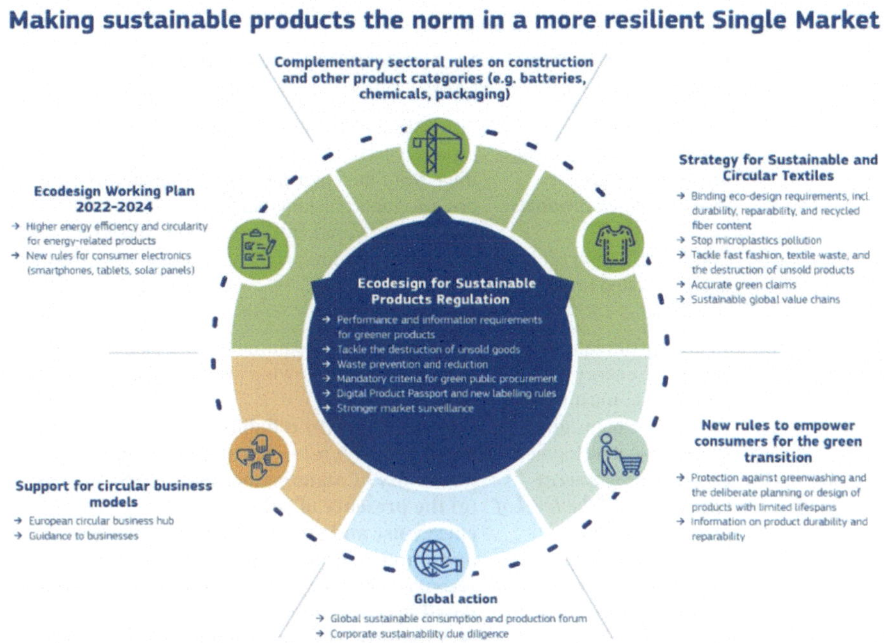

Fig. 4 EU's policy framework of 2022 to make sustainable products the norm [41]

the Ecodesign Directive 2009/125/EC as the cornerstone of the EU's ecodesign framework.

The ESPR is not the only initiative by the Commission to improve circularity, but it does sit at the centre of the Commission's policy framework introduced in the early 2020's aimed at making sustainable products the norm.

Compared to the previous Ecodesign Directives, the ESPR extends the ecodesign framework in two ways, namely by expanding the scope from energy-related products to virtually all physical products (with a few exceptions such as—but not limited to—food, medicinal products, or veterinary medicinal products) and by expanding the scope of possible ecodesign measures beyond energy related requirements—see Table 5.

Similar to the Ecodesign Directives, product-specific regulations will need to be introduced to establish the product-specific requirements related to the various product aspects covered by the ESPR. Typically, these requirements may relate to one or both of the following:

- Requirements related to performance, including durability, availability of spare parts, and minimum recycled content.
- Requirements related to information, such as energy efficiency information, information on material composition, or maintenance information. They will mainly be made available through the introduction of the digital product passport or,

Table 5 ESPR versus previous ecodesign directives

	Directive 2009/125/EC	Ecodesign for Sustainable Products Regulation (2024/1781)
Scope	"… establishes …Community Ecodesign requirements for ***energy-related products*** …." (Art 1) For the definition of energy-related products, see above	"… setting of Ecodesign requirements that ***products*** have to comply with …, with the aim of improving the environmental sustainability …in order to make sustainable products the norm and to reduce the overall carbon footprint and environmental footprint …." (Art 1)
Requirements foreseen	… setting of requirements which the energy-related products … must fulfil …. It contributes to sustainable development by ***increasing energy efficiency and the level of protection of the environment, while … increasing the security of the energy supply***. (Art 1)	… the Ecodesign requirements… shall be such as to improve the following product aspects ("product aspects") …: **(a) durability; (b) reliability; (c) reusability; (d) upgradability; (e) repairability; (f) the possibility of maintenance and refurbishment; (g) the presence of substances of concern; (h) energy use and energy efficiency; (i) water use and water efficiency; (j) resource use and resource efficiency; (k) recycled content; (l) the possibility of remanufacturing; (m) recyclability; (n) the possibility of the recovery of materials; (o) environmental impacts, including carbon footprint and environmental footprint; (p) expected generation of waste** (Art 5)

for products with energy labels, via the European Product Registry for Energy Labelling (EPREL).

In a deviation from the traditional product-specific approach, the ESPR also explicitly enables the Commission to adopt "horizontal" ecodesign requirements which may be set for those product groups that display one or more similarities which allow these products to be improved based on common requirements (note that ecodesign regulation 1275/2008 on energy consumption in "off" and various "standby" modes, amended in 2013 to include "networked standby" and replaced by Regulation 2023/826 was already designed as a horizontal piece of legislation, covering a range of electronic products).

In addition to the above, a digital product passport (DPP) is also introduced by the ESPR, supporting green public procurement requirements to become mandatory and creating a framework avoiding the destruction of unsold consumer products.

Particularly, the digital product passport, or DPP, has the potential to become a significant innovation for providing product information, as a DPP will likely become a requirement for products entering the EU market within future ecodesign regulations.

As explained by the Commission, the DPP is "*a digital identity card …, which will store … information to support products' sustainability, promote their circularity*

…. This information will be accessible electronically, … to make more informed decisions related to sustainability, circularity and regulatory compliance. It will also allow … automatic checks on the existence and authenticity of the DPPs …" [42]. The Commission will need to develop secondary legislation in the years ahead to specify the details concerning the DPPs data requirements, technical design, and operation before it will become fully operational.

Economic operators will also need to take necessary measures to prevent the need to destroy unsold consumer goods. In first step, the disclosure of information on the number and weight of unsold consumer goods discarded per year will be mandated. In a subsequent step, the destruction of certain consumer products such as apparel and clothing accessories or footwear will be prohibited.

The impact of the ESPR will reside in the implementation of the framework by the European Commission in the years ahead. Several pieces of additional regulations are needed to introduce the DPP and establish the framework to prevent the destruction of unsold consumer goods. As with the previous Ecodesign Directives, the main impact resides in the product-specific regulations—or in case of the ESPR also possible horizontal regulations—established under the new framework. The first ESPR working plan adopted in April 2025 provides a first indication of how the ESPR will be implemented.

4.1 ESPR and Energy Labelling Working Plan 2025–2030

This working plan is the first to apply the increased scope of the ESPR to specific products targeting "*a balance between the ESPR's potential positive environmental impacts, … and the need to simplify regulations*".

As stated by the Commission, "*the ESPR builds on the approach … under the EU's current Ecodesign and Energy Labelling Framework …. It is estimated that the current Ecodesign and energy labelling requirements achieved a 12% reduction in final energy consumption in 2023. …, avoiding 145 million tonnes of CO2 emissions that year…. The ESPR aims to replicate this success on a broader scale while maintaining the close synergies achieved with the Energy Labelling Framework.*"

Following a prioritisation exercise, "*the first working plan will include four final products, two intermediate products and two legal acts …, in addition to a list of substantial work … carried over from the last Ecodesign and energy labelling working plan*".

New products included:

Textiles/apparel	Tyres
Furniture	Mattresses

Intermediate products:

Iron and Steel	Aluminium

Horizontal requirements:

Repairability (including scoring)	Recycled content and recyclability of electrical and electronic equipment

ICT products are not specifically mentioned but could potentially be covered by the horizontal requirements.

Besides the new product categories introduced above, the first ESPR working plan also elaborates on the transition from the previous Ecodesign Directive to the ESPR framework.

The Ecodesign and Energy Labelling working plan for 2022–2024 covered 35 products for which considerable progress has been made. For nineteen of these products, a "*transition period until 31 December 2026*" applies. This means that the revision and adoption of these measures will continue under the previous framework, with a deadline of 31 December 2026 by which these measures need to be adopted. For the remaining sixteen products and due to efficiency reasons, these sixteem products will be addressed by the 2025–2030 working plan and will thus fall under the scope of the ESPR.

Product groups covered by the transition period falling under the previous Ecodesign Directive:

Photovoltaic panels	Water pumps
Space and combination heaters	Industrial fans
Water heaters	Circulators
Solid fuel local space heaters	External power supplies
Air conditioners including air-to-air heat pumps and comfort fans	Computers
Solid fuel boilers	Servers and data storage products
Air heating and cooling products	Power transformers
Ventilation units	Equipment for professional refrigeration
Vacuum cleaners	Equipment for imaging
Cooking appliances	

Product groups carried forward and included in the ESPR 2025–2030 Working Plan:

Low-temperature emitters	Refrigerating appliances (including household fridges and freezers)
Displays	Refrigerating appliances with a sales function

(continued)

(continued)

EV chargers	Light sources and (only for ecodesign) separate control gears
Household dishwashers	Welding equipment
Household washing machines and household washer–dryers	Mobile phones and tablets
Professional laundry appliances	Local space heaters
Professional dishwashers	Tumble dryers
Electric motors and variable speed drives	Standby and off mode consumption

5 Next Steps and Conclusions

The importance and impact of the EU's ecodesign framework on the environment, consumers, and businesses can hardly be overstated, as it is establishing "*production and consumption patterns …, including climate, environmental, energy, resource-use and biodiversity targets, …*" [1].

A significant amount of product regulations has been introduced in the past two decades, and the ESPR will further develop these regulations while also introducing new product groups and new concepts such as the DPP or horizontal ecodesign requirements. As the scope expands, ecodesign requirements will affect an increasing number of businesses, impacting product design (e.g. repairability), product testing (e.g. to establish energy efficiency or durability ratings), product maintenance (e.g. availability of spare parts), material composition (e.g. requirements on recycled content), energy efficiency (e.g. energy labelling or minimum performance standards), and introducing a potentially substantial amount of information and reporting requirements (e.g. the introduction of the DPP). The exact impact on businesses will depend on the type products in scope, the proposed requirements, and the type of business affected (e.g. a potentially significant impact on small and medium sized enterprises).

The European Commission will equally face an ambitious amount of work to develop, draft, and update ecodesign measures as it needs to keep on top of ever changing technologies, address potential trade-offs between product aspects within each product category, carefully assess product scope for each product specific regulation, and avoid potential overlap between horizontal ecodesign requirements and product-specific regulations. In addition, the ecodesign framework is not the only policy initiative aimed at improving the circular economy. Alignment will be needed as the ESPR sits within a wider framework of regulations such as market access regulations (e.g. the New Legislative Framework), consumer legislation (e.g. the "right to repair" Directive 2024/1799), or other product legislation (e.g. the EU Regulation concerning batteries and waste batteries 2023/1542).

Finally, the implementation of the ecodesign framework requires a significant number of technical details on measurement techniques or methodologies as well as in certain cases the development of specific tools (e.g. software tools that run certain workloads that allow consistent measurement of energy usage across products within a certain product category). As such, standardisation bodies face a significant amount of work, and the Commission needs to facilitate the development and maintenance of software applications supporting the ecodesign requirements.

That being said, the environmental imperative to collectively reduce our environmental footprint remains paramount, and ecodesign requirements have proven to effectively reduce "*the environmental, energy and climate impacts ... and in improving circularity*" [43].

If the Commission manages to strike the right balances, maximise the environmental benefits of the ESPR framework while steering clear of excessive or avoidable burdens on businesses, the ESPR framework can indeed, as the Commission states, "*contribute to ... EU policy goals related to the environment, energy, climate, consumer protection, competitiveness, resilience and the single market*" [43].

Acknowledgements The authors would like to sincerely thank Dr. Christophe Gaie for his invaluable support and insightful contributions throughout the development of this book chapter. His expertise, guidance, and constructive feedback greatly enhanced the quality and clarity of this chapter.

Disclaimer Note that the authors have assembled the information above to the best of their knowledge as available at the time when the present book was written. Any interpretation is the personal opinion of the authors. Also, the regulation framework can change and evolve any time. The only legally binding text is the one provided in the documents formally published in the European Union Official Journal.

References

1. Regulation (EU) 2024/1781 of the European parliament and of the council of 13 June 2024 establishing a framework for the setting of ecodesign requirements for sustainable products, amending Directive (EU) 2020/1828 and Regulation (EU) 2023/1542 and repealing Directive 2009/125/EC (Text with EEA relevance). https://eur-lex.europa.eu/eli/reg/2024/1781/oj/eng
2. Directive 2009/125/EC of the European parliament and of the council of 21 October 2009 establishing a framework for the setting of ecodesign requirements for energy-related products (recast) (Text with EEA relevance). https://eur-lex.europa.eu/eli/dir/2009/125/oj/eng
3. Regulation (EU) 2017/1369 of the European parliament and of the council of 4 July 2017 setting a framework for energy labelling and repealing directive 2010/30/EU (Text with EEA relevance). https://eur-lex.europa.eu/eli/reg/2017/1369/oj/eng
4. Hübner R (2012) Ecodesign: reach, limits and challenges 20 years of eco-design—time for a critical reflection. Renate Hübner. https://www.researchgate.net/publication/265633795_Ecodesign_reach_limits_and_challenges_20_years_of_Ecodesign_-_time_for_a_critical_reflection
5. Green Paper on Integrated Product Policy, European Commission. https://eur-lex.europa.eu/legal-content/EN/TXT/PDF/?uri=CELEX:52001DC0068

6. Communication from the commission to the council and the European parliament—integrated product policy—building on environmental life-cycle thinking, European Commission, 2003. https://eur-lex.europa.eu/legal-content/EN/TXT/PDF/?uri=CELEX:52003DC0302
7. Establishment of the working plan 2012–2014 under the ecodesign directive, Euopean commission working document, 2012. https://ec.europa.eu/docsroom/documents/9952/attachments/1/translations/en/renditions/pdf
8. Understanding the energy label, European Commission. https://energy-efficient-products.ec.europa.eu/Ecodesign-and-energy-label/understanding-energy-label_en
9. New Legislative Framework, European Commission. https://single-market-economy.ec.europa.eu/single-market/goods/new-legislative-framework_en
10. Mueck M, Gaie C (2025) European digital regulations. Springer, Berlin
11. Harmonised Standards, European Commission. https://single-market-economy.ec.europa.eu/single-market/european-standards/harmonised-standards_en
12. Janahi NA, Durugbo CM, Al-Jayyousi OR (2021) Eco-innovation strategy in manufacturing: a systematic review. Clean Eng Technol 5:100343. https://doi.org/10.1016/j.clet.2021.100343
13. Sartal A, Bellas R, Mejías AM, García-Collado A (2014) The sustainable manufacturing concept, evolution and opportunities within Industry 40: a literature review. Adv Mech Eng 12(5):25232. https://doi.org/10.1177/1687814020925232
14. Alderman A (2019) Knowing the link between product regulations and product standards can put you ahead of the competition. In: PSMA update (Q2 2019). Power sources manufacturers association. https://www.psma.com/HTML/newsletter/Q2_2019/page16.html
15. Holt HR (1994) A first step in electronic ecodesign. Electronics and the environment, 1994. ISEE 1994, proceedings, 1994 IEEE international symposium on IEEE, pp 191–194
16. Roy R (1994) The evolution of ecodesign. Technovation 14(6):363–380. https://doi.org/10.1016/0166-4972(94)90016-7
17. Tu J-C, Hsu F-L (1999) The ecodesign strategy on product research and development from the life-cycle design. In: Proceedings first international symposium on environmentally conscious design and inverse manufacturing. Tokyo, pp 351–356. https://doi.org/10.1109/ECODIM.1999.747637
18. Aoe T (2006) Eco-efficiency and ecodesign in electrical and electronic products. J Clean Prod 15(15):1406–1414. https://doi.org/10.1016/j.jclepro.2006.06.004
19. Lofthouse V (2006) Ecodesign tools for designers: defining the requirements. J Clean Prod 14(15–16):1386–1395. https://doi.org/10.1016/j.jclepro.2005.11.013
20. Karlsson R, Luttropp C (2006) EcoDesign: what's happening? An overview of the subject area of EcoDesign and of the papers in this special issue. J Clean Prod 14(15–16):1291–1298. https://doi.org/10.1016/j.jclepro.2005.11.010
21. Vallet F, Eynard B, Millet D, Mahut SG, Tyl B, Bertoluci G (2012) Using eco-design tools: an overview of experts' practices. Des Stud 34(3):345–377. https://doi.org/10.1016/j.destud.2012.10.001
22. Pigosso DCA, McAloone TC, Rozenfeld H (2015) Characterization of the state-of-the-art and identification of main trends for ecodesign tools and methods: classifying three decades of research and implementation. Indian Instit Sci J 94(4):405–427
23. Rossi M, Germani M, Zamagni A (2016) Review of ecodesign methods and tools: barriers and strategies for an effective implementation in industrial companies. J Clean Prod 129:361–373. https://doi.org/10.1016/j.jclepro.2016.04.051
24. Bundgaard AM, Mosgaard MA, Remmen A (2017) From energy efficiency towards resource efficiency within the ecodesign directive. J Clean Prod 144:358–374. https://doi.org/10.1016/j.jclepro.2016.12.144
25. Landi D, Capitanelli A, Germani M (2017) Ecodesign and energy labelling: the role of virtual prototyping. Proced CIRP 61:87–92. https://doi.org/10.1016/j.procir.2016.11.144
26. Kim H, Cluzel F, Leroy Y, Yannou B, Yannou-Le Bris G (2020) Research perspectives in ecodesign. Des Sci 6:e7. https://doi.org/10.1017/dsj.2020.5
27. Schäfer M, Löwer M (2021) Ecodesign—a review of reviews. Sustainability 13(1):315. https://doi.org/10.3390/su13010315

28. Bhamra T, Hernandez RJ (2021) Thirty years of design for sustainability: an evolution of research, policy and practice. Des Sci 7:e2. https://doi.org/10.1017/dsj.2021.2
29. Overview of existing EU Ecodesign, Energy labelling and tyre labelling measures (Sept. 2023). European Commission. https://commission.europa.eu/system/files/2023-09/Summary%20overview%20of%20ED-EL%20measures%20Sept23%20v2-%20for%20web.pdf
30. Directive 2005/32/EC of the European parliament and of the council of 6 July 2005 establishing a framework for the setting of ecodesign requirements for energy-using products and amending council directive 92/42/EEC and directives 96/57/EC and 2000/55/EC of the European parliament and of the council, 2005. https://eur-lex.europa.eu/eli/dir/2005/32/oj/eng
31. Methodology for the ecodesign of energy-related products (MEErP), European Commission. https://energy-efficient-products.ec.europa.eu/methodology-meerp_en
32. Ecodesign and Energy Labelling Working Plan 2009–2011, European Commission. https://eur-lex.europa.eu/legal-content/EN/TXT/HTML/?uri=CELEX:52008DC0660
33. Ecodesign and Energy Labelling Working Plan 2012–2014, European Commission. https://ec.europa.eu/docsroom/documents/9952/attachments/1/translations/en/renditions/pdf
34. Ecodesign and Energy Labelling Working Plan 2016–2019, European Commission. https://eur-lex.europa.eu/legal-content/EN/TXT/PDF/?uri=CELEX:52016DC0773
35. Ecodesign and Energy Labelling Working Plan 2022–2024, European Commission. https://energy.ec.europa.eu/publications/Ecodesign-and-energy-labelling-working-plan-2022-2024_en
36. ESPR and Energy Labelling Working Plan 2025–2030, European Commission. ESPR and Energy Labelling Working Plan 2025–2030|Environment
37. Commission rolls out plan to boost circular and efficient products in the EU, European Commission. https://ec.europa.eu/commission/presscorner/detail/en/ip_25_1071. Accessed on 16 April 2025
38. Commission Regulation (EU) 2023/1670 of 16 June 2023 laying down ecodesign requirements for smartphones, mobile phones other than smartphones, cordless phones and slate tablets pursuant to directive 2009/125/EC of the European parliament and of the council and amending commission regulation (EU) 2023/826 (Text with EEA relevance). https://eur-lex.europa.eu/eli/reg/2023/1670/oj/eng
39. Commission Delegated Regulation (EU) 2023/1669 of 16 June 2023 supplementing regulation (EU) 2017/1369 of the European parliament and of the council with regard to the energy labelling of smartphones and slate tablets (Text with EEA relevance). https://eur-lex.europa.eu/eli/reg_del/2023/1669/oj/eng
40. Communication from the Commission to the European parliament, the council, the European economic and social committee and the committee of the regions a new circular economy action plan for a cleaner and more competitive Europe, 2020. https://eur-lex.europa.eu/legal-content/EN/TXT/?qid=1583933814386&uri=COM:2020:98:FIN
41. More environmentally sustainable and circular products, European Commission. https://ec.europa.eu/newsroom/growth/items/769584/
42. Ecodesign for sustainable products regulation, European Commission. https://commission.europa.eu/energy-climate-change-environment/standards-tools-and-labels/products-labelling-rules-and-requirements/Ecodesign-sustainable-products-regulation_en
43. Ecodesign for sustainable products and energy labelling working plan 2025–2030, European Commission, 2025. https://eur-lex.europa.eu/legal-content/EN/TXT/PDF/?uri=CELEX:52025DC0187

Chapter 4
Standardisation Supporting Sustainability Objectives

Markus Mueck, Beniamino Gorini, Kristof Claesen, and Stéphane Du Boispéan

Abstract In separate chapters of the present book, a detailed introduction is given to the European Green Deal as a major framework under which a number of legislations have been initiated and maintained, and the related *Ecodesign for Sustainable Products Regulation* and *Energy Labelling Regulation* are managed. In the present chapter, we complement this overview by providing a summary of available standards, and specifically harmonised standards, supporting the implementation of the Ecodesign and Energy Labelling Legislations. While there are currently no references of harmonised standards available which would be providing presumption of conformity to the ecodesign and energy labelling framework directives, there are such harmonised standards published in the Official Journal in the context of related implementing measures. The understanding of the related framework is essential especially for manufacturers for related products to ensure that all requirements are met for accessing the European Union Single Market.

Keywords Regulation · Ecodesign · Circular economy · Market access · *European* Union single market

M. Mueck (✉)
Intel Deutschland GmbH, Neubiberg, Belgium
e-mail: markus.muck@gmail.com

B. Gorini
Nokia Networks, Massy, France
e-mail: Beniamino.Gorini@nokia.com

K. Claesen · S. Du Boispéan
Intel Corporation SA/NV, Brussels, Belgium
e-mail: kristof.claesen@intel.com

C. Gaie and M. Mueck (eds.), *Digital Sustainability: Legislations, Standards, Solutions and Related Case Studies*, Advances in Sustainability Science and Technology,
https://doi.org/10.1007/978-981-92-0458-8_4

1 Introduction

To address the topic of standardisation for sustainability, it is appropriate to focus on legislations relating to the European Green Deal and especially the *Ecodesign for Sustainable Products Regulation* [1] (repealing the prior Directive 2009/125/EC [2]) and the *Energy Labelling Regulation* [3].

As discussed by the European Commission [4, 5], "*there are no references of Harmonised Standards ... published in the Official Journal (OJ) of the European Union*" which would provide presumption of conformity to "*the Ecodesign and energy labelling framework directives*". However, there are such harmonised standards available and published *in the Official Journal (OJ) of the European Union* in support of **implementing measures** for ecodesign and energy labelling.

Indeed, both the *Ecodesign for Sustainable Products Regulation* and *Energy Labelling Regulation* are aligned with the New Legislative Framework (NLF) [6, 7]. This alignment provides a unique opportunity for all stakeholders to contribute to the implementation process of the Regulation as illustrated in Fig. 1: While the legislation itself sets high-level principles, it also established the fact that the European Standardisation Organisations (ESOs) are tasked with writing the standards that implement the essential requirements of harmonised legislations. Once the harmonised standards are referenced, they are published in the Official Journal (OJ) of the European Union and listed in a related database [8]. Further information and a more detailed introduction to the NLF are available in [7].

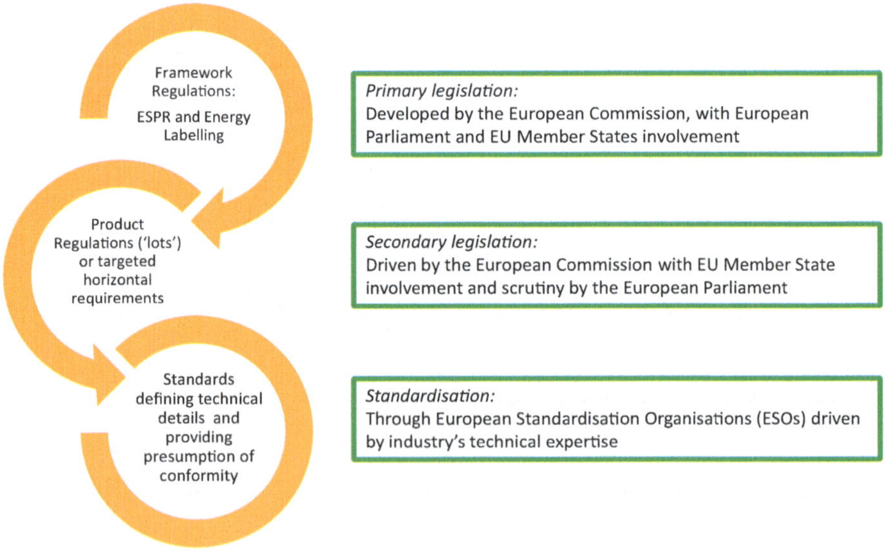

Fig. 1 High-level illustration on the layers of the ecodesign regulatory framework and relationships between primary legislations, secondary legislations, and standardisation

By following the standard, a presumption of conformity with the harmonised legislations exists. In other words, harmonised standards serve as a "passport to the market," reducing legal uncertainty and ensuring consistent application of technical requirements across all EU Member States. A harmonised standard is thus a key tool for manufacturers and other stakeholders to ensure compliance of related products with the essential requirements of the legislations; only if such compliance is demonstrated, access to the European Union Single Market is granted.

In the present chapter, we will thus focus on available harmonised standards in support of ecodesign and energy labelling **implementing measures**.

2 State of the Art

2.1 Role of Standardisation in Sustainability Policy

Many recent publications now frame the idea of standardisation not simply as a form of technical coordination but also as a way to promote sustainability. Indeed, standardisation can provide strategic policy tools that align market actions with public goals. Academic research on the economics of standardisation has typically addressed two roles for standardisation: first, reducing transaction costs, and second, facilitating innovation via interoperability and common knowledge bases [9]. This idea is also supported by empirical studies that outline the link between research and innovation. Indeed, standards facilitate the emergence of new and sustainability-related technologies, as they favour stable markets. This also reduces uncertainty for firms and regulators alike [10]. Consequently, standards can be seen as instruments that facilitate the development of policies, technological advancements, and market implementation.

The interaction between mandatory regulations and voluntary standards is a key theme in the sustainability literature. Many times, voluntary standards, such as ISO management system or product standards, are developed prior to or complement legislation by turning broad sustainability objectives into operationalised practices that can be implemented [11]. There are multiple studies showing how certification programs and third-party conformity assessment entities contribute to sustainable development goals (SDGs) by disseminating best practices; they also allow for monitoring of progress and lend credibility to businesses along global supply chains [11, 12]. In many situations where formal regulations are disjointed or respond slowly, it is not uncommon for voluntarily adopted standards to function as quasi-regulatory bodies, thereby facilitating enforcement and compliance with national laws, while also facilitating and creating policy coherence on an international basis [13].

Standards are being recognised as mechanisms for transforming overarching sustainable development ambitions into concrete and measurable criteria through research that correlates ISO standards with a variety of SDGs (e.g. ensuring food security, environmentally sounds production practices, and protecting our ecosystem).

Such research demonstrates how overall policy goals can be effectively integrated into routine organisational processes and associated performance measurements using a common set of standards [12, 13], particularly in conjunction with wider European (and worldwide) policy objectives. The European policy includes recognition of standardisation as a major driver of long-term sustainability, as well as an important element contributing to achieving regulatory outcomes, and fostering the continued expansion of the internal markets across Europe [14]. The European Commission's long-term strategy suggests that standards should be used as tools for promoting the implementation and compliance of sustainability policies across Member States through the establishment of a common framework to enhance the consistency, transparency, and enforceability of those policies, thereby reinforcing their position as an integral aspect of sustainability governance and not simply as technical adjuncts to it.

2.2 Harmonised Standards Under the New Legislative Framework

According to Decision 768/2008/EC, the Coordinated Approach to New Legislation (NLF) framework creates a common regulatory model that defines principles for key aspects of regulations, such as essential requirements, conformity assessments, and market monitoring to create uniformity across sectors. In du Boispéan et al. [15], highlight that this framework is a "toolbox" of regulation that ensures consistency of regulation across sectors while allowing for variation through the voluntary nature of standardisation.

The NLF's presumption of conformity system matches essential requirements established by legislators with harmonised standards produced by European Standardization Organizations (ESOs) like CEN, CENELEC, and ETSI. By showing compliance with harmonised standards, economic operators are entitled to a presumption that they comply with the EU law and facilitate their access to the common market. The presumption of conformity mechanism is increasingly being used to achieve sustainability objectives, especially under the European Green Deal. Additionally, recent regulations have expanded the scope of these essential requirements, such as the Ecodesign for Sustainable Products Regulation. These now cover long-term durability, ease of repair, ease of recycling, and environmental performance over the complete life cycle of the product.

The value of standardisation extends far beyond mere compliance with harmonised standards. As Blind [16] indicates that the use of standards can reduce the effects of coordination failure and mitigate information asymmetries between the involved parties. In addition, standardisation will lead to a decrease in uncertainty, which tends to hinder transformation that is moving towards mission-oriented objectives, with examples being the transition to a more circular and lower carbon economy. By establishing a common technical language and developing stable expectations, standards allow for increased investment into sustainable technologies and increase the interoperability of those technologies, thereby enhancing the likelihood of their diffusion. In addition, standards provide support for the gradual removal of existing legacy technologies.

As digital technologies become more fully integrated into how sustainability is regulated, the role of harmonised standards continues to expand. For example, the digital product passport demonstrates the need for standards that will provide interoperability, data quality, and tracking of sustainability information throughout a value chain. According to Micklitz [17], the evolution of standards has placed them at the intersection of technical design and governance for the public interest. Therefore, harmonised standards have developed into quasi-regulatory tools that influence how environmental performance, consumer rights, and transparency are delivered in practice. This raises issues of legitimacy, accountability, and stakeholder involvement whenever some public governance decision that involves significant social implications is delegated to a private standarisation organisation.

Incorporating feedback from institutional actors, these issues have been acknowledged and reflected in the European Telecommunication Standards Institute's (ETSI) position papers [18]. The papers highlight the necessity for a balanced standardisation framework that fosters both innovation and sustainability. They also emphasise the importance of consistency with EU regulatory policy and the protection of human rights in all areas of life. Additionally, reviews of the legal framework for the determination of harmonised EU standards demonstrate the significance of procedural safeguards within the Scheme. They stress the need for transparency of procedure and substantial participation by public interest stakeholders in reinstating the legitimacy of the New Approach Framework (NAF) [19].

2.3 Standardisation in Support of Ecodesign and Energy Labelling

Standardisation plays a central role to make EU's ecodesign and energy labelling frameworks applicable. It offers the technical bridge between regulatory objectives and measurable product requirements. Early ecodesign measures focused mainly on energy efficiency in-use. As a consequence, they reflected the priority of climate and energy security. Dalhammar [20] noticed in that the original stages of ecodesign industry participation with ecodesign standards were initially based on energy-in-use metrics. Indeed, they were the easiest metrics to quantify due to the existence of harmonised testing processes. In the New Legislative Framework, the European Commission establish mandates upon European Standardisation Organisation (ESOs) to develop harmonised standards. It provides guidance for the application of essential energy performance requirements through reproducible laboratory testing methods. The use of standardisation at this met subjected more towards comparability, market surveillance, and harmonising the EU internal market.

The scope of ecodesign has evolved significantly over time. What initially focused on improving energy efficiency has gradually broadened to address multiple dimensions of performance and the wider objectives of the circular economy. They now include durability, reparability, recyclability, and the presence of critical raw materials. This shift has substantially increased the technical complexity of standardisation. As demonstrated by Talens-Peiró et al. [21], translating policy ambitions such as "improved durability" into enforceable requirements requires the development of precise test procedures. These now concern cycle testing protocols, disassembly assessments, and material composition analyses. Standards therefore function as conversion mechanisms, transforming qualitative sustainability goals into quantifiable indicators. These indicators can be verified through conformity assessment.

Material efficiency requirements have continued to become further integrated under the Ecodesign Directive, highlighting tensions and coordination issues between these regulatory frameworks. Bundgaard and Huulgaard [22] indicate that Ecodesign and Energy Labelling Regulation (2017/1369) requirements need to be aligned. Traditionally, energy labels communicated to consumers in an easy-to-understand format for energy performance but now ecodesign is considering many life-cycle impacts. As more performance measures are applied, the relationship between product requirements (minimum performance thresholds) and consumer information instruments (comparative labels) becomes more complicated. This regulatory misalignment increases the need for coordinated standardisation processes to enable the connection of evolving metrics across instruments.

There is a wider movement towards circular economy goals that can also be seen in the progressive revisions that are being made in ecodesign implementing measures. Product regulation has gradually changed from being solely about energy efficiency to a combination of energy efficiency and life-cycle performance, as well as resource efficiency, according to Barkhausen et al. [23]. In addition, Cordella et al. [24] show how EU Ecolabel criteria are increasingly being developed to include material efficiency criteria; this will help to further reinforce the inclusion of circularity in product standards. In these newly expanded standards, they provide not just a means of measuring energy, but they will also provide a framework of systemic sustainability standards that cut across production, use, and end-of-life stages. Digitalisation will provide another source of complexity and opportunity.

The digital product passport (DPP) introduced through the Ecodesign for Sustainable Product Regulation is changing the way product data is created, verified, and shared in the supply chain. Numerous studies, including Adisorn et al. [25], Psarommatis and May [26], and Voulgaridis et al. [27], indicate that in order to successfully implement a digital product passport, it is necessary to have interoperable data structures with common reporting formats and secure protocols for transferring that data. From this perspective, standardisation pertains to not only the physical testing of the product, but also applicable to the governance of data, the traceability of products, and the infrastructure related to the interoperability of systems. Therefore, the development of standards is key to creating the digital circular economy (DCE) ecosystem by ensuring that sustainability data is comparable, verifiable, and available to all participants within and across geographical borders.

2.4 Positioning of the Present Chapter

The available research and policy analyses highlight the increasing significance of harmonised standards in facilitating the implementation of Ecodesign and Energy Labelling regulations. Studies have focused on understanding standards' governance role; how they contribute to attaining material efficiency requirements; and their expanded role in relation to achieving broader digitalisation and circular economic goals. Nevertheless, available literature on this topic reveals that the various implementing measures under Ecodesign and Energy Labelling regulations do not yet have a coherent means of linking them to their respective harmonised standards that operationalise them.

Available literature tends to either focus on regulatory progression, discuss material efficiency criteria, or focus solely on digital tools (e.g. digital product passport) separate from one another. As a result, we still do not have a comprehensive/integrated overview that can demonstrate how the various implementing measures under Ecodesign and Energy Labelling relate to their associated harmonised standards. This fragmentation obstructs our ability to measure coherence, find overlap or gaps in coverage, and determine if/how effectively sustainability goals translate to technical specifications.

This chapter will address this gap by producing a systematic mapping of existing Ecodesign/Energy Labelling implementing measures with their relevant harmonised standards. This chapter also enhances our understanding of how sustainability requirements are put into practice and provides a means to understand how standardisation serves as a key tool in implementing EU product policy.

3 Harmonised Standards in Support of Ecodesign and Energy Labelling Implementing Measures

Applicable implementing measures for ecodesign and energy labelling are discussed by the European Commission [4, 5, 28]. A summary of available Directives/Regulations is given in Tables 1, 2, 3, 4, 5, and 6. The tables provide an overview of the applicable regulatory landscape.

In the following subsections, available harmonised standards under the above-mentioned applications are summarised and discussed.

Table 1 Ecodesign and energy labelling directives/regulations (part 1)

Application	Further details on Application	Ecodesign	Energy Labelling
Circuators	Circulators and glandless circulators integrated in products	(EC) 641/2009, (EU) 622/2012	
Computers	Computers and computer servers	(EU) 617/2013	
Computers & small servers	Enterprise servers and data storage products	(EU) 2019/424	
Dishwashers	Household dishwashers	(EU) 2019/2022	(EU) 2019/2017, EPREL
Domestic ovens, hobs & range hoods	Domestic ovens, hobs and range hoods	(EU) 66/2014, CR (EC) 1275/2008, (EU) 2023/826 (Hobs)	(EU) 65/2014, EPREL

Table 2 Ecodesign and energy labelling directives/regulations (part 2)

Application	Further details on Application	Ecodesign	Energy Labelling
Electronic displays	Displays including televisions	(EU) 2019/2021	(EU) 2019/2013
Electric motors	Electric motors	(EU) 2019/1781	
Fans	Industrial fans driven by motors	(EU) 327/2011, (EU) 2024/1834	
Game Consoles	Voluntary agreement	VA 2015-2022, (EU) 2023/826	
Lighting	Light sources	(EU) 2019/2020	(EU) 2019/2015, EPREL

Table 3 Ecodesign and energy labelling directives/regulations (part 3)

Application	Further details on Application	Ecodesign	Energy Labelling
Heating and cooling appliances	Air conditioners and comfort fans	(EU) 206/2012	(EU) 626/2011, (EC) 2002/31
	Hot-water boilers	92/42/EEC	
	Water heaters and hot water storage tanks	(EU) 814/2013	(EU) 812/2013
	Space heaters	(EU) 813/2013	(EU) 811/2013
	Local space heaters, Solid fuel local space heaters	(EU) 2015/1188 (EU) 2015/1185, (EU) 2024/1103	(EU) 2015/1186
	Solid fuel boilers	(EU) 2015/1189	(EU) 2015/1187
	Air heating products, cooling products, high temperature process chillers and fan coil units	(EU) 2016/2281	(EU) 2019/2015, EPREL

Table 4 Ecodesign and energy labelling directives/regulations (part 4)

Application	Further details on Application	Ecodesign	Energy Labelling
Power supplies	External power supplies	(EU) 2019/1782	
Refrigerating appliances	Professional refrigerated storage cabinets	(EU) 2015/1095	(EU) 2015/1094, EPREL
	Household refrigerating appliances	(EU) 2019/2019	(EU) 2019/2016, EPREL
	Refrigerating appliances with a direct sales function	(EU) 2019/2024	(EU) 2019/2018
Set-top boxes	Simple set-top boxes	(EC) 107/2009	

3.1 *Circulators*

In the context of circulators, the following Regulations are available [4, 5]:

- Commission Regulation (EC) No 641/2009 (glandless standalone circulators and glandless circulators integrated in products), see http://eur-lex.europa.eu/legal-content/EN/TXT/?uri=CELEX:32009R0641&locale=en
- Commission Regulation (EU) No 622/2012 of 11 July 2012 amending Regulation (EC) No 641/2009 (glandless standalone circulators and glandless circulators

Table 5 Ecodesign and energy labelling directives/regulations (part 5)

Application	Further details on Application	Ecodesign	Energy Labelling
Smartphones and Tablets	smartphones, mobile phones other than smartphones, cordless phones and slate tablets	(EU) 2023/1670	(EU) 2023/1669
Standby and off mode	Electric power consumption standby and off mode of electrical and electronic household and office equipment	(EC) 1275/2008, replaced by 2023/826 in 2023, (EC) 801/2013	
Transformers	Small, medium and large power transformers	(EU) 548/2014, (EU) 2019/1783	
Tumble driers	Household tumble driers	(EU) 932/2012, (EU) 2023/2533	(EU) 392/2012, (EU) 2023/2534
Tyres			(EU) 2020/740

Table 6 Ecodesign and energy labelling directives/regulations (part 6)

Application	Further details on Application	Ecodesign	Energy Labelling
Vacuum cleaners	Vacuum cleaners	(EU) 666/2013	
Ventilation	Ventilation units	(EU) 1253/2014	(EU) 1254/2014
Washing machines and washer-dryers	Household washing machines and washer-dryers	(EU) 2019/2023	(EU) 2019/2014, EPREL
Water pumps	Water pumps	(EU) 547/2012	
Welding equipment	Welding equipment	(EU) 2019/1784	

integrated in products), see http://eur-lex.europa.eu/legal-content/EN/TXT/?qid=1492615669686&uri=CELEX:32012R0622

Related titles and references of harmonised standards under Directive 641/2009 for eco-design—circulators are summarised in Table 7. These harmonised standards provide presumption of conformity to the underlying legislations and are thus a useful tool for stakeholders, specifically ensuring harmonised testing methods.

Ecodesign requirements for circulators are further discussed in [29]. Circulators are designed to create a circulation through a closed loop system, as for example in heating systems where hot water is transported. The ecodesign regulation includes requirements for two types of circulators: standalone and boiler-integrated devices. The annual energy savings are expected to amount to 14 TWh by 2030 in the EU

Table 7 Related titles and references of harmonised standards under directive 641/2009 for eco-design—circulators

ESO (European Standardisation Organisation)	Reference and title of the Harmonised Standard (and reference document)	First publication in the Official Journal (OJ)
CEN	EN 16297-1:2012: Pumps - Rotodynamic pumps - Glandless circulators - Part 1: General requirements and procedures for testing and calculation of energy efficiency index (EEI)	1st publication
CEN	EN 16297-2:2012: Pumps - Rotodynamic pumps - Glandless circulators - Part 2: Calculation of energy efficiency index (EEI) for standalone circulators	1st publication
CEN	EN 16297-3:2012: Pumps - Rotodynamic pumps - Glandless circulators - Part 3: Energy efficiency index (EEI) for circulators integrated in products	1st publication

corresponding to a 1.000.000 tonnes reduction of CO_2 emissions. As of 2020, 137 M circulators were deployed in the EU, leading to a total fluid energy output of 20.8 TWh. In average, each circulator moved ~30.000 L of water per day on 208 days of the year. 24.4 TWh of electrical energy were needed corresponding to 1% of the 2020 total energy consumption of the EU.

Standardising testing procedures and establishing efficiency indices are two important directions that contribute to transforming energy efficiency goals into actual results. This may be achieved by utilising the regulatory framework governing circulators. Indeed, the standardisation of testing procedures establishes a "common framework" for all manufacturers that provides a way to compare their performance. This makes it much easier for governments to monitor the marketplace and enforce compliance. The gradual decrease in energy-saving performance thresholds has been one of the primary drivers of technological development. This illustrates how the standardisation of testing procedures can both prompt transformation of the marketplace and deliver substantial reductions in energy use and CO_2 emissions.

3.2 Computers

In the context of computers, the following Regulation is available [4, 5]:

- Commission Regulation (EU) No 617/2013 of 26 June 2013 (ecodesign requirements for computers and computer servers), see http://eur-lex.europa.eu/legal-content/EN/TXT/?uri=CELEX%3A32013R0617

Related harmonised standards are currently not available which would provide presumption of conformity to the respective legislations. Rather, temporary guidance is provided by transitional measurement methods; those support the maintenance of consistent compliance in the mean time.

Such transitional methods of measurement available are as follows:

- Commission communication in the framework of the implementation of Commission Regulation (EU) No 617/2013 (ecodesign requirements for computers and computer servers), see http://eur-lex.europa.eu/legal-content/EN/TXT/?uri=uriserv:OJ.C_.2014.110.01.0108.01.ENG

As discussed in [30], it can be considered that the year 1990 marks the approximate start of public usage of the Internet in Europe, when 55 M PCs were deployed. Among those, 97% were desktops. The number of deployed PCs reached 349 M PCs by 2020 with 24% being desktops and 72% laptops, notebooks, or tablet computers. Approx. 2/3rd were used in homes—89% of the EU households had a notebook or tablet and 28% a desktop computer. Thin clients and workstations represented approx. 4% of the total and were mainly deployed in offices. The energy consumption per PC per year was reduced by approx. **65%** to ~100 kWh/year for desktops (40% switched off, 15% sleep mode, and 45% idle mode) and to ~29 kWh/year for notebook (25% switched off, 39% sleep mode, and 36% idle mode) between the years1990 and 2000. The 2020 aggregate electricity consumption amounted to 19 TWh for PCs in the EU versus 31 TWh in 2010. The 2020 average PCs related consumption per EU household amounted to 57 kWh.

Regulating fast-evolving digital technologies is challenging because there are no harmonised computer standards yet. Transitional measurement methods may provide temporary certainty for legal purposes but are not considered as stable or have the same ability to presume conformity as harmonised standards do. This ongoing conflict highlights the difference between how quickly innovations occur and the relatively slower nature of standardisation processes. The reliance on electricity for digital products will increase, making it necessary to develop adaptable, technology-neutral standards so that regulatory consistency and performance are achieved over time.

3.3 Computers and Small Servers

In the context of computers and small servers, the following Regulation is available [4, 5]:

- Commission Regulation (EU) 2019/424 of 15 March 2019 (ecodesign requirements for servers and data storage products), see https://eur-lex.europa.eu/legal-content/EN/TXT/?uri=CELEX:32019R0424

No information is published by the European Commission on harmonised standards which would provide presumption of conformity to the respective legislations [4, 5]. It is understood that this lack of harmonised standards is mainly due to the dynamicity of the sector, characterised by short innovation cycles. In such a context, it is challenging to develop formal standards.

Energy consumption of servers and data products in the EU are discussed in [31]:

- 1990: 16 M computer servers were deployed in the EU, and 24% of them were used for cloud services;
- 2020: 17 M computer servers were deployed, and 51% of them were used for cloud services; 3.1 M data storage products (versus 2.4 M in 2010) were deployed;
- 2030 (projected): 23 M computer servers will be deployed, and 56% used for cloud services. 3.8 M data storage products will be deployed.

A key focus of European Regulation is on PSUs (internal power supply units). 26.5 TWh of electrical energy were used by computer servers and 15.5 TWh by data storage products in total in the EU per year in the 2010–2030 period. The corresponding average PSU efficiency amounted to 71% in 1990 (77% for data storage products) and to 86% in 2020 (91% for data storage products). A further increase to 92% for servers and 95% for data storage products is expected by 2030.

The rise of digital infrastructures, specifically servers and data storage products, has led to the development of a regulatory framework within European energy policy. Although harmonised standards have not yet been established, the focus on PSU efficiency and performance metrics highlights the positive effects of regulation on technological optimisation. Efforts towards standardisation will likely focus on incorporating system efficiency, including data centre interaction and cloud architecture. This outlines the relevance of aligning digital transformation with sustainability objectives.

3.4 Dishwashers

In the context of dishwashers, the following Regulations are available [4, 5]:

- Commission Regulation (EU) 2019/2022 of 1 October 2019 (household dishwashers), see https://eur-lex.europa.eu/legal-content/EN/TXT/?uri=uriserv%3AOJ.L_.2019.315.01.0267.01.ENG&toc=OJ%3AL%3A2019%3A315%3ATOC
- Commission Delegated Regulation (EU) 2019/2017 of 11 March 2019 (energy labelling of household dishwashers), see https://eur-lex.europa.eu/legal-content/EN/TXT/?uri=uriserv%3AOJ.L_.2019.315.01.0134.01.ENG&toc=OJ%3AL%3A2019%3A315%3ATOC

Related titles and references of harmonised standards in support of Regulation 2019/2022 and Delegated Regulation 2019/2017 on dishwashers are summarised in Table 8. The EN 60436 series ensures that performance claims for dishwashers, including energy and water efficiency, are measured in a consistent and comparable way across the European Union.

Table 8 Harmonised standards in support of regulation 2019/2022 and delegated regulation 2019/2017 on dishwashers

ESO (European Standardisation Organisation)	Reference number of the standard	Title of the standard	OJ reference for publication in OJ
CENELEC	EN 60436:2020, EN 60436:2020/AC:2020-06, EN 60436:2020/A11:2020	Electric dishwashers for household use - Methods for measuring the performance	Restrictions apply, see [4,8]
CENELEC	EN 60436:2020, EN 60436:2020/AC:2020-06, EN 60436:2020/A11:202 EN 60436:2020/A12:2022	Electric dishwashers for household use - Methods for measuring the performance	OJ L 224 - 12/09/2023
CENELEC	EN 60436:2020, EN 60436:2020/AC:2020-06, EN 60436:2020/A11:2020	Electric dishwashers for household use - Methods for measuring the performance	Restrictions apply, see [4,8]
CENELEC	EN 60436:2020, EN 60436:2020/AC:2020-06, EN 60436:2020/A11:2020, EN 60436:2020/A12:2022	Electric dishwashers for household use - Methods for measuring the performance	OJ L 224 - 12/09/2023

The energy efficiency of dishwashers is discussed in [31]: Current and projected usage is as follows:

- 2020: 97 M dishwashers are deployed in the EU using 214 M m^3 of water;
- 2030 (projected): 126 M dishwashers will be deployed in the EU.

The corresponding energy and water consumption is as follows:

- 1990: 310 kWh/year of average electricity consumption and 6.7 m^3/year of water per dishwasher;
- 2020: 174 kWh/year (versus estimated 262 kWh/year without regulation) of average electricity consumption and 2.2 m^3/year (versus estimated 5.3 m^3/year without regulation) of water per dishwasher;
- 2030: 159 kWh/year of average electricity consumption and 1.9 m^3/year of water per dishwasher.

It is estimated that the related regulations have led to a saving of a total of 7.3 TWh of electricity and 295 M m^3 of water in 2020. The projected savings in 2030 amount to 10 TWh of electricity and 421 M m^3 of water.

The labelling of dishwashing machines provides an interesting example of how harmonised standards dictate the development of ecodesign regulations and energy efficiency. More specifically, the EN 60436 series of standards provides a means of accurately measuring the performance of dishwashers. This allows manufacturers to supply clear information to consumers, while also providing a level playing field for manufacturers to compete with one another. As a result of this ecodesign measure, there has been a substantial decrease in electricity and water usage through regulatory thresholds and standardised methods of testing. Further, by including water efficiency in ecodesign, it becomes apparent that ecodesign has a far wider impact on the environment than purely the energy consumption dimension.

3.5 Domestic Ovens, Hobs, and Range Hoods

In the context of domestic ovens, hobs, and range hoods, the following Regulations are available [4, 5]:

- Commission Delegated Regulation (EU) No 65/2014 of 1 October 2013 (energy labelling of domestic ovens and range hoods), see http://eur-lex.europa.eu/legal-content/EN/TXT/?uri=CELEX:32014R0065
- Commission Regulation (EU) No 66/2014 of 14 January 2014 (ecodesign requirements for domestic ovens, hobs, and range hoods), see http://eur-lex.europa.eu/legal-content/EN/TXT/?uri=CELEX:32014R0066

Available harmonised standards in support of Regulations for domestic ovens, hobs, and range hoods are summarised in Table 9. These harmonised standards provide presumption of conformity to the underlying legislations and are thus a useful tool for stakeholders, specifically ensuring harmonised testing methods.

Energy efficiency of domestic ovens, hobs, and range hoods is discussed in [32–34]. 240 M ovens were deployed in the EU in 2020 with the share of electric ovens evolving from 76% (1990) to 84% (2020) and projected to 87% (2030).

Due to regulation initiatives, the standby power has dropped considerably from 3 to 0.5 W which has contributed to an overall efficiency improvement by 27% in 2020 versus the estimates without regulation. On average, a newly sold electric oven had an energy consumption of 84 kWh/year as of 2020. The overall savings in terms of primary energy amounted to 12.5 TWh in 2020 (with 48% efficiency of electricity generation) and are projected to 12.7 TWh in 2030. Regulations thus led to a 24% saving.

Also mainly due to regulations, range hoods have become more efficient by 17% in 2020 versus the estimates without regulation. The projection in 2030 amounts an efficiency improvement by 28%. The energy consumption of new range hoods amounts to 110 kWh/year in 2020.

Concerning hobs, 233 M such products were deployed in Europe in 2020. Electric hobs represent a share of 43% (1990) which further evolved towards 66% (2020) and is projected to reach 71% (2030).

The previous paragraphs outline how harmonised standards have been fundamental in providing reliable measurements of energy consumption and standby performance for this product category. The ongoing improvement in efficiency, specifically the decrease in standby power, is a direct result of a series of regulations. The current trend towards greater electrification will also increase the need for harmonised standards, as cooking appliances become more integrated into smart home systems and require careful choices by consumers.

Table 9 Harmonised standards in support of regulations for domestic ovens, hobs, and range hoods

ESO (European Standardisation Organisation)	Reference and title of the Harmonised Standard (and reference document)	First publication in the Official Journal (OJ)
CEN	EN 30-2-1:2015: Domestic cooking appliances burning gas - Part 2-1: Rational use of energy - General	12/04/2017
CEN	EN 15181:2017 (new): Measuring method of the energy consumption of gas fired ovens	1st publication
CENELEC	EN 60350-1:2013: Household electric cooking appliances - Part 1: Ranges, ovens, steam ovens and grills - Methods for measuring performance IEC 60350-1:2011 (Modified), EN 60350-1:2013/A11:2014	10/07/2015
CENELEC	EN 60350-1:2016: Household electric cooking appliances - Part 1: Ranges, ovens, steam ovens and grills - Methods for measuring Performance IEC 60350-1:2016 (Modified)	12/04/2017
CENELEC	EN 60350-2:2013: Household electric cooking appliances - Part 2: Hobs - Methods for measuring performance, IEC 60350-2:2011 (Modified), EN 60350-2:2013/A11:2014	12/04/2017
CENELEC	EN 61591:1997: Household range hoods and other cooking fume extractors - Methods for measuring performance, IEC 61591:1997, EN 61591:1997/A1:2006, IEC 61591:1997/A1:2005, EN 61591:1997/A2:2011, IEC 61591:1997/A2:2010	10/07/2015
CENELEC	EN 61591:1997/A11:2014 EN 61591:1997/A12:2015	14/10/2016

3.6 Electronic Displays

In the context of electronic displays, including televisions, the following Regulations are available [4, 5]:

- Commission Regulation (EU) 2019/2021 of 1 October 2019 (ecodesign requirements for electronic displays), see https://eur-lex.europa.eu/eli/reg/2019/2021/oj
- Commission Delegated Regulation (EU) 2019/2013 of 11 March 2019 (energy labelling of electronic displays), see https://eur-lex.europa.eu/eli/reg_del/2019/2013/oj

Related harmonised standards are currently not available which would provide presumption of conformity to the respective legislations. Consequently, regulatory texts and transitional methods need to be used for compliance until harmonised standards are available.

Energy efficiency of electronic displays is discussed in [35]. As of 2020, there were 522 M electronic displays deployed in the EU with an overall display area of 242 km^2 and a total view time of 817B hours (93,000,000 years corresponding to an overage on-time of 8.5 h/day for household displays). Among those, 80% were televisions, 16% were monitors, and 4% signage displays. In 2030, the display area is estimated to increase to 398 km^2.

The consumption of televisions is evolving as follows:

- 1990: television consumption amounted to 8.8 W/dm^2;
- 2020: television consumption amounted to 1.0 W/dm^2 (versus an estimated 1.59 W/dm^2 without regulation) with the average screen size of televisions increasing from 10 dm^2 (1990) to 51 dm^2 (2020);
- 2030: television consumption is projected to amount to 0.35 W/dm^2.

The electronic display industry illustrates the regulatory complexity in the absence of harmonised standards. Manufacturers will have no presumption of conformity and must rely on direct regulatory provisions and transitional methods to meet compliance requirements, making compliance more complex than before. However, the significant improvements in efficiency over the last several decades show that regulatory requirements are effective. Future standardisation efforts can enhance legal certainty and facilitate the integration of new requirements, such as repairability, material efficiency, and circular economy.

3.7 Electric Motors

In the context of electric motors the following Regulations are available [4, 5]:

- Commission Regulation (EU) 2019/1781 of 1 October 2019 (ecodesign requirements for electric motors and variable-speed drives), see https://eur-lex.europa.eu/eli/reg/2019/1781/oj

Related publications in the Official Journal (under repealed Regulation (EC) No 640/2009) are available as follows:

- Commission communication in the framework of the implementation of the Commission Regulation (EC) No 640/2009 of 22 July 2009 (ecodesign requirements for electric motors), see http://eur-lex.europa.eu/LexUriServ/LexUriServ.do?uri=OJ:C:2012:394:0020:0021:EN:PDF

Energy efficiency of electric motors is discussed in [36]. As of 2020, 380 M electric motors (covered by the ecodesign regulation) were deployed in Europe with power levels ranging from 0.12 to 1000 kW and amounting to a total mechanical output of 1177 TWh. A total of 1326 TWh of electricity was used (53% of the total electricity consumption of the EU), leading to an average efficiency of 88.8%. This represents an improvement from the 1990 levels (85.7% of efficiency).

The regulations led to an estimated saving of 52 TWh of electricity (2020) and projected 107 TWh (2030). As of 2020, those savings correspond to 2.2% of the total electricity consumption of the EU.

The previous paragraphs outline that electric motors constitute one of the most energy-intensive product groups covered by ecodesign legislation. This makes regulation in this domain particularly impactful. A key approach is to set standards that maximise the application of optimised performance. This is achieved by defining precise efficiency classifications and testing procedures across a wide power range, thereby adding consistency to the enforcement of minimum performance levels for electric motors. Given the large share of electricity consumed by electric motors, it is critical that continued standardisation efforts take place in both variable-speed drive technology and system integration to support the achievement of long-term decarbonisation targets.

3.8 Fans

In the context of fans, the following Regulation is available [4, 5]:

- Commission regulation (EU) No 327/2011 of 30 March 2011 (ecodesign requirements for fans driven by motors with an electric input power between 125 W and 500 kW), see http://eur-lex.europa.eu/legal-content/EN/TXT/?uri=CELEX:32011R0327&locale=en

Related harmonised standards are currently not available which would provide presumption of conformity to the respective legislations. Consequently, regulatory provisions need to be used for conformity assessment, which possibly increase complexity for stakeholders.

Energy efficiency of industrial fans is discussed in [37]. As of 2020, a total of 214 M regulated industrial fans were deployed in the EU, with power levels ranging from 0.125 to 500 kW and consuming a total of 281 TWh of electricity. While they correspond to only 5% of the 4B fans in the EU, their electricity consumption amounts to 80% of the total. Such fans generated an output airflow energy amounting to 125 TWh, leading to an average efficiency of 44.5%.

Defining clear performance metrics is essential for energy-intensive industrial applications. Indeed, the metrics established by regulatory frameworks have already led to measurable increases in efficiency. The lack of formalised standards may, however, create complexity in conformity assessments and restrict comparability between businesses operating within the internal market. Thus, the development of harmonised testing procedures in this industry provides additional opportunities for improving energy efficiency in industrial systems.

3.9 Heating and Cooling Appliances

Heating and cooling appliances are divided into the following sub-categories which are assessed separately in the following subsections:

- Air conditioners and comfort fans;
- Hot water boilers;
- Water heaters and hot water storage tanks;
- Space heaters;
- Local space heaters, solid fuel local space heaters;
- Solid fuel boilers;
- Air heating products, cooling products, high-temperature process chillers, and fan coil units.

This classification is used for tailored requirements and testing standards which are designed to meet the needs of each product group.

3.9.1 Air Conditioners and Comfort Fans

In the context of air conditioners and comfort fans, the following Regulations are available [4, 5]:

- Commission Regulation (EU) No 206/2012 of 6 March 2012 (ecodesign requirements for air conditioners and comfort fans), see http://eur-lex.europa.eu/legal-content/EN/TXT/?uri=CELEX:32012R0206&locale=en
- Commission Delegated Regulation (EU) No 626/2011 of 4 May 2011 (energy labelling of air conditioners), see http://eur-lex.europa.eu/legal-content/EN/TXT/?uri=CELEX:32011R0626&locale=en

Titles and references of harmonised standards under Directive 206/2012 for 626/2011—ecodesign-energy labelling—air conditioners are summarised in Table 10. These standards are used as a basis for evaluating efficiency, sound power levels, and seasonal performance.

Energy efficiency of air conditioners and comfort fans are discussed in [38]. As of 2020, 57 M RACs (room air conditioners) are deployed in the EU (86% fixed installed and 14% portable) providing 78 TWh of cooling (requiring 15 TWh of electricity) and 112 TWh of heating (requiring 29 TWh of electricity). 47% of fixed RACs were also heating-capable. The total electricity consumption (44 TWh) corresponds to 1.8% of the total electricity consumption of the EU as of 2020, which is 2384 TWh. The applicable regulations are estimated to lead to savings of 22 TWh of electricity in 2020 (approx. 0.9% of the total electricity consumption of the EU), increasing to 31 TWh in 2030.

The example of air conditioning shows how harmonised standards help to uniformly define seasonal efficiency metrics, sound power levels, and test conditions that reflect typical usage. Given the rapid increase expected in cooling demand

Table 10 Titles and references of harmonised standards under directive 206/2012 for 626/2011—eco-design—energy labelling—air conditioners

ESO (European Standardisation Organisation)	Reference and title of the Harmonised Standard (and reference document)	First publication in the Official Journal (OJ)
CEN	EN 12102-1:2017 (new): Air conditioners, liquid chilling packages, heat pumps, process chillers and dehumidifiers with electrically driven compressors - Determination of the sound power level - Part 1: Air conditioners, liquid chilling packages, heat pumps for space heating and cooling, dehumidifiers and process chillers	1st publication
CEN	EN 12900:2013: Refrigerant compressors - Rating conditions, tolerances and presentation of manufacturer's performance data	11/04/2014
CEN	EN 14511-2:2013: Air conditioners, liquid chilling packages and heat pumps with electrically driven compressors for space heating and cooling - Part 2: Test conditions	11/04/2014
CEN	EN 14511-3:2013: Air conditioners, liquid chilling packages and heat pumps with electrically driven compressors for space heating and cooling - Part 3: Test methods	11/04/2014
CEN	EN 14825:2016 (new): Air conditioners, liquid chilling packages and heat pumps, with electrically driven compressors, for space heating and cooling - Testing and rating at part load conditions and calculation of seasonal performance	1st publication
CEN	EN 15218:2013: Air conditioners and liquid chilling packages with evaporatively cooled condenser and with electrically driven compressors for space cooling - Terms, definitions, test conditions, test methods and requirements	11/04/2014

across Europe, particularly under climate change scenarios, standardised seasonal performance indicators, such as SEER and SCOP methodologies [39], are critical for ensuring comparability and preventing regulatory gaps. Introducing such harmonised measurement frameworks will presumably provide significant energy-saving potential. These measures are essential for the transition and decarbonisation of sectors where load impacts are becoming increasingly significant.

3.9.2 Hot Water Boilers

In the context of hot water boilers, the following Directive is available [4, 5]:

- Council Directive 92/42/EEC of 21 May 1992 on efficiency requirements for new hot water boilers fired with liquid or gaseous fuels, see http://eur-lex.europa.eu/legal-content/EN/TXT/?uri=CELEX:31992L0042&locale=en

Related harmonised standards are currently not available which would provide presumption of conformity to the respective legislations.

Transitional measurement methods following the regulations can be technical ground for both compliance and business needs. This will provide benefits to the industry and economy and generate positive environmental effects. In fact, water heating systems have become more integrated into smart grid and renewable energy systems. Therefore, they are likely to play an even larger role in improving sustainability.

3.9.3 Water Heaters and Hot Water Storage Tanks

In the context of water heaters and hot water storage tanks, the following Regulations are available [4, 5]:

- Commission Regulation (EU) No 814/2013 of 2 August 2013 (ecodesign requirements for water heaters and hot water storage tanks), see http://eur-lex.europa.eu/legal-content/EN/TXT/?uri=CELEX:32013R0814
- Commission Delegated Regulation (EU) No 812/2013 of 18 February 2013 (energy labelling of water heaters, hot water storage tanks, and packages of water heater and solar device), see http://eur-lex.europa.eu/legal-content/EN/TXT/?uri=CELEX%3A32013R0812

Related harmonised standards are currently not available which would provide presumption of conformity to the respective legislations.

However, there are transitional methods of measurement available as follows:

- Commission communication in the framework of the implementation of Commission Regulation (EU) No 814/2013 (ecodesign requirements for water heaters and hot water storage tanks) and of Commission Delegated Regulation (EU) No 812/2013 (energy labelling of water heaters, hot water storage tanks, and packages of water heater and solar device), see http://eur-lex.europa.eu/legal-content/EN/TXT/?uri=uriserv:OJ.C_.2014.207.01.0022.01.ENG

Energy efficiency of water heaters is discussed in [40]. As of 2020, 228 M water heaters were deployed in the EU (60% water heaters (mainly electric) and 40% space and water heaters (mainly gas)) to produce 5.2B m^3 of 60 °C sanitary hot water. The efficiency of water heaters evolves over time as follows:

- 1990: 42% of average efficiency;
- 2020: 66% of average efficiency (estimated to be 55% without regulation);
- 2030: estimated to 71% due to an increased usage of solar and heat pump technology.

As of 2020, the applicable regulations are estimated to save 34 TWh of primary energy. By 2030, it is expected to increase to 73 TWh.

In the European Union today, heating and storing heat is one of the largest energy-consuming domains. The efficiency gains made with this type of appliance through ecodesign regulations and standardised measurement are much larger than the reductions that would have happened before ecodesign regulations were put into place. Given that many of these appliances last a long time, it will take a fair amount of time for the full regulatory benefit of these standards to occur; this reinforces the need for stable, long-term standards that will take future efficiency standards into account. Harmonisation of standards is critical to ensuring that technological advances will yield real and measurable reductions in energy consumption, particularly for heating through heat pumps and hybrid systems.

3.9.4 Space Heaters

In the context of space heaters the following Regulations are available [4, 5]:

- Commission Regulation (EU) No 813/2013 of 2 August 2013 (ecodesign requirements for space heaters and combination heaters), see http://eur-lex.europa.eu/legal-content/EN/TXT/?uri=CELEX:32013R0813
- Commission Delegated Regulation (EU) No 811/2013 of 18 February 2013 (energy labelling of space heaters, combination heaters, packages of space heater, temperature control and solar device, and packages of combination heater, temperature control, and solar device), see http://eur-lex.europa.eu/legal-content/EN/TXT/?uri=CELEX%3A32013R0811

Harmonised standards are addressed in the following Publications in the Official Journal:

- Commission Implementing Decision (EU) 2024/2909 of 22 November 2024, see https://eur-lex.europa.eu/eli/dec_impl/2024/2909/oj

Also, there is a publication of titles and references of transitional methods of measurement as follows:

- Commission communication in the framework of the implementation of Commission Regulation (EU) No 813/2013 (ecodesign requirements for space heaters and combination heaters) and of Commission Delegated Regulation (EU) No 811/2013 (energy labelling of space heaters, combination heaters, packages of space heater, temperature control and solar device, and packages of combination heater, temperature control, and solar device), see http://eur-lex.europa.eu/legal-content/EN/TXT/?uri=uriserv:OJ.C_.2014.207.01.0002.01.ENG

Energy efficiency of space heaters is discussed in [41]. As of 2020, 10 M central hydronic space heaters were deployed in the EU (77% based on gas, 14% based on oil, 6.5% based on electric heat pumps, 2.4% based on electric resistance, 0.8% based on solar-assisted combis, 0.5% based on hybrid gas/electric technology, and

0.1% based on micro combined heat and power). In 2020, those provided 951 TWh of heat. The efficiency of such space heaters evolved over time as follows:

- 1990: 46% of average primary energy efficiency;
- 2020: 82% of average primary energy efficiency (estimated to be 63% without regulation);
- 2030: estimated 96% of average primary energy efficiency.

The long lifetime and thus product replacement cycle is a specific challenge, and it typically takes 20 years or more to have all central hydronic space heaters replaced by a more efficient generation.

As of 2020, the applicable regulation led to savings of 136 TWh of primary energy which is expected to increase to 306 TWh in 2030.

Using energy to heat a space has a huge carbon footprint. The efficiency gains made with this type of appliance through ecodesign regulations and standardised measurement are much larger than the reductions that would have happened before ecodesign regulations were put into place. Regulatory intervention has significantly improved energy performance and accelerated technological progress. The availability of older heating systems has made it hard for new energy-efficient heating systems to be adopted. However, there must still be stable and future-thinking standards that help make this transition possible and provide measurable and verifiable efficiency improvements.

3.9.5 Local Space Heaters, Solid Fuel Local Space Heaters

In the context of local space heaters and solid fuel local space heaters, the following Regulations are available [4, 5]:

- Commission Regulation (EU) 2015/1185 of 24 April 2015 (ecodesign requirements for solid fuel local space heaters), see http://eur-lex.europa.eu/legal-content/EN/TXT/?uri=uriserv:OJ.L_.2015.193.01.0001.01.ENG
- Commission Regulation (EU) 2015/1188 of 28 April 2015 (ecodesign requirements for local space heaters), see http://eur-lex.europa.eu/legal-content/EN/TXT/?uri=uriserv:OJ.L_.2015.193.01.0076.01.ENG
- Commission Delegated Regulation (EU) 2015/1186 of 24 April 2015 (energy labelling of local space heaters), see http://eur-lex.europa.eu/legal-content/EN/TXT/?uri=uriserv%3AOJ.L_.2015.193.01.0020.01.ENG

Related harmonised standards are currently not available which would provide presumption of conformity to the respective legislations.

However, transitional methods of measurement are available as follows:

- Commission communication in the framework of the implementation of Commission Regulation (EU) 2015/1188 (ecodesign requirements for local space heaters), of the implementation of Commission Regulation (EU) 2015/1185 (ecodesign requirements for solid fuel local space heaters) and of the implementation of

Commission Delegated Regulation (EU) 2015/1186 (energy labelling of local space heaters), see http://eur-lex.europa.eu/legal-content/EN/TXT/?uri=uriserv:OJ.C_.2017.076.01.0004.01.ENG&toc=OJ:C:2017:076:TOC

Energy efficiency of local space heaters is discussed in [42]. As of 2020, 33 M local space heaters (LSHs) were deployed in the EU (78.8% electric, 20.5% solid fuel-based, 0.6% gaseous fuel-based, and 0.1% liquid fuel-based) producing 245 TWh/year of heat. The efficiency of local space heaters evolved over time as follows:

- 2020: 77.9% of electric efficiency (electric local space heaters) and 67.8% of primary efficiency (non-electric local space heaters);
- 2030: estimated to be 87.8% of electric efficiency (electric local space heaters) and 75.7% of primary efficiency (non-electric local space heaters).

The applicable regulations are estimated to achieve 16.8 TWh of electricity and 18 TWh of fuel savings as of 2030.

In the area of solid fuel technology, transitional measurement methods remain widely used due to the lack of clear guidelines and the complexity of measuring performance and emissions. Measurable efficiency gains have been supported through the routine use of these methods. While there are measurable efficiency improvements, particulate emissions and their effects on air quality are equally important when discussing sustainability. The establishment of harmonised standards will provide a higher degree of comparability between different types of solid fuels, thereby enhancing the integration of climate and air quality objectives within the regulatory framework.

3.9.6 Solid Fuel Boilers

In the context of solid fuel boilers, the following Regulations are available [4, 5]:

- Commission Regulation (EU) 2015/1189 of 28 April 2015 (ecodesign requirements for solid fuel boilers), see http://eur-lex.europa.eu/legal-content/EN/TXT/?uri=uriserv:OJ.L_.2015.193.01.0100.01.ENG
- Commission Delegated Regulation (EU) 2015/1187 of 27 April 2015 (energy labelling of solid fuel boilers and packages of a solid fuel boiler, supplementary heaters, temperature controls, and solar devices), see http://eur-lex.europa.eu/legal-content/EN/TXT/?uri=CELEX%3A32015R1187

Harmonised standards are addressed in the following Publications in the Official Journal:

- Commission Implementing Decision (EU) 2024/564 of 14 February 2024 on harmonised standards for solid fuel boilers and packages of a solid fuel boiler, supplementary heaters, temperature controls, and solar devices, see https://eur-lex.europa.eu/eli/dec_impl/2024/564/oj

Energy efficiency of solid fuel boilers are discussed in [43]. As of 2020, 9 M solid fuel boilers (SFBs) were deployed in the EU (66% using wood logs, pellets, or chips and 34% using coal). In 2020, those were producing 156 TWh of heat (6.6% of the EU demand for space heating). The efficiency of solid fuel boilers evolved over time as follows:

- 1990: 41.4% of average efficiency;
- 2020: 75.1% of average efficiency (estimated to be 71.5% without regulation);
- 2030: estimated to 78% of average efficiency.

In 2020, 4.4 TWh of primary energy was saved through the applicable regulations which is estimated to reach 10.6 TWh of savings in 2030.

With the introduction of common standards in this area, the link between ecodesign requirements and presumption of conformity is strengthened. This leads to greater legal certainty and market transparency. Energy efficiency and emissions will continue to be equally important for all solid fuel technologies. The continued work being done around standards will be especially important as technologies associated with utilising biomass for heating are examined with respect to overall decarbonisation strategies. Importantly, care should be taken regarding any indirect increase in utilisation of a technology as the result of improvements made to its efficiency; as this could offset some or all of the overall environmental benefits of its use.

3.9.7 Air Heating Products, Cooling Products, High-Temperature Process Chillers, and Fan Coil Units

In the context of air heating products, cooling products, high-temperature process chillers, and fan coil units, the following Regulations are available [4, 5]:

- Commission Regulation (EU) 2016/2281 of 30 November 2016 (ecodesign requirements for air heating products, cooling products, high-temperature process chillers, and fan coil units), see http://eur-lex.europa.eu/legal-content/EN/TXT/?qid=1500298084425&uri=CELEX:32016R2281

Harmonised standards are addressed in the following Publications in the Official Journal:

- Commission Implementing Decision (EU) 2024/2638 of 9 October 2024 on the harmonised standards drafted in support of Regulation (EU) 2016/2281 (ecodesign requirements for air heating products, cooling products, high temperature process chillers, and fan coil units), see https://eur-lex.europa.eu/eli/dec_impl/2024/2638/oj

Also, there is a publication available of titles and references of transitional methods of measurement as follows:

- Commission communication in the framework of the implementation of Commission Regulation (EU) 2016/2281 (ecodesign requirements for air heating products, cooling products, high-temperature process chillers, and fan coil units), see http://eur-lex.europa.eu/legal-content/EN/TXT/?uri=uriserv:OJ.C_.2017.229.01.0001.01.ENG&toc=OJ:C:2017:229:TOC

Energy efficiency of air heating and cooling products is discussed in [44]. As of 2020, a total of 10 M central air heating and cooling devices were deployed in the EU requiring a total of 798 TWh/year of cooling (75% for process chillers, 15% for comfort chillers, and 10% for central air conditioning) and 211 TWh/year of heating (64% for central air conditioning and 36% for air heaters). As of 2030, applicable regulations are estimated to save 18.3 TWh of electricity and 15.8 TWh of fuel, taken together the savings correspond to 0.31% of the total EU energy consumption as of 2021 which is 10922 TWh.

This heterogeneous product group provides various examples of the difficulties associated with regulating equipment used in residential, commercial, and industrial settings. The interaction between harmonised standards and transitional solutions underlines an ever-evolving technical landscape in which system-level performance increasingly matters. More specifically, as these products interact with other industrial energy systems, future standardisation efforts should address integrated performance metrics. Thus, they will enable the alignment of equipment efficiency with overall system optimisation and sector coupling strategies.

3.10 Lighting

In the context of lighting, the following Regulations are available [4, 5]:

- Commission Regulation (EU) 2019/2020 of 1 October 2019 (ecodesign requirements for light sources and separate control gears), see https://eur-lex.europa.eu/eli/reg/2019/2020/oj
- Commission Delegated Regulation (EU) 2019/2015 of 11 March 2019 (energy labelling of light sources), see https://eur-lex.europa.eu/eli/reg_del/2019/2015/oj

No information on available harmonised standards is provided by the European Commission [4, 5].

Energy efficiency of light sources is discussed in [45]. As of 2020, 11B light sources were deployed in the EU (41% based on LEDs, 28% based on compact fluorescent, 19% based on linear fluorescent, 11% based on incandescent, and 1% based on high-intensity discharge). With the total electrically produced light increasing by 2.8× since 1990, a total of 13 T lumens of light output was produced in 2020. The efficiency evolves over time as follows:

- 1990: 17.5 lm/W of average efficiency;
- 2020: 85 lm/W of average efficiency (estimated to be 38 lm/w without regulation);
- 2030: estimated to be 165 lm/W of average efficiency.

As of 2020, the applicable regulations are estimated to save 81 TWh of electricity, which will reach 96 TWh by 2030.

The lighting sector is one of the most successful models of market transformation through ecodesign and energy labelling policies. Although limited public information is available relating to harmonised standards for lighting, the rapid market transformation from conventional to LED is an example of how effectively the signalling effect of regulatory efficiency thresholds, coupled with consumer-facing labelling schemes, function to drive consumer behaviour. In addition, standardised performance metrics for lighting are critical in establishing comparability and driving innovation, thereby directly contributing to significant savings in EU electricity consumption and greenhouse gas emissions [46].

3.11 Power Supplies

In the context of power supplies, the following Regulations are available [4, 5]:

- Commission Regulation (EC) No 278/2009 of 6 April 2009 (ecodesign requirements for no-load condition electric power consumption and average active efficiency of external power supplies), see http://eur-lex.europa.eu/legal-content/EN/TXT/?uri=CELEX:32009R0278&locale=en

The following publications are available in the Official Journal:

- Commission communication in the framework of the implementation of the Commission Regulation (EC) No 278/2009 of 6 April 2009 (ecodesign requirements for no-load condition electric power consumption and average active efficiency of external power supplies), see http://eur-lex.europa.eu/legal-content/EN/TXT/PDF/?uri=uriserv:OJ.C_.2015.120.01.0002.01.ENG

Titles and references of harmonised standards under Directive 278/2009 for Ecodesign—Power Supplies are summarised in Table 11.

The benefits of performance standards can be seen in external power supplies. By establishing a set of efficiency targets and supporting the measures with harmonised measurement standards, significant cumulative energy savings have been realised across billions of small devices [47]. The regulation of no-load consumption and average active efficiency limits both active and passive energy losses, emphasising the need for a complete test methodology. The continued standardisation of electronic devices and chargers will help to prevent rebound effects and keep pace with digitalisation.

Energy efficiency of external power supplies is discussed in [48]. As of 2020, 1.7B EPS (external power supplies) were deployed in the EU (75% for residential use). Those were used to convert 48 TWh/year of electricity. The efficiency of external power supplies evolves over time as follows:

- 1990: 68% of average efficiency;

Table 11 Titles and references of harmonised standards under directive 278/2009 for eco-design—power supplies

ESO (European Standardisation Organisation)	Reference and title of the Harmonised Standard (and reference document)	First publication in the Official Journal (OJ)
CENELEC	EN 50563:2011: External a.c. - d.c. and a.c. - a.c. power supplies – Determination of no-load power and average efficiency of active modes	07/05/2013
CENELEC	EN 50563:2011/A1:2013 (new)	1st publication

- 2020: 82% of average efficiency (estimated to be 75% without regulation).

As of 2020, the applicable regulations led to 5 TWh of electricity savings, which is estimated to reach 5.4 TWh in 2030.

3.12 *Refrigerating Appliances*

Heating and cooling appliances are divided into the following sub-categories which are assessed separately in the following subsections:

- Professional refrigerated storage cabinets;
- Household refrigerating appliances;
- Refrigerating appliances with a direct sales function.

Harmonised standards establish a common technical and base for measuring performance and carrying out market surveillance within refrigeration, be it professional, domestic, or direct-sale type of refrigeration equipment. After 30 years, the significant reduction of electrical energy consumption due to standardising performance testing methods and efficiencies illustrates the value of these two areas in fostering progressive technical improvements over time [49, 50]. The refrigeration sector is especially critical, due to its ongoing operation and relatively high energy density, which places a fundamental role for standardisation in the continuing pursuit of energy savings.

3.12.1 Professional Refrigerated Storage Cabinets

In the context of professional refrigerated storage cabinets, the following Regulations are available [4, 5]:

- Commission Delegated Regulation (EU) 2015/1094 of 5 May 2015 (energy labelling of professional refrigerated storage cabinets), see http://eur-lex.europa.eu/legal-content/EN/TXT/?uri=CELEX%3A32015R1094
- Commission Regulation (EU) 2015/1095 of 5 May 2015 (ecodesign requirements for professional refrigerated storage cabinets, blast cabinets, condensing units, and process chillers), see http://eur-lex.europa.eu/legal-content/EN/TXT/?uri=CELEX%3A32015R1095

The following Publications are available in the Official Journal (Commission Regulation (EU) 2015/1095):

- Commission Implementing Decision (EU) 2024/1589 of 4 June 2024 on the Harmonised Standards for condensing units for refrigeration, see https://eur-lex.europa.eu/eli/dec_impl/2024/1589/oj

Energy efficiency of professional refrigerated storage cabinets is discussed in [51]. As of 2020, 8 M professional refrigerators and freezers were deployed in the EU (40% storage cabinets, 59% condensing units, and 1% large medium- or low-temperature process chillers). The applicable regulations are estimated to save 5.3 TWh in 2020, which is estimated to reach 14.1 TWh in 2030.

3.12.2 Household Refrigerating Appliances

In the context of household refrigerating appliances, the following Regulations are available [4, 5]:

- Commission Regulation (EC) No 643/2009 of 22 July 2009 (ecodesign requirements for household refrigerating appliances), see http://eur-lex.europa.eu/legal-content/EN/TXT/?uri=CELEX:32009R0643&locale=en
- Commission Delegated Regulation (EU) No 1060/2010 of 28 September 2010 (energy labelling of household refrigerating appliances), see http://eur-lex.europa.eu/legal-content/EN/TXT/?uri=CELEX:32010R1060&locale=en

There are the following Publications available of titles and references of transitional methods of measurement:

- Commission communication in the framework of the implementation of Commission delegated Regulation (EU) No 1060/2010 (energy labelling of household refrigerating appliances), see http://eur-lex.europa.eu/legal-content/EN/TXT/?uri=uriserv:OJ.C_.2011.049.01.0006.01.ENG

Also, there are the following Publications in the Official Journal:

Table 12 Titles and references of harmonised standards under directive 643/2009 for 1060/2010 ecodesign—energy labelling—refrigerating appliances

ESO (European Standardisation Organisation)	Reference and title of the Harmonised Standard (and reference document)	First publication in the Official Journal (OJ)
CENELEC	EN 62552:2013: Household refrigerating appliances - Characteristics and test methods IEC 62552:2007 (Modified)	1st publication

- Commission communication in the framework of the implementation of the Commission Regulation (EC) No 643/2009 of 22 July 2009 (ecodesign requirements for household refrigerating appliances) and of Commission Delegated Regulation (EU) No 1060/2010 of 28 September 2010 (energy labelling of household refrigerating appliances), see http://eur-lex.europa.eu/LexUriServ/LexUriServ.do?uri=OJ:C:2014:022:0032:0033:EN:PDF

Titles and references of harmonised standards under Directive 643/2009 for 1060/2010 Ecodesign—Energy Labelling—Refrigerating Appliances are summarised in Table 12.

Energy efficiency of fridges and freezers are discussed in [52]. As of 2020, 258 M household refrigerators and freezers are deployed in the EU, cooling 76.6 M m^3 (net volume) at −18 °C (22%) and +5 °C (78%). The efficiency of refrigerator–freezers evolves over time as follows:

- 1990: 477 KWh/year of average electricity consumption;
- 2020: 181 KWh/year of average electricity consumption (estimated to be 417 KWh/year without regulation);
- 2030: estimated to reach 114 kWh/year.

As of 2020, the applicable regulations saved 55 TWh of electricity which is estimated to reach 75 TWh in 2030.

In 1990, the average electricity consumption of sold refrigerator–freezers was 477 kWh/a. Without measures, in 2020 this would have been 417 kWh/a, but due to ecodesign and energy labelling it was reduced to 181 kWh/a (**−56%**). In 2030 a further decrease to **114 kWh/a** is expected.

3.12.3 Refrigerating Appliances with a Direct Sales Function

In the context of refrigerating appliances with a direct sales function, the following Regulations are available [4, 5]:

- Commission Regulation (EU) 2019/2024 of 1 October 2019 (ecodesign requirements for refrigerating appliances with a direct sales function), see http://data.europa.eu/eli/reg/2019/2024/oj
- Commission Delegated Regulation (EU) 2019/2018 of 11 March 2019 (energy labelling of refrigerating appliances with a direct sales function), see http://data.europa.eu/eli/reg_del/2019/2018/oj

The following Publications in the Official Journal are available:

- Commission Implementing Decision (EU) 2024/2397 of 11 September 2024 on a harmonised standard for refrigerating appliances with a direct sales function, see https://eur-lex.europa.eu/eli/dec_impl/2024/2397/oj
- Commission Implementing Decision (EU) 2025/533 of 25 March 2025 on a harmonised standard for refrigerating appliances with a direct sales function, see https://eur-lex.europa.eu/eli/dec_impl/2025/533/oj

Energy efficiency of commercial refrigerators is discussed in [53]. As of 2020, 14.4 M commercial refrigerators (refrigerating appliances with direct sales function) were deployed in the EU, cooling 17.4 M m^3 (net volume) used for beverage coolers (20%), spiral vending machines (3%), ice cream freezers (5%), and supermarket display fridges (72%). The efficiency of commercial refrigerators evolves over time as follows:

- 1990: 6531 kWh/year of average electricity consumption for commercial fridges;
- 2030: estimated to be 2373 kWh/year of average electricity consumption (estimated to be 4161 kWh/year without regulation).

The applicable regulations are leading to estimated 16 TWh savings of electricity by 2030, reaching 23 TWh by 2040.

3.13 Set-Top Boxes

In the context of set-top boxes, the following Regulation is available [4, 5]:

- Commission Regulation (EC) No 107/2009 of 4 February 2009 (ecodesign requirements for simple set-top boxes), see http://eur-lex.europa.eu/legal-content/EN/TXT/?uri=CELEX:32009R0107&locale=en

Related harmonised standards are currently not available which would provide presumption of conformity to the respective legislations.

In the absence of harmonised standard, basic ecodesign requirements (such as the standby mode) have been helpful in reducing the power used by continuously powered devices or those that sit idle [54]. As the industry of audiovisual services moves towards streamed content (Netflix, Amazon Prime, HBO …) future regulatory approaches will need to become more directive towards network consumption [55].

Harmonised standards would enable to uniformise solutions among similar states or regions (Europe, North America, and Asia).

3.14 Standby and off Mode

In the context of standby and off mode, the following Regulations are available [4, 5]:

- Commission Regulation (EC) No 1275/2008 of 17 December 2008 (ecodesign requirements for standby and off mode electric power consumption of electrical and electronic household and office equipment), see http://eur-lex.europa.eu/legal-content/EN/TXT/?uri=CELEX:32008R1275&locale=en
- Commission Regulation (EU) No 801/2013 of 22 August 2013 (ecodesign requirements for standby, off mode electric power consumption of electrical and electronic household and office equipment, and televisions), see http://eur-lex.europa.eu/legal-content/EN/TXT/?qid=1470402070642&uri=CELEX:32013R0801

There are the following Publications in the Official Journal:

- Commission communication in the framework of the implementation of the Commission Regulation (EC) No 1275/2008 of 17 December 2008 (ecodesign requirements for standby and off mode electric power consumption of electrical and electronic household and office equipment), see http://eur-lex.europa.eu/legal-content/EN/TXT/PDF/?uri=uriserv:OJ.C_.2016.460.01.0001.01.ENG

Summary list of titles and references of harmonised standards under Directive 1275/2008 for Ecodesign—Standby and Off Mode are summarised in Table 13.

Energy efficiency effects through standby and off mode is discussed in [56]. As of 2020, 4B products were consuming electricity while being in off mode or standby mode in the EU. 3B of such devices were regulated for standby only. Examples include the following:

- 57% for diverse small appliances;
- 29% for sound, media, and network equipment;
- 5% for coffee makers;
- 2% for electric toothbrushes;
- And more.

The regulations for the remaining 1B products are covering standby and active use including.

- Washing machines;
- Dishwashers;
- Laundry dryers;
- Ovens;
- Hobs;

Table 13 Titles and references of harmonised standards under directive 1275/2008 for eco-design—standby and off mode

ESO (European Standardisation Organisation)	Reference and title of the Harmonised Standard (and reference document)	First publication in the Official Journal (OJ)
CENELEC	EN 50242:2016 (new): Electric dishwashers for household use - Methods for measuring the performance, IEC 60436:2004 (Modified) IEC 60436:2004/A1:2009 (Modified) + A1:2009 (Modified), IEC 60436:2004/A2:2012 (Modified)	1st publication
CENELEC	EN 50564:2011: Electrical and electronic household and office equipment - Measurement of low power consumption IEC 62301:2011 (Modified)	20/12/2012
CENELEC	EN 60350-1:2013 (new): Household electric cooking appliances - Part 1: Ranges, ovens, steam ovens and grills - Methods for measuring performance IEC 60350-1:2011 (Modified)	1st publication
CENELEC	EN 60350-1:2013/A11:2014 (new)	1st publication
CENELEC	EN 60350-2:2013 (new): Household electric cooking appliances - Part 2: Hobs - Methods for measuring performance IEC 60350-2:2011 (Modified)	1st publication
CENELEC	EN 60350-2:2013/A11:2014 (new)	1st publication

- Printers;
- Fixed phones;
- Game consoles;
- And more.

The efficiency of these products evolves over time as follows:

- 2020: 1.9 W of average standby power (estimated to be 3 W without regulation);
- 2030: estimated to be 1.8 W.

As of 2020, the regulations are estimated to achieve 29 TWh of savings of electricity, reaching 35 TWh in 2030.

Standby and off mode energy consumption is a cross-cutting source of wasted power, which makes it a horizontal approach to regulation. In this context, harmonised standards are important because consistency of measurement is critical for regulating billions of dissimilar products. The cumulative savings achieved, notably in IoT, illustrate how small efficiency gains for individual products can result in large system-wide effects [57], thus reinforcing the strategic benefit of establishing uniform standards for dealing with diffused energy usage patterns.

3.15 Transformers

In the context of transformers, the following Regulation is available [4, 5]:

- Commission Regulation (EU) No 548/2014 of 21 May 2014 (small, medium, and large power transformers), see http://eur-lex.europa.eu/legal-content/EN/TXT/?qid=1440506867412&uri=CELEX:32014R0548

  There are the following Publications in the Official Journal:

- Commission communication in the framework of the implementation of the Commission Regulation (EU) No 548/2014 of 21 May 2014 (small, medium, and large power transformers), see http://eur-lex.europa.eu/legal-content/EN/TXT/PDF/?uri=uriserv:OJ.C_.2016.416.01.0012.01.ENG

Summary list of titles and references of harmonised standards under Directive 548/2014 for Small, Medium, and Large Power Transformers are summarised in Table 14.

Energy efficiency of power transformers is discussed in [58]. As of 2020, 4.4 M utility transformers were deployed in the EU (63% for distribution of electricity to households, 32% for distribution of electricity to industry, 3% for attaching energy sources to the networking including wind, solar, and hydro, and 2% for high-voltage transmission networks). The aggregate loss in the transformers amounted to 105 TWh in 2020, corresponding to 4% of the overall electricity consumption in the EU (2384 TWh in 2020). The efficiency of transformers evolves over time as follows:

- 1990: 22 MWh of average annual electricity loss;

Table 14 Titles and references of harmonised standards under directive 548/2014 for small, medium, and large power transformers

ESO (European Standardisation Organisation)	Reference and title of the Harmonised Standard (and reference document)	First publication in the Official Journal (OJ)
CENELEC	EN 50588-1:2015: Medium power transformers 50 Hz, with highest voltage for equipment not exceeding 36 kV - Part 1: General requirements	11/09/2015
CENELEC	EN 50588-1:2015/A1:2016 (new)	1st publication
CENELEC	EN 50629:2015: Energy performance of large power transformers (Um > 36 kV or Sr ≥ 40 MVA)	11/09/2015
CENELEC	EN 50629:2015/A1:2016 (new)	1st publication

- 2020: 23 MWh of average annual electricity loss (estimated to be 28 MWh without regulation).

The regulations are estimated to lead to 5.7 TWh of electricity savings in 2020, increasing to 17 TWh in 2030.

Power transformers have a central function in the electric power transmission and distribution system. As a consequence, any minor increase in their operational performance can have a large total savings impact [59]. The establishment of consistent global measurement standards has allowed the development of standard testing protocols for loss measurement and classification of efficiency. As the demand for electricity continues to rise, and as the injection of renewable energy continues to proliferate, standardisation is expected to reduce system losses and to support the decarbonisation of the infrastructure system.

3.16 Tumble Driers

In the context of tumble driers the following Regulations are available [4, 5]:

- Commission Regulation (EU) No 932/2012 of 3 October 2012 (ecodesign requirements for household tumble driers), see http://eur-lex.europa.eu/legal-content/EN/TXT/?uri=CELEX:32012R0932&locale=en%22
- Commission Delegated Regulation (EU) No 392/2012 of 1 March 2012 (energy labelling of household tumble driers), see http://eur-lex.europa.eu/legal-content/EN/TXT/?uri=CELEX:32012R0392&locale=en

There are the following Publications in the Official Journal:

- Corrigendum to Commission Communication in the framework of the implementation of Commission Regulation (EU) No 932/2012 (ecodesign requirements for household tumble driers) and of Commission Delegated Regulation (EU) No 392/2012 (energy labelling of household tumble driers), see https://eur-lex.europa.eu/legal-content/EN/TXT/?uri=uriserv:OJ.C_.2017.080.01.0008.01.ENG&toc=OJ:C:2017:080:TOC
- Commission Communication in the framework of the implementation of Commission Regulation (EU) No 932/2012 of 3 October 2012 (ecodesign requirements for household tumble driers) and of Commission Delegated Regulation (EU) No 392/2012 of 1 March 2012 (energy labelling of household tumble driers), see https://eur-lex.europa.eu/legal-content/EN/TXT/?uri=uriserv:OJ.C_.2014.149.01.0001.01.ENG&toc=OJ:C:2014:149:TOC

Summary list of titles and references of harmonised standards under Directive 932/2012 for 392/2012—Ecodesign—Energy Labelling—Tumble Driers are available in Table 15.

Energy efficiency of tumble dryers is discussed in [60]. As of 2020, 50 M laundry dryers were deployed in the EU, representing a considerable increase from 19 M

Table 15 Titles and references of harmonised standards under directive 932/2012 for 392/2012—eco-design—energy labelling—tumble driers

ESO (European Standardisation Organisation)	Reference and title of the Harmonised Standard (and reference document)	First publication in the Official Journal (OJ)
CENELEC	EN 61121:2013: Tumble dryers for household use - Methods for measuring the performance IEC 61121:2012 (Modified)	03/12/2013

dryers in 1990 and estimated to reach 69 M dryers in 2030. The efficiency of laundry dryers evolves over time as follows:

- 1990: 270 kWh/year of average electricity consumption;
- 2020: 150 kWh/year of average electricity consumption (estimated to be 295 kWh/year without regulation);
- 2030: estimated to be 101 kWh/year.

The applicable regulations have led to a 3.8 TWh of electricity saving in 2020, reaching 10.2 TWh by 2030.

In the tumble dryer industry, the use of harmonised standards helps support both the requirements for ecodesign and the consumer-focused energy labelling schemes. The development of heat pump dryers is a case of innovation driven by ever-increasing efficiency targets [61, 62]. Standardised testing methods allow for assurance and reliability for consumers when comparing products and are also evidenced by the cumulative savings from regulatory action on high energy consuming appliances in the home.

3.17 Vacuum Cleaners

In the context of vacuum cleaners, the following Regulations are available [4, 5]:

- Commission Delegated Regulation (EU) No 665/2013 on the energy labelling of vacuum cleaners was annulled by the General Court of the European Union, see https://curia.europa.eu/jcms/upload/docs/application/pdf/2018-11/cp180168en.pdf
- Commission Regulation (EU) No 666/2013 of 8 July 2013 (ecodesign requirements for vacuum cleaners), see http://eur-lex.europa.eu/legal-content/EN/TXT/?uri=CELEX:32013R0666&locale=en

The publication of references of harmonised standards under Commission Delegated Regulation (EU) No 665/2013 is not valid anymore because the Regulation was annulled by the General Court of the European Union. Rather, the following Publications in the Official Journal are available:

- Commission communication in the framework of the implementation of Commission Delegated Regulation (EU) No 665/2013 (energy labelling of vacuum cleaners) and of Commission Regulation (EU) No 666/2013 (ecodesign requirements for vacuum cleaners), see http://eur-lex.europa.eu/legal-content/EN/TXT/PDF/?uri=uriserv:OJ.C_.2017.267.01.0004.01.ENG

Energy efficiency of vacuum cleaners is discussed in [63]. As of 2020, 271 M vacuum cleaners were deployed in the EU including 45 M cordless and robots. 94% of the vacuum cleaners are used in the residential sector. Their efficiency evolves over time as detailed below:

- 1990: 65 kWh/year of the standardised (50 h/year) electricity consumption;
- 2020: 35 kWh/year of the standardised (50 h/year) electricity consumption (expected to be 105 kWh/year without regulation).

As of 2020, the applicable regulations have led to 11.7 TWh of electricity saving, reaching 14.6 TWh by 2030.

In addition to quantified savings, vacuum cleaners demonstrate how ecodesign has transformed both innovation as well as competition in the marketplace through the implementation of power limitations and minimum performance requirements. As a result, marketing has changed from featuring wattage to promoting suction, airflow, and filtration performance. This type of regulation has led to increased technological improvements while providing consumers with increased performance. Additionally, the repeal of the energy labelling regulation emphasises the need for rigorous and representative testing methods. It is also important to define their end-of-life to limit the ecological impact [64]. Future harmonised standards will be vital in keeping up with new usage trends; therefore, updates must be made particularly for cordless and robotic models [65] in order to maintain the long-term energy efficiency improvements achieved in the internal market.

3.18 Ventilation

In the context of ventilation, the following Regulations are available [4, 5]:

- Commission Regulation (EU) No 1253/2014 of 7 July 2014 (ecodesign requirements for ventilation units), see http://eur-lex.europa.eu/legal-content/EN/TXT/?uri=uriserv:OJ.L_.2014.337.01.0008.01.ENG
- Commission Delegated Regulation (EU) No 1254/2014 of 11 July 2014 (energy labelling of residential ventilation units), see http://eur-lex.europa.eu/legal-content/EN/TXT/?uri=uriserv%3AOJ.L_.2014.337.01.0027.01.ENG

Related harmonised standards are currently not available which would provide presumption of conformity to the respective legislations.

However, there is a publication of titles and references of transitional methods of measurement:

- Commission communication in the framework of the implementation of Commission Regulation (EU) No 1253/2014 (ecodesign requirements for ventilation units) and of the implementation of Commission Delegated Regulation (EU) No 1254/2014 (energy labelling of residential ventilation units), see http://eur-lex.europa.eu/legal-content/EN/TXT/?uri=uriserv:OJ.C_.2016.416.01.0016.01.ENG&toc=OJ:C:2016:416:TOC

Energy efficiency of ventilation units is discussed in [66]. As of 2020, 28 M ventilation units (continuously operating, i.e. excluding intermittently operating exhaust fans as used in toilets, bathrooms, or similar) were deployed in the EU. 20 M ventilation units were operated in residential households and 8 M in a non-residential environment. Air ventilation in the heating seasons amounts to 56% of the total use, with ventilation units recovering 100 TWh of heat as of 2020, corresponding to 4% of the overall EU head demand (2360 TWh). This has required 43 TWh of electricity in 2020.

In 2020, residential ventilation units have consumed 7.7 TWh of electricity in 2020 (estimated to be 8.4 TWh without regulation). In the non-residential environment, the 2020 consumption is 35 TWh (estimated to be 36.7 TWh without regulation). The applicable regulations have led to estimated 2.5 TWh of electricity savings in 2020, reaching an estimated 8.4 TWh by 2030.

The importance of system-level efficiency regarding heat recovery and building performance is increasingly being shown through ventilation units. Though no harmonised standard providing presumption of conformity exists, transitional approaches may allow compliance verification to occur. The interrelationship between the variables of ventilation, indoor air quality, and heating demand suggests that future standardisation activities may offer a growing opportunity for the interaction of energy efficiency with health and comfort, further supporting sustainability's multi-dimensional nature [67].

3.19 Washing Machines and Washer–Dryers

In the context of washing machines and washer–dryers, the following Regulations are available [4, 5]:

- Commission Regulation (EU) 2019/2023 of 1 October 2019 (ecodesign requirements for household washing machines and household washer–dryers), see https://eur-lex.europa.eu/legal-content/EN/TXT/?uri=uriserv%3AOJ.L_.2019.315.01.0285.01.ENG
- Commission Delegated Regulation (EU) 2019/2014 of 11 March 2019 (energy labelling of household washing machines and household washer–dryers), see https://eur-lex.europa.eu/legal-content/EN/TXT/?uri=uriserv%3AOJ.L_.2019.315.01.0029.01.ENG

There are the following Publications in the Official Journal available:

- Commission Implementing Decision (EU) 2021/936 of 3 June 2021 on the harmonised standards for household washing machines and household washer–dryers, see https://eur-lex.europa.eu/legal-content/EN/TXT/?uri=uriserv%3AOJ.L_.2021.204.01.0042.01.ENG
- Amendment of 15 January 2025 to Implementing Decision (EU) 2021/936, see https://eur-lex.europa.eu/eli/dec_impl/2025/72/oj

Energy efficiency of washing machines and washer–dryers is discussed in [68, 69], respectively. As of 2020, 175 M washing machines and washer–dryers were deployed in the EU washing 121B kilos of laundry and using 1161B litres of water. The efficiency of these machines has evolved over time as follows:

- 1990: 350 kWh/year of average electricity consumptions and 22 m^3/year water consumption for washing machines operated in average at 56 °C;
- 2020: 96 kWh/year of average electricity consumptions (estimated to be 159 kWh/year without regulation) and 6.3 m^3/year water consumption (estimated to be 13 m^3/year without regulation) for washing machines operated in average at 40 °C.

The applicable regulations have led to estimated 14.1 TWh of electricity and 1212B L of water savings in 2020, reaching estimated 12.4 TWh of electricity (a decrease) and 1464B L of water savings in 2020.

The washing machine sector has been responsible for an impressive reduction in energy and water use resulting from EU policy implementations linked to harmonised standards and the combined implementation of energy and water efficiency regulations. This was facilitated by the implementation of ecodesign and energy labels [70].

Significant reductions in both electricity and water consumption have resulted from the development of methodologies for measuring performance, which has directly impacted design features such as lower operating temperatures and the use of new drum technologies. Competition between countries in different regions has facilitated the emergence of environmental innovation [71]. Finally, the incorporation of durability and repairability into recent regulatory requirements extends the scope of sustainability beyond a sole focus on operational efficiency.

3.20 Water Pumps

In the context of water pumps, the following Regulations are available [4, 5]:

- Commission Regulation (EU) No 547/2012 of 25 June 2012 (ecodesign requirements for water pumps), see http://eur-lex.europa.eu/legal-content/EN/TXT/?uri=CELEX:32012R0547&locale=en

Related harmonised standards are currently not available which would provide presumption of conformity to the respective legislations.

Still, the following transitional methods of measurement are available:

- Commission Regulation (EU) No 547/2012 (ecodesign requirements for water pumps), see http://eur-lex.europa.eu/legal-content/EN/TXT/?uri=CELEX:32012R0547&locale=en

Energy efficiency of water pumps is discussed in [72]. As of 2020, 14.5 M water pumps under applicable regulations were deployed in the EU (39% end-suction pumps, 16% vertical multistage, and 45% submergible multistage). In 2020, these water pumps have generated a total fluid energy output amounting to 96 TWh using 179 TWh of electricity (7.5% of the total EU electricity consumption in 2020) and implying an average efficiency of 54%. The applicable regulations have led to estimated 3.3 TWh of electricity savings in 2020, reaching estimated 3.7 TWh in 2030.

As the use of electric pumps is a significant part of all European Union electrical consumption, there is high potential for affecting energy use through enhanced pump performance. Research highlights the need for improved energy efficiency indicators and methodologies to better select and operate pumping equipment, supporting targeted efficiency gains beyond current practice [73]. Additionally, studies on hydraulic systems controlled by variable-speed drives show that matching pump speed and system demand through adjustable drives can substantially reduce energy consumption and improve overall system efficiency, reinforcing the importance of variable-speed and system-level approaches in future standardisation efforts [74]. Presently, EU approved standards that set minimum performance levels have not been established, but guidelines for determining minimum pump performance through measurement of the minimum efficiency index (MEI) are available to improve pump design. Because of the close relationship between pumps and hydraulic systems, future attention in standardisation will be towards developing hydraulic systems that use variable-speed drives and operate at higher system-level efficiency.

3.21 Welding Equipment

In the context of welding equipment, the following Regulations are available [4, 5]:

- Commission Regulation (EU) 2019/1784 of 1 October 2019 (ecodesign requirements for welding equipment), see https://eur-lex.europa.eu/legal-content/EN/TXT/?uri=uriserv%3AOJ.L_.2019.272.01.0121.01.ENG&toc=OJ%3AL%3A2019%3A272%3ATOC

Related harmonised standards are currently not available which would provide presumption of conformity to the respective legislations.

Energy efficiency of welding equipment is discussed in [75]. As of 2020, 3 M welding units were deployed in the EU (47% manual metal arc technology, 14% tungsten inert gas, 35% metal active or inert gas, and 4% plasma arc cutting). These units were used for an average of 440 h per year and at 3.4 kW output power, while they were put in idle state for 1560 h per year. In total, 4.6 TWh/year of energy was produced in 2020 consuming 6.4 TWh/year of electricity. By 2030, the applicable regulations are expected to lead to an efficiency improvement from 76.9 to 87.5%, while idle state power is estimated to be reduced from 79 to 50 W.

The applicable regulations are leading to estimated 0.87 TWh of electricity savings in 2030.

Ecodesign regulations extend beyond just products into industrial categories with their incorporation into welding apparatus; therefore, they encompass challenges associated with sustainability throughout various joining methods: there are many studies that show how various types of welding differ dramatically in terms of energy consumption and environmental burden depending on whether the process is an automated or manual manner; therefore generally speaking, using manual methods can be a greater consumer of resources than using automated/hybrid (i.e. the total amount of energy consumed over the entire life cycle of the welding operation) based upon available data [76].

In addition, as noted by the results from sustainability assessments conducted on the sample of the welding operations studied, reducing electricity usage, material input(s), and emissions throughout the full life cycle of the welding operation can provide a significant reduction in the environmental impact of the welding operation and support better overall sustainability for their manufacturing decisions [77]. Overall, this demonstrates the importance for harmonisation between test procedures and performance metrics to establish equal basis for comparison within energy efficiency improvements associated with manufacturing processes as industrial electrification continues to expand.

4 Upcoming Harmonised Standards

Currently available harmonised standards in support of ecodesign and energy labelling directives/regulations are listed above. However, the work is continuously ongoing. In addition to the work on which only CEN/CENELEC was involved, the following activities are currently in preparation, in coordination with CEN/CENELEC, by the European Telecommunications Standards Institute (ETSI) in its Technical Committee Environmental Engineering (TC EE):

- ETSI maintains activities in relation to *Commission Regulation (EU) 2023/826 of 17 April 2023 laying down ecodesign requirements for off mode, standby mode, and networked standby energy consumption of electrical and electronic household and office equipment* [78]. This is a horizontal measure that covers a myriad of electronic and electrical products. ETSI had published a European Norm in the past, ETSI EN 303 423 [79] that was prepared under the former mandate M/544 [80] and could be used to demonstrate compliance to the EU Regulation (EC) No 1275/2008 [81] amended by Regulation (EU) 801/2013 [82] now superseded by EU Regulation 2023/826 [78]. As soon as the Standardisation Request related to (EU) 2023/826 will be published, ETSI will initiate the creation of a harmonised standard for the "interconnecting equipment" on the basis of ETSI EN 303 423 [79] while CEN/CENELEC will work on the standard for the "edge equipment" following the basis of the standard EN 50643 [83] that was published for the superseded EU Regulation (EC) No 1275/2008 [81] amended by Regulation (EU) 801/2013 [82].

- ETSI maintains activities in relation to *Commission Regulation (EU) 2019/424 of 15 March 2019 laying down ecodesign requirements for servers and data storage products* [84]. The related Standardisation Request had expired, and as soon as a renewed version is available, ETSI will proceed with the creation of harmonised standards on the basis of the following available European Norms:
  - ETSI EN 303 800-2: Secure data deletion functionality [85];
  - ETSI EN 303 800-3: Availability of firmware and of security updates to firmware [86];
  - ETSI EN 303 800-5: Product disassembly and disassembly instruction [87];
  - ETSI EN 303 470: Energy efficiency measurement methodology and metrics for servers [88];
  - ETSI EN 303 804: Energy efficiency metrics and measurement methods for data storage equipment [89].
- ETSI maintains activities in relation to *Commission Delegated Regulation (EU) 2023/1669 of 16 June 2023 supplementing Regulation (EU) 2017/1369 of the European Parliament and of the Council with regard to the energy labelling of smartphones and slate tablets* [90]. The corresponding standardization request had been published [91], and all three ESOs, i.e. ETSI, CEN, and CENELEC, have agreed on a work split. Notably, ETSI has already started the work on the following requests by Annex 1 of the Standardisation Request [91]:
  - European standard[s] providing detailed technical specifications of the measurement methods and calculations for the preparation for reuse of smartphones, mobile phones other than smartphones, cordless phones, and slate tablets. This will be published as EN 304 236;
  - European standard[s] providing detailed technical specifications of the measurement methods and calculations for the design for reliability of smartphones, mobile phones other than smartphones, and slate tablets. This will be published as EN 304 238;
  - European standard providing detailed technical specifications of the measurement methods and calculations for the standardised physical dimensions of rechargeable batteries of cordless phones. This will be published as EN 304 235;
  - European standard providing detailed technical specifications of the measurement methods and calculations for the energy efficiency index (EEI) of smartphones and slate tablets. This will be published as EN 304 237.
- Starting in 2024, the European Union is implementing a new regulation requiring nearly all products sold in the EU to feature a Digital Product Passport (DPP) [92]. This initiative is part of the *Regulation (EU) 2024/1781 of the European*

Parliament and of the Council of 13 June 2024 establishing a framework for the setting of Ecodesign requirements for sustainable products, amending Directive (EU) 2020/1828 and Regulation (EU) 2023/1542 and repealing Directive 2009/125/EC [1]. CEN/CENELEC and ETSI have accepted the standardisation request for the standards on DPP, and the agreement among these SDOs is that CEN/CENELEC Joint Technical Committee (JTC) 24 will publish the relevant standards but the work is done in cooperation with ETSI. Eight harmonised standards will be produced by end 2026 under the present standardisation request, and they are on:

- Unique identifiers;
- Data carriers;
- Access rights management;
- Interoperability;
- Data processing and exchange;
- Data storage and archiving;
- Data authentication and integrity;
- Application programming interface for life-cycle management.

5 Next Steps and Conclusions

In the present chapter, harmonised standards and other key publications have been listed which are applicable to ecodesign and energy labelling implementing measures. Harmonised standards provide presumption of conformity with the related legislations and are thus a key tool for manufacturers to fulfil all requirements to place products onto the European Union Single Market. Beyond those deliverables, the reader is recommended to check for other supporting documents which are developed by the European Standardisation Organisations (ESOs), i.e. ETSI, CEN, and CENELEC.

This context is of course highly dynamic and new harmonised standards and other deliverables may be published or withdrawn at any time. Indeed, for a number of product categories, no harmonised standards are available or are in the process to be developed. The future availability of related deliverables needs to be closely monitored.

In the future, EU goals for sustainability and energy efficiency may introduce new product categories or stricter performance requirements. Companies, and more broadly any type of manufacturer, should adopt flexible design approaches and use digital tools to track compliance in real time, allowing them to adapt quickly to new or revised regulations.

6 Disclaimer

Note that the authors have assembled the information above to the best of their knowledge as available at the time when the present book was written. Any interpretation is the personal opinion of the authors. Also, the regulation framework can change and evolve any time. The only legally binding text is the one provided in the documents formally published in the European Union Official Journal.

Acknowledgements The authors would like to sincerely thank Dr. Christophe Gaie for his invaluable support and insightful contributions throughout the development of this book chapter. His expertise, guidance, and constructive feedback greatly enhanced the quality and clarity of this chapter.

References

1. Regulation (EU) 2024/1781 of the European Parliament and of the Council of 13 June 2024 establishing a framework for the setting of Ecodesign requirements for sustainable products, amending Directive (EU) 2020/1828 and Regulation (EU) 2023/1542 and repealing Directive 2009/125/EC (Text with EEA relevance). https://eur-lex.europa.eu/eli/reg/2024/1781/oj/eng
2. Directive 2009/125/EC of the European Parliament and of the Council of 21 October 2009 establishing a framework for the setting of Ecodesign requirements for energy-related products (recast) (Text with EEA relevance). https://eur-lex.europa.eu/eli/dir/2009/125/oj/eng
3. Regulation (EU) 2017/1369 of the European Parliament and of the Council of 4 July 2017 setting a framework for energy labelling and repealing Directive 2010/30/EU (Text with EEA relevance). https://eur-lex.europa.eu/eli/reg/2017/1369/oj/eng
4. Harmonised Standards for Ecodesign and Energy Labelling, European Commission. https://single-market-economy.ec.europa.eu/single-market/goods/european-standards/harmonised-standards/Ecodesign_en
5. All the products covered by the Directive 2009/125/EC and Regulation (EU) 2017/1369, European Commission. https://energy-efficient-products.ec.europa.eu/product-list_en
6. New Legislative Framework, European Commission. https://single-market-economy.ec.europa.eu/single-market/goods/new-legislative-framework_en
7. Mueck M, Gaie C (2025) European digital regulations. Springer, Berlin
8. Harmonised Standards, European Commission. https://single-market-economy.ec.europa.eu/single-market/european-standards/harmonised-standards_en
9. Blind K (2016) The impact of standardisation and standards on innovation. In: Edler J, Cunningham P, Gök A, Shapira P (eds) Handbook of innovation policy impact. Edward Elgar Publishing Ltd., New York, pp 423–449
10. Blind K, Gauch S (2009) Research and standardisation in nanotechnology: evidence from Germany. J Technol Transf 34:320–342. https://doi.org/10.1007/s10961-008-9089-8
11. Ikram M, Zhang Q, Sroufe R, Ferasso M (2021) Contribution of certification bodies and sustainability standards to sustainable development goals: an integrated grey systems approach. Sustain Prod Consump 28:326–345. https://doi.org/10.1016/j.spc.2021.05.019
12. Zhao X, Castka P, Searcy C (2020) ISO standards: a platform for achieving sustainable development goal 2. Sustainability 12(22):9332. https://doi.org/10.3390/su12229332
13. Ceko E (2023) On the relationship between ISO standards and sustainable development. Problemy Ekorozwoju 18(2):148–158

14. Communication from the commission to the European parliament, the council and the european economic and social committee a strategic vision for European standards: Moving forward to enhance and accelerate the sustainable growth of the European economy by 2020 (Text with EEA relevance). https://eur-lex.europa.eu/legal-content/EN/TXT/?uri=celex:52011DC0311
15. du Boispéan S, Mueck M, Gaie C (2025) Introduction to the European new legislative framework. In: Mueck M, Gaie C (eds) European digital regulations: intelligent systems reference library, vol 265. Springer, Cham
16. Blind K (2026) Standardization as an innovation policy instrument. In: Schulz-Schaeffer I, Windeler A, Blättel-Mink B (eds) Handbook of innovation. Springer, Cham
17. Micklitz HW (2023) The role of standards in future EU digital policy legislation. In: A consumer perspective. https://anec.eu/images/Publications/other-publications/2023/ANEC-DIGITAL-2023-G-138.pdf
18. ETSI (2026) ETSI's statement paper No. 2. In: ETSI's statement on European standardisation and the European commission's forthcoming European product act, 1st edn. pp 1–3. https://www.etsi.org/images/files/ETSIPositionPapers/ETSI-Statement-Paper-No2.pdf
19. Eliantonio M, Volpato A (2022) The European system of harmonised standards. In: Legal opinion for ECOS. https://doi.org/10.2139/ssrn.4055292
20. Dalhammar C (2015) Industry attitudes towards ecodesign standards for improved resource efficiency. J Clean Prod 123:155–166. https://doi.org/10.1016/j.jclepro.2015.12.035
21. Talens-Peiró L, Ardente F, Mathieux F (2016) Analysis of material efficiency of energy-related products for the development of EU ecolabel criteria: analysis of product groups: personal computers and electronic displays. In: EUR 27793 EN. Publications Office of the European Union. https://doi.org/10.2788/642541
22. Bundgaard AM, Huulgaard RD (2022) The role of standards in support of material efficiency requirements under the ecodesign directive. J Clean Product 385:135599. https://doi.org/10.1016/j.jclepro.2022.135599
23. Barkhausen R, Durand A, Fick K (2022) Review and analysis of ecodesign directive implementing measures: product regulations shifting from energy efficiency towards a circular economy. Sustainability 14(16):10318. https://doi.org/10.3390/su141610318
24. Cordella M, Alfieri F, Sanfelix J et al (2020) Improving material efficiency in the life cycle of products: a review of EU Ecolabel criteriaThe. Int J Life Cycle Assess 25:921–935. https://doi.org/10.1007/s11367-019-01608-8
25. Adisorn T, Tholen L, Götz T (2021) Towards a digital product passport fit for contributing to a circular economy. Energies 14(8):2289. https://doi.org/10.3390/en14082289
26. Psarommatis F, May G (2024) Digital product passport: a pathway to circularity and sustainability in modern manufacturing. Sustainability 16(1):396. https://doi.org/10.3390/su16010396
27. Voulgaridis K, Lagkas T, Angelopoulos CM et al (2024) Digital product passports as enablers of digital circular economy: a framework based on technological perspective. Telecommun Syst 85:699–715. https://doi.org/10.1007/s11235-024-01104-x
28. List of energy efficient products regulations and related Impact Assessments by product group, European Commission. https://energy-efficient-products.ec.europa.eu/policy-making-0/list-energy-efficient-products-regulations-and-related-impact-assessments-product-group_en
29. Circulators—Ecodesign requirements apply to this product. European Commission. https://energy-efficient-products.ec.europa.eu/product-list/circulators_en
30. Computers—ecodesign requirements apply to this product. European Commission. https://energy-efficient-products.ec.europa.eu/product-list/computers_en
31. Servers and Data Storage Products—ecodesign requirements apply to these products. European Commission, https://energy-efficient-products.ec.europa.eu/product-list/servers-and-data-storage-products_en
32. Domestic Ovens—energy labelling and ecodesign requirements apply to this product. European Commission. https://energy-efficient-products.ec.europa.eu/product-list/domestic-ovens_en
33. Range Hoods—energy labelling and ecodesign requirements apply to this product. European Commission. https://energy-efficient-products.ec.europa.eu/product-list/range-hoods_en

34. Hobs—ecodesign requirements apply to this product. European Commission. https://energy-efficient-products.ec.europa.eu/product-list/hobs_en
35. Electronic Displays—energy labelling and ecodesign requirements apply to this product. European Commission. https://energy-efficient-products.ec.europa.eu/product-list/electronic-displays_en
36. Electric Motors and variable speed drives—ecodesign requirements apply to these products. European Commission. https://energy-efficient-products.ec.europa.eu/product-list/electric-motors_en
37. Industrial Fans—ecodesign requirements apply to this product. European Commission. https://energy-efficient-products.ec.europa.eu/product-list/industrial-fans_en
38. Air Conditioners and Comfort Fans—energy labelling and ecodesign requirements apply to this product. European Commission. https://energy-efficient-products.ec.europa.eu/product-list/air-conditioners-and-comfort-fans_en
39. Erginer V, Aydoğdu G (2025) Experimental evaluation of SEER and SCOP deviations in inverter air conditioners under standard test modes and real-life user settings. Energy Build 350:116631. https://doi.org/10.1016/j.enbuild.2025.116631
40. Water Heaters—energy labelling and ecodesign requirements apply to this product. European Commission. https://energy-efficient-products.ec.europa.eu/product-list/water-heaters_en
41. Space Heaters—energy labelling and ecodesign requirements apply to this product. European Commission. https://energy-efficient-products.ec.europa.eu/product-list/space-heaters_en
42. Local Space Heaters—energy labelling and ecodesign requirements apply to this product. European Commission. https://energy-efficient-products.ec.europa.eu/product-list/local-space-heaters_en
43. Solid Fuel Boilers—energy labelling and ecodesign requirements apply to this product. European Commission. https://energy-efficient-products.ec.europa.eu/product-list/solid-fuel-boilers_en
44. Air Heating and Cooling Products—ecodesign requirements apply to this product. European Commission. https://energy-efficient-products.ec.europa.eu/product-list/air-heating-and-cooling-products_en
45. Light Sources—energy labelling and ecodesign requirements apply to this product. European Commission. https://energy-efficient-products.ec.europa.eu/product-list/light-sources_en
46. Singh D, Basu C, Meinhardt-Wollweber M, Roth B (2015) LEDs for energy efficient greenhouse lighting. Renew Sustain Energy Rev 49:139–147. https://doi.org/10.1016/j.rser.2015.04.117
47. International Telecommunication Union (ITU) and Global e-Sustainability Initiative (GeSI) (2012) An energy-aware survey on ICT device power supplies. In: ITU-T study group 5. https://www.itu.int/dms_pub/itu-t/oth/4B/01/T4B010000070002PDFE.pdf
48. External Power Supplies—ecodesign requirements apply to this product. European Commission. https://energy-efficient-products.ec.europa.eu/product-list/external-power-supplies_en
49. Calm JM (2008) The next generation of refrigerants: historical review, considerations, and outlook. Int J Refrig 31(7):1123–1133. https://doi.org/10.1016/j.ijrefrig.2008.01.013
50. McLinden MO, Huber ML (2020) (R)evolution of refrigerants. J Chem Eng Data 65(9):4176–4193. https://doi.org/10.1021/acs.jced.0c00338
51. Professional Refrigerated Storage Cabinets—energy labelling and ecodesign requirements apply to this product. European Commission. https://energy-efficient-products.ec.europa.eu/product-list/professional-refrigerated-storage-cabinets_en
52. Fridges and Freezers—energy labelling and ecodesign requirements apply to this product. European Commission. https://energy-efficient-products.ec.europa.eu/product-list/fridges-and-freezers_en
53. Commercial Refrigerators—energy labelling and ecodesign requirements apply to this product. European Commission. https://energy-efficient-products.ec.europa.eu/product-list/commercial-refrigerators_en
54. Lee W, Lee Y-T, Choi M (2015) Energy saving system for set-top boxes with passive standby mode. In: Proceedings of the 2015 17th international conference on advanced communication technology (ICACT), PyeongChang, Korea (South), pp 553–555. https://doi.org/10.1109/ICACT.2015.7224856

55. Shehabi A, Walker B, Masanet E (2014) The energy and greenhouse-gas implications of internet video streaming in the United States. Environ Res Lett 9(5):054007. https://doi.org/10.1088/1748-9326/9/5/054007
56. Standby, networked standby and off mode. European Commission. https://energy-efficient-products.ec.europa.eu/product-list/standby-networked-standby-and-mode_en
57. Wang W, Su J, Hicks Z, Campbell B (2020) The standby energy of smart devices: problems, progress, and potential. In: Proceedings of the 2020 IEEE/ACM fifth international conference on internet-of-things design and implementation (IoTDI), Sydney, pp 164–175. https://doi.org/10.1109/IoTDI49375.2020.00024
58. Power Transformers—ecodesign requirements apply to this product. European Commission. https://energy-efficient-products.ec.europa.eu/product-list/power-transformers_en
59. Baggini A, Bua F (2015) Power transformers energy efficiency programs: a critical review. In: Proceedings of the 2015 IEEE 15th international conference on environment and electrical engineering (EEEIC), Rome, pp 1961–1965. https://doi.org/10.1109/EEEIC.2015.7165474
60. Tumble Dryers—energy labelling and ecodesign requirements apply to this product. European Commission. https://energy-efficient-products.ec.europa.eu/product-list/tumble-dryers_en
61. Fil BE, Garimella S (2021) The state of the art in energy saving techniques for garment/textile drying. Drying Technol 40(11):2235–2250. https://doi.org/10.1080/07373937.2021.1938599
62. Graban A, Groll E, Braun J, Ziviani D (2021) Review of heat pump tumble clothes dryer energy efficiency and related technologies. Int Refrig Air Condit Confer 41:2222
63. Vacuum Cleaners—ecodesign requirements apply to this product. European Commission. https://energy-efficient-products.ec.europa.eu/product-list/vacuum-cleaners_en
64. Pérez-Belis V, Bakker C, Juan P, Bovea MD (2017) Environmental performance of alternative end-of-life scenarios for electrical and electronic equipment: a case study for vacuum cleaners. J Clean Prod 159:158–170. https://doi.org/10.1016/j.jclepro.2017.05.032
65. Asafa T, Afonja T, Olaniyan E, Alade H (2018) Development of a vacuum cleaner robot. Alex Eng J 57(4):2911–2920. https://doi.org/10.1016/j.aej.2018.07.005
66. Ventilation Units—energy labelling and ecodesign requirements apply to this product. European Commission, https://energy-efficient-products.ec.europa.eu/product-list/ventilation-units_en
67. Giama E (2022) Review on ventilation systems for building applications in terms of energy efficiency and environmental impact assessment. Energies 15(1):98. https://doi.org/10.3390/en15010098
68. Washing Machines—energy labelling and ecodesign requirements apply to this product. European Commission. https://energy-efficient-products.ec.europa.eu/product-list/washing-machines_en
69. Washer Dryers—energy labelling and ecodesign requirements apply to this product. European Commission. https://energy-efficient-products.ec.europa.eu/product-list/washer-dryers_en
70. Boyano-Larriba A, Cordella M, Espinosa-Martinez M, Villanueva-Krzyzaniak A, Graulich K, Rüdinauer I, Alborzi F, Hook I, Stamminger R (2017) Ecodesign and energy label for household washing machines and washer dryers. In: EUR 28809 EN, publications office of the European Union, Luxembourg
71. Kim J, Park Y, Yun C et al (2015) Comparison of environmental and economic impacts caused by the washing machine operation of various regions. Energ Effi 8:905–918. https://doi.org/10.1007/s12053-015-9333-7
72. Water Pumps—ecodesign requirements apply to this product. European Commission. https://energy-efficient-products.ec.europa.eu/product-list/water-pumps_en
73. Ivchenko O, Andrusiak V, Kondus V, Pavlenko I, Petrenko S, Krupińska A, Włodarczak S, Matuszak M, Ochowiak M (2023) Energy efficiency indicator of pumping equipment usage. Energies 16(15):5820. https://doi.org/10.3390/en16155820
74. Malrait F, Jebai AK, Ejjabraoui K (2017) Power conversion optimization for hydraulic systems controlled by variable speed drives. J Process Control 74:133–146. https://doi.org/10.1016/j.jprocont.2017.11.002

75. Welding Equipment—ecodesign requirements apply to this product. European Commission. https://energy-efficient-products.ec.europa.eu/product-list/welding-equipment_en
76. Schischke K, Nissen NF, Lang K (2014) Welding equipment under the energy-related products directive. J Ind Ecol 18(4):517–528. https://doi.org/10.1111/jiec.12164
77. Doncheva E, Djokikj J, Avramov N, Krstevska A, Petreski M (2023) Sustainability and application of life cycle assessment in welded structures. Ss. Cyril and Methodius University, Faculty Of mechanical Engineering, Skopje, North Macedonia. https://stumejournals.com/journals/innovations/2023/1/3
78. Commission Regulation (EU) 2023/826 of 17 April 2023 laying down Ecodesign requirements for off mode, standby mode, and networked standby energy consumption of electrical and electronic household and office equipment pursuant to Directive 2009/125/EC of the European Parliament and of the Council and repealing Commission Regulations (EC) No 1275/2008 and (EC) No 107/2009 (Text with EEA relevance). https://eur-lex.europa.eu/eli/reg/2023/826/oj/eng
79. ETSI EN 303 423 V1.3.1 (2021-07), Environmental Engineering (EE); Electrical and electronic household and office equipment; Measurement of networked standby power consumption of Interconnecting equipment. https://www.etsi.org/deliver/etsi_en/303400_303499/303423/01.03.01_60/en_303423v010301p.pdf
80. C(2015)9468—Standardisation request M/544: commission implementing decision of 5.1.2016 on a standardisation request to the European standardisation organisations as regards Ecodesign requirements for networked standby in support of Regulation (EC) No 1275/2008 and Regulation (EC) No 642/2009. https://ec.europa.eu/growth/tools-databases/enorm/mandate/544_en
81. Commission Regulation (EC) No 1275/2008 of 17 December 2008 implementing Directive 2005/32/EC of the European parliament and of the council with regard to ecodesign requirements for standby and off mode electric power consumption of electrical and electronic household and office equipment (text with EEA relevance). https://eur-lex.europa.eu/eli/reg/2008/1275/oj/eng
82. Commission Regulation (EU) No 801/2013 of 22 August 2013 amending regulation (EC) No 1275/2008 with regard to ecodesign requirements for standby, off mode electric power consumption of electrical and electronic household and office equipment, and amending regulation (EC) No 642/2009 with regard to ecodesign requirements for televisions Text with EEA relevance. https://eur-lex.europa.eu/eli/reg/2013/801/oj/eng
83. EN 50643:2018-07: electrical and electronic household and office equipment—measurement of networked standby power consumption of edge equipment. https://www.dinmedia.de/de/norm/din-en-50643/282633462
84. Commission Regulation (EU) 2019/424 of 15 March 2019 laying down Ecodesign requirements for servers and data storage products pursuant to Directive 2009/125/EC of the European Parliament and of the Council and amending Commission Regulation (EU) No 617/2013 (Text with EEA relevance.). https://eur-lex.europa.eu/eli/reg/2019/424/oj/eng
85. ETSI EN 303 800-2 V1.1.1 (2025-06), Environmental Engineering (EE); Assessment of material efficiency of ICT network infrastructure goods (circular economy); Part 2: Server and data storage product secure data deletion functionality. https://www.etsi.org/deliver/etsi_en/303800_303899/30380002/01.01.01_60/en_30380002v010101p.pdf
86. ETSI EN 303 800-3 V1.1.1 (2025-02), Environmental Engineering (EE); Assessment of material efficiency of ICT network infrastructure goods (circular economy); Part 3: Server and data storage product availability of firmware and of security updates to firmware. https://www.etsi.org/deliver/etsi_en/303800_303899/30380003/01.01.01_60/en_30380003v010101p.pdf
87. ETSI EN 303 800-5 V1.1.1 (2024-08), Environmental Engineering (EE); Assessment of material efficiency of ICT network infrastructure goods (circular economy); Part 5: Server and data storage product disassembly and disassembly instruction. https://www.etsi.org/deliver/etsi_en/303800_303899/30380005/01.01.01_60/en_30380005v010101p.pdf
88. ETSI EN 303 470 V1.1.1 (2019-03)—UPDATE IN PREPARATION, Environmental Engineering (EE); Energy Efficiency measurement methodology and metrics for servers. https://www.etsi.org/deliver/etsi_en/303400_303499/303470/01.01.01_60/en_303470v010101p.pdf

89. ETSI EN 303 804 V1.1.1 (2025-06), Environmental Engineering (EE); Energy efficiency metrics and measurement methods for data storage equipment. https://www.etsi.org/deliver/etsi_en/303800_303899/303804/01.01.01_60/en_303804v010101p.pdf
90. Commission Delegated Regulation (EU) 2023/1669 of 16 June 2023 supplementing Regulation (EU) 2017/1369 of the European Parliament and of the Council with regard to the energy labelling of smartphones and slate tablets (Text with EEA relevance). https://eur-lex.europa.eu/eli/reg_del/2023/1669/oj/eng
91. Commission implementing decision of 3.4.2025 on a standardisation request to the European standardisation organisations as regards ecodesign requirements for smartphones, mobile phones other than smartphones, cordless phones and slate tablets in support of Regulation (EU) 2023/1670 and Delegated Regulation (EU) 2023/1669, M/608—C(2025)1927. https://ec.europa.eu/growth/tools-databases/enorm/mandate/608_en
92. EU's digital product passport: advancing transparency and sustainability. European Commission. https://data.europa.eu/en/news-events/news/eus-digital-product-passport-advancing-transparency-and-sustainability

Chapter 5
Overview of Digital Architectures to Foster Sustainability

Christophe Gaie and Laurent Denis

Abstract This chapter will provide a holistic overview on available architectural strategies and approaches in the context of digital systems to support sustainability objectives. These are indeed key requirements on multiple levels—government legislations impose more and more stringent sustainability requirements, manufacturers desire to position themselves as key innovators in the field of sustainability, and customers expect that manufacturers design products in accordance with green design principles. Example strategies include shutting down of temporarily unused parts of a component or system, adapting the performance to the actual needs while minimizing expenses in terms of energy consumption, and applying semantic and goal-oriented communication principles to minimize the over-the-air capacity requirements, etc.

Keywords Architecture · Sustainability · Green IT · Design · Performance · Energy consumption

1 Introduction to Digital Architecture

The traditional definition of architecture as "the art or practice of designing and making buildings"[1] holds utmost importance by ensuring structures are well designed, notably for safety, usability, comfort, cost-effectiveness, durability, and esthetics. Extending this architectural lens to IT systems is now crucial because these systems form the foundation of modern society, with almost every process and service relying on digital solutions.

[1] Definition of the term "architecture" given by the Cambridge Dictionary https://dictionary.cambridge.org/dictionary/english/architecture.

C. Gaie · L. Denis
French Prime Minister Office, Paris, France

C. Gaie (✉)
Laboratoire Magellan, Université Jean Moulin—Lyon 3, Lyon, France
e-mail: christophe.gaie@gmail.com

C. Gaie and M. Mueck (eds.), *Digital Sustainability: Legislations, Standards, Solutions and Related Case Studies*, Advances in Sustainability Science and Technology,
https://doi.org/10.1007/978-981-92-0458-8_5

Applying architectural thinking to digital systems brings significant advantages, fostering well-structured, scalable, and maintainable solutions. It enables better management of complexity, enhances reliability and security, and promotes efficient resource utilization. Furthermore, a conscious architectural approach allows for the integration of sustainability principles. This includes designing energy-efficient systems, optimizing hardware usage, reducing digital waste, and considering the long-term environmental impact of IT infrastructure and software development practices. By embedding sustainability into the architecture of digital systems, we can build evolvable solutions that are able to adjust to future technological evolutions more efficiently.

The use of architectural thinking for digital systems provides many advantages that can lead to better organized, scalable, and maintainable solutions. It facilitates managing various levels of complexity, integrating reliability and security, and optimizing resource utilization. In addition, digital architecture should now consider the recent evolution toward promoting sustainability. As a consequence, the principles set by modern architecture should seek to maximize the effectiveness of energy consumption, reduce the use of mineral resources, and decrease digital waste. This will increase the longevity of infrastructural IT technology and software development processes, while also protecting the environment. This evolution means that sustainability must be integrated into digital architecture from the beginning of projects, so that systems can evolve and adapt more effectively to changes.

In the digital domain, the roots of architecture lie in the necessity to organize and manage a growing number of systems with increasing complexity. Early computers first focused on physical architecture, aiming to optimally organize the hardware components of the central processing unit (CPU), memory, and input/output devices. The initial work of John von Neumann in the 1940s defined the storage of both data and instructions in a single memory space [1], a foundational architectural principle that persists in many systems today.

Then, software architecture emerged to specify and organize the growing number of components in a software system, as well as the relationships and interactions between those components. It introduced notions such as modularity, layering, and client–server systems in order to build scalable, maintainable, and robust solutions. This evolution continued with the emergence of networking and distributed systems, which required architectures to manage communications, distributed data, and resource management across interconnected computers [2].

Figure 1 shows the main elements involved in designing a sustainable digital architecture.

This chapter aims to explain why sustainability is a key concern in the architecture of digital systems and is structured as follows. Section 5.2 presents the historical evolution of digital architecture, tracing its development from early physical configurations to contemporary enterprise-wide strategies. Section 5.3 identifies and discusses the fundamental architectural concepts that contribute to sustainability, such as flexibility, modularity, scalability, transparency, and standardization. Section 5.4 focuses on specific strategies and tools that foster sustainability within IT architectures, highlighting innovations at the hardware, software, and enterprise

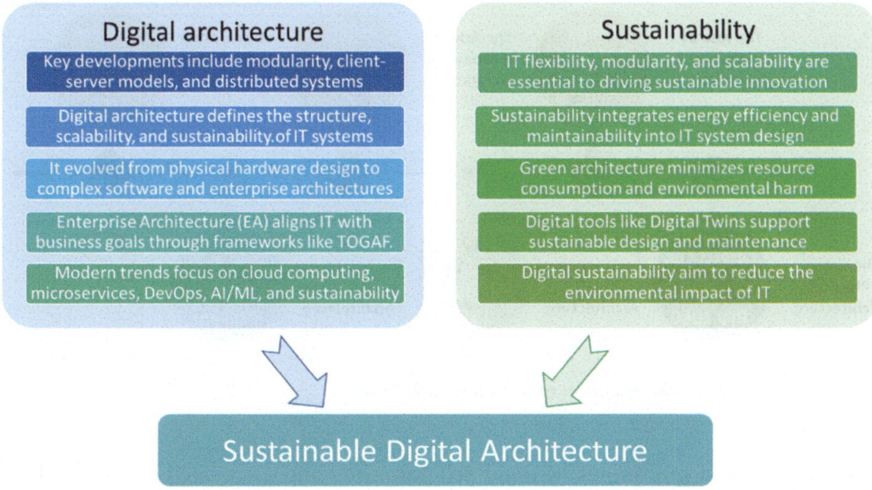

Fig. 1 Designing a sustainable digital architecture

levels. Finally, Sect. 5.5 concludes by summarizing the key insights of this chapter and emphasizing the growing importance of integrating sustainability as a foundational element in future digital architectures.

2 The Evolution of Digital Architecture

The evolution of digital architecture mirrors the increasing complexity and sophistication of computing and information technology, progressing from a focus on physical hardware to encompassing intricate software systems and enterprise-wide strategies. This evolution is illustrated in Fig. 2.

The origin of digital architecture can be traced back to what is called "Physical Architecture," which existed in the pre-1960s. During this time, the primary focus was on the physical arrangement of hardware components within individual computers. Key characteristics of this era included direct manipulation of hardware, limited software, and machine-specific programming. Two important milestones of this period were the development of early computing devices like vacuum tubes and transistors, which served as fundamental building blocks, and the Von Neumann Architecture in the 1940s, which proposed storing both data and instructions in a single memory space—a foundational principle that persists in many systems today [3].

Between the 1960s and 1980s, architecture expanded from hardware to software. This period focused on organizing software components and their connections in order to manage the complex imbrications between software systems. The emergence of operating systems, programming languages, and modular design played an important role in defining architectural principles. More precisely, the 1960s and

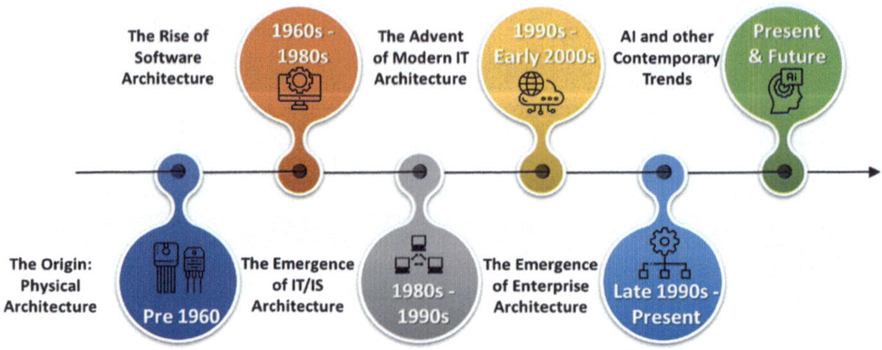

Fig. 2 Main phases in the evolution of digital architecture

1970s represented milestones of fixed programming and associated cultural values around modularity [4, 5]. Waterfall model methodologies that appeared during this time were early, systematic approaches to building software [6]. Additionally, the client–server model emerged in the late 1970s and 1980s, decentralizing processing across client and server. This continued evolution and acknowledgment of significant patterns and styles associated with software design allowed for what many consider the "golden age" of software architecture. This contributed to defining and structuring the central principles of architecture and building their robustness over time [7].

In the 1980s and 1990s, the "Emergence of IT/IS Architecture" marked a significant transition in which IT systems were expected to align with business needs and adopt an enterprise focus on information flow and data management [8]. This domain became known as Information Systems (IS) Architecture. Several characteristics defined this period, including the strong introduction of database management technology, the rise of network capabilities (e.g., local area networks and early wide area networks), and the development of local application frameworks. Moreover, it was during this period that relational models became widespread, building on contemporaneous work of the late 1970s to create a standard method for storing and retrieving data [9]. This period also saw the creation of foundational frameworks for IS Architecture, notably the Information Systems Architecture (ISA) framework [10].

In the 1990s and early 2000s, the "Advent of Modern IT Architecture" led to a focus on integration, interoperability, and the scalability of distributed systems as a result of the internet and globalization. This period was characterized by the widespread adoption of three-tier architecture, rapid growth of web technologies, component-based development, and the introduction of early Enterprise Application Integration (EAI) solutions. Significant milestones of this period included the rise of the internet and web technologies around 1995, which disrupted the way applications were delivered and integrated [11, 12], as well as the broad use of object-oriented architecture and design, leading to improved reusability and maintainability [13, 14].

Finally, early EAI solutions addressed the need to connect disparate applications across the enterprise [15].

The "Emergence of Enterprise Architecture" from the late 1990s to the present represents a shift toward taking a holistic, enterprise-wide view of IT. This involved aligning IT strategy with business strategy and structuring all dimensions of IT, including business, data, application, and technology. Significant events from this period include John Zachman's Framework, one of the first to formalize the concepts of Enterprise Architecture in 1987 [10]. Then, at the end of the 1980s, the term itself became more established, leading to the formation of a range of more complete frameworks that exist today, such as The Open Group Architecture Framework (TOGAF) and the Federal Enterprise Architecture Framework (FEAF). These frameworks contributed to the democratization of Enterprise Architecture and accelerated its adoption in organizations. Thanks to this work, Enterprise Architecture is now recognized as a domain that provides strategic value for organizations [16].

Finally, "Contemporary Trends in IT Architecture," which are present today, are characterized by a focus on agility, cloud computing, microservices, DevOps, data-centricity, security by design, and sustainability. Enterprise Architecture continues to evolve to incorporate these trends. Key characteristics include cloud-native architectures, containerization, API-led connectivity, event-driven architectures, AI/ML integration, DevSecOps practices, and Green IT initiatives. Important milestones include the widespread adoption of cloud computing with IaaS, PaaS, and SaaS models [17]; the rise of microservices architecture, which breaks down monolithic applications into smaller, independent services [18, 19]; the adoption of DevOps and Agile methodologies, which integrate development and operations for faster delivery [20]; and an increasing emphasis on data governance and analytics, which leverages data as a strategic asset [21]. This development reflects a constant need to manage complexity, increase efficiency, align information technology (IT) with business objectives, and adapt to technological change. Today, Enterprise Architecture is the most developed form of architecture, as it provides a strategic framework for viewing the entire information technology ecosystem of an organization.

Built upon the evolution of software architecture to address the increasing complexity of organizations, Enterprise Architecture is a holistic discipline [2, 22]. It offers a comprehensive framework to structure the different architectural dimensions of IT systems and align them with business objectives and strategies. It extends beyond individual systems or software components to include the entire IT landscape of the organization, considering business architecture, data architecture, application architecture, and technology architecture. By identifying clear relationships and dependencies between domains, it improves decision-making, IT governance, IT resource allocation, and overall business agility. Importantly, EA enables a reduction of business risks by integrating human resources, governance, technology, and processes [23]. As a consequence, it also reinforces the resilience of organizations in continuous digital transformation, especially given the increase in cybersecurity threats. Finally, EA aligns IT investments with business objectives to build a modern digital ecosystem that can adapt to market evolutions and technological innovations.

Contemporary IT Architecture primarily aims to better address agility, facilitate the adoption of cloud computing, and integrate the practices of DevOps, alongside the development of modular microservices. Furthermore, there is a significant emphasis on managing data and knowledge, incorporating security by design from the outset, and a growing commitment to sustainability in IT operations [24]. Key characteristics defining today's IT architectures include the prevalence of cloud-native architectures, which fully leverage the capabilities of cloud platforms, and the widespread use of containerization, exemplified by technologies like Docker and Kubernetes. Moreover, the usage of APIs is now standard, enabling seamless integration between disparate systems and services [25], while event-driven architectures are gaining prominence for their ability to react in real time to changes and events within the system. The integration of artificial intelligence (AI) and machine learning (ML) is rapidly growing, aiming to provide task automation, improve prediction, and enhance decision-making capabilities within IT systems. Significant improvements were made to accelerate the usage of cloud computing across various models like Infrastructure as a Service (IaaS), Platform as a Service (PaaS), and Software as a Service (SaaS) [26]. This offers a profound transformation in the IT architecture of organizations, facilitating the provision and scaling of IT resources.

3 The Emergence of Sustainability

The rapid evolution of digital architecture is increasingly converging with the critical imperative of sustainability. Once a peripheral thought, this concept has become a fundamental pillar of modern IT, acknowledging the significant environmental footprint of digital technologies. It aims to tackle the problem of massive energy consumption by data centers [27], as well as the proliferation of electronic waste, whose environmental impact can no longer be overlooked. This evolution is depicted in Fig. 3.

To overcome these challenges, governments have settled stringent regulations. Manufacturers have also invested in innovation to gain competitive advantages and optimize their image by complying with growing expectations from consumers who prioritize eco-friendly practices in their purchasing decisions. This shift also reflects a broader understanding of "Green Architecture," or "sustainable architecture," which emphasizes minimizing resource consumption and environmental harm [28]. This is now directly applicable to digital infrastructure and the creation of sustainable digital infrastructures [29]. Moreover, Enterprise Architecture processes are increasingly being leveraged to align IT strategies and functions with an organization's overarching sustainability goals, fostering IT flexibility and collaborative innovation to create social and economic value within a sustainable business ecosystem [30].

During this foundational phase, the concept of "Green IT" emerged as a set of principles and practices aimed at making IT operations environmentally sustainable [31]. Academic discourse began to highlight the environmental impact across the entire Information and Communication Technology (ICT) ecosystem, from the

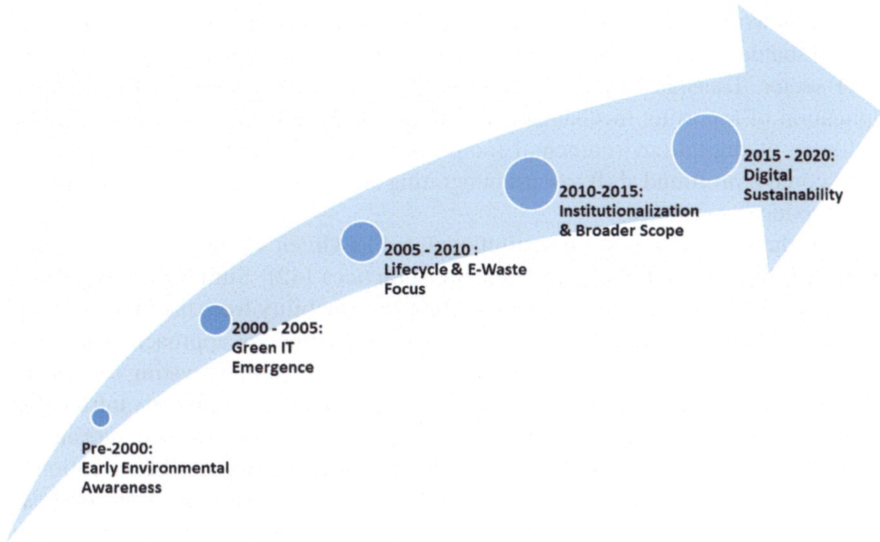

Fig. 3 Evolution toward digital sustainability

energy consumption of data centers and networks to the environmental footprint associated with manufacturing, use, and disposal of digital products [32]. Researchers identified key objectives for Green IT, emphasizing not only energy conservation but also broader considerations like responsible manufacturing, waste management, and the design of environmentally sound software and hardware [33]. This early understanding was crucial in acknowledging the growing power consumption of ICT, including the increasing energy demands of mobile and fixed networks, which was becoming a significant concern [34]. The goal was to establish an ecological information technology approach that recognized the inherent conflict between IT growth and its environmental consequences, preparing a future where sustainability would be fully integrated into the IT landscape [35, 36].

Then, the extent of Green IT expanded to the entire life cycle of IT products. To this end, it addresses the manufacturing process, aiming to significantly reduce energy consumption and environmental impacts, as detailed in [37]. It then considers ethical impacts in the supply chain, such as social equity, environmental protection, and economic feasibility. The importance of responsible practices in ICT manufacturing was also outlined in [38]. Moreover, the strategic design of products for modularity, repairability, and upgradability was recognized as essential to extending product life and minimizing resource consumption [39]. Finally, responsible end-of-life disposal and recycling became a central aspect to consider, particularly for addressing the growing challenge of e-waste management [40].

This era also marked the emergence of rigorous analytical tools like Lifecycle Assessment (LCA) for IT products, providing deeper insights into environmental impacts across all stages of product development. Alongside this, sustainable supply chain frameworks gained prominence [38], complemented by the adoption of international standards such as ISO 14001 for environmental management systems within the IT sector. Transparency and accountability were further reinforced through the application of reporting mechanisms like the GRI Guidelines, which promoted the disclosure of digital environmental footprints [41]. This comprehensive approach underscored a profound shift toward integrating sustainability across every phase of an IT product's existence.

Then, these principles were institutionalized as Green IT evolved toward more structured, organization-wide sustainability practices [42]. Simultaneously, the IT community became conscious that integrating sustainability from the initial conception stage of digital products and services is the most efficient approach. This advocates for incentivizing IT services to engage resources toward innovating for sustainable solutions [43]. This period also witnessed governments increasingly introducing regulations specifically targeting energy consumption in digital infrastructure and the management of e-waste, with international frameworks like the Basel Convention on the Control of Transboundary Movements of Hazardous Wastes and Their Disposal strengthening controls, including those on e-waste [44]. At the same time, consumer expectations for eco-friendly digital products increased, significantly influencing manufacturers to adopt more sustainable practices.

At the end of the 2010s and in the early 2020s, there was a shift from Green IT toward digital sustainability. This broader concept integrates the alignment of digital technologies with comprehensive environmental and social goals at a strategic level [45]. Among the new concepts adopted, the authors underline the importance of achieving "Sustainable by Design" (SBD) principles, which promote energy-efficient algorithms [46]. Other important concepts include building lean and scalable digital infrastructures and applying data-minimization techniques to reduce storage and computational footprints. Moreover, the emergence of artificial intelligence enables the optimization of energy grids and the development of robust blockchain systems to ensure transparent and more sustainable supply chains [47]. The use of AI also facilitates big data analytics for advanced climate modeling [48]. Altogether, these developments confirm that during this period, the focus shifted toward designing digital infrastructure with sustainability as a core principle [49].

Then, following this evolution, digital sustainability has become a fundamental pillar of modern IT. It became fully integrated into organizational strategies [45] and recognized as crucial for achieving global environmental goals, such as those outlined in the UN Sustainable Development Goals (SDGs). Today, this integration drives innovation toward a sustainable business ecosystem, emphasizing holistic, lifecycle-based digital design, meticulous measurement, tracking, and reporting of IT's environmental impact [50], and leveraging digital transformation as an enabler for broader sustainability goals. The Global E-waste Monitor 2024 outlines that the growth of e-waste makes it more urgent than ever to increase efforts toward collecting, recycling, and building sustainable digital practices and initiatives [51].

4 IT Architecture Concepts that Contribute to Sustainability

The evolution of digital architecture has been marked by significant shifts, yet it remains guided by core principles that ensure its continuous adaptation and effectiveness. These foundational concepts are crucial for understanding how modern IT systems are designed, built, and managed to meet evolving business requirements and integrate technological advancements. Therefore, this section elaborates on the importance of key concepts such as: IT flexibility, modularity, scalability, transparency, and standardization. They are illustrated in Fig. 4.

IT flexibility refers to the capacity of IT systems and architectures to readily and cost-effectively respond to new requirements, new technologies, and new business models. It outlines the transformation from rigid, monolithic structures, like legacy taxation systems [52], toward adaptable architectures that can support changes without

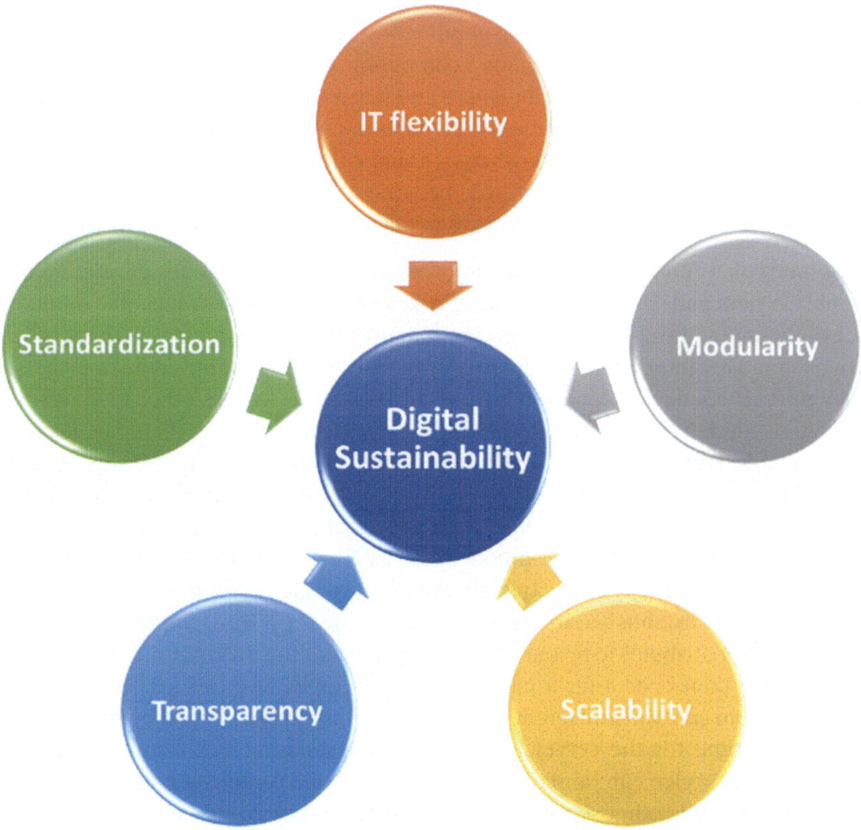

Fig. 4 Five concepts that contribute to digital sustainability

requiring major rewrites. Achieving IT flexibility requires loose coupling, defined interfaces, and a careful approach to cloud services that allows businesses to operate efficiently, adapt quickly, and implement new features based on changing requirements. Application Programming Interfaces (APIs), as in [53], depict an adaptable architecture that responds to rapid technological change by separating the front end (simple and stable) from the back end (possibly complex and changing). As a consequence, IT flexibility is a key enabler of digital sustainability, allowing efficient adaptation to change without significant rework.

Achieving IT flexibility often involves the adoption of loosely coupled components, well-defined interfaces, and the strategic embrace of cloud services. For instance, the move toward cloud-native applications, characterized by their modularity and composability, directly supports greener IT practices by enabling more efficient resource utilization and reducing energy consumption [54]. Furthermore, advanced software technologies facilitated by flexible IT architectures, such as sophisticated virtualization and containerization within cloud data centers, are proven to significantly enhance energy efficiency, thereby minimizing the environmental footprint of digital operations [55]. Thus, IT flexibility is a strategic imperative for organizations committed to building genuinely sustainable digital ecosystems.

These principles are central to cloud computing, allowing organizations to provision new services on demand, only where and when they are needed. Achieving this level of flexibility is not automatic for IT departments; it requires careful planning from the very beginning of service design and often involves more sophisticated and complex architectures. In addition, because each cloud provider offers unique services, teams need specialized skills to manage and optimize them effectively.

Modularity is a fundamental principle of modern architectural design that takes a complex system and splits it into smaller, independent, and interchangeable modules. Each of these modules performs a specific function and communicates with the others through standard interfaces. Modularity simplifies development, maintenance, and debugging, as changes to one module are less likely to impact other modules. An example of modularity is the microservices architectural style, which enables each service to be deployed and scaled independently and to use its own technology stack. Microservices provide flexibility and resilience to the overall system. Service-Oriented Architecture (SOA) is another architectural style that embodies modularity and is critically important as a component of sustainable digital infrastructures [29].

From a sustainability perspective, modularity provides several advantages, many of which are well established in software engineering [56]. Most importantly, modularity directly contributes to resource efficiency by enabling the right sizing of components. Rather than executing a large monolithic application on a single server, a modular system can scale different services according to their actual demand, which reduces idle time for the server and lowers energy usage. For example, the user authentication service can be scaled to use more resources during hours when users are logging in, while other services can remain at a lower resource level. By allowing older, obsolete, or inefficient components to be updated or replaced without requiring changes to the entire application, modularity reduces the need for continuous large-scale remaking of systems. This is important because digital sustainability and technical debt are strongly related to the remaking or replacement of complete systems [57].

Scalability describes an IT system's ability to manage rising workloads. It is vital for handling variations in user demand, data size, and transactional load. This requires architectures capable of distributing requests across multiple machines or environments, benefiting from cloud computing infrastructure, and implementing strategies like load balancing and auto-scaling for sustained performance and availability. Digital transformation and data growth continuously generate an increasing need for IT resources, which is facilitated by developing scalable architectures. While scalability supports this growth, it does not inherently guarantee sustainability. Indeed, data centers face challenges in terms of both performance and energy efficiency [58]. Moreover, edge computing (or fog computing) can provide low-latency, energy-efficient services by bringing computation closer to the data, but it often extends rather than changes existing cloud models, without any inherent green principles.

The publication of Carrillo et al. details a scalable, sustainable architecture, inspired by the circular economy, focusing on rethinking computation, optimizing deployment, and regenerating energy. This aligns with broader concerns about IT's environmental impact. Cloud computing's high energy consumption leads to increased operational costs and carbon emissions [59]. Therefore, energy-efficient solutions, based on deep analysis of power efficiency and existing literature, are vital to minimize cloud computing's environmental footprint.

Transparency in IT architecture ensures that the workings of a system are clear, understandable, and easily auditable, both for technical teams and for stakeholders with business interests. This principle promotes visibility into system components, data flows, and operational processes, which is vital for effective governance, troubleshooting, and compliance. Transparent architectures foster trust, facilitate collaborative development, and enable better decision-making by providing clear insights into how IT assets support business functions and where potential improvements or risks might lie.

From a sustainability perspective, transparency is necessary for achieving IT operations that respect the environment. Indeed, it enables precise monitoring and auditing of resource consumption, allowing organizations to identify and optimize energy inefficiencies and carbon footprints. This extends to consumer awareness, where digital transparency plays a crucial role in educating consumers about product origins and sustainability [60]. Digital platforms, including blockchain-enabled systems, mobile apps, and e-commerce sites, provide a medium for brands to communicate supply chain transparency. This influences consumer trust, decision-making, and the adoption of sustainable products. However, the effectiveness of these platforms leans on the clarity, accessibility, and credibility of the information, with challenges like greenwashing and information overload potentially reduces efforts and trust [60].

The broader context of environmental governance also highlights that while transparency is often seen as a solution for accountability and effectiveness, its design and efficacy in different policy contexts remain contested [61]. The transformative potential of transparency is being re-examined. The impacts of real-time transparency enabled by novel digital technologies in the sustainability domain require further analysis. These insights underscore that IT transparency is not only about internal system clarity but also extends to critical external dimensions of sustainability, ethical consumption, and effective environmental governance.

Standardization provides consistency, interoperability, and efficiency in every information system considered. The objective of standards (as well as norms or protocols) is to define rules that ensure uniformity within a domain and, consequently, facilitate integration, reduce complexity, and minimize technical debt. Standardization can be applied to many elements, from data formatting and communication protocols to development frameworks and deployment processes. It helps establish a consistent and repeatable IT environment, which leads to greater component reusability and reduced training requirements, while creating a cost-effective and resilient digital ecosystem.

Standardization contributes significantly to sustainability as it reduces the environmental footprint of digital technologies through design for energy efficiency. Moreover, it promotes interoperable solutions that prevent monopolistic behaviors by large tech companies (e.g., Google, Apple, Amazon, Microsoft, Broadcom, Tesla, and Nvidia). It also often extends hardware longevity through common norms like USB-C. The road to a "Digital Green Deal" starts with cross-sectoral green digitalization policies that establish coherence between sustainability and digital policies [62]. Standardization offers a process that ensures digital technologies can be developed and used in environmentally compatible ways, rather than have their productivity, energy, or resource demands exacerbated.

Furthermore, standardization is critical for effective AI governance, which is essential for ensuring that artificial intelligence contributes positively to societal goals, including the UN Sustainable Development Goals (SDGs) [63]. AI governance involves establishing shared principles, mechanisms for certification, and robust data protection. Standardized approaches to data use are fundamental to promoting AI ethics and sustainability. This global approach to AI governance, facilitated by standardization, aims to provide citizen-centered and societally beneficial outcomes.

5 Strategies and Tools for Contributing to Sustainability

5.1 Different Strategies for Contributing to Sustainability

To ensure the sustainability of IT architectures, it is essential to consider the entire chain involved in delivering services to users. Beyond the software aspects discussed in Sect. 5.3, this also includes data centers, the hardware they contain, and the communication systems that support any digital service.

Data Centers:

- State-of-the-art buildings capable of hosting a high density of machines while optimizing the resources needed by servers, both in terms of power consumption and heat dissipation.
- Reuse of waste heat through connections with facilities that require heating (such as swimming pools).
- Innovative approaches, such as experimenting with submerged data centers that use the ocean to cool machines instead of relying on large amounts of fresh water (e.g., Microsoft's *Project Natick*, or the *Yellow Sea experiment* by the Chinese company Highlander).

Hardware Innovations to Match Power Consumption with Actual Usage:

- Deactivating inactive components (*Power Gating*).
- Adjusting voltage and frequency based on workload (*Dynamic Voltage and Frequency Scaling—DVFS*).
- Managing power in sleep or idle states (*Low-power sleep modes*).

Chip Design Innovations to Reduce Energy Consumption:

- In parallel with Moore's Law, energy consumption becomes a critical challenge as it grows increasingly difficult to integrate more components on a given silicon surface.
- Using AI to optimize the layout of blocks within a chip. Google has achieved very good results in this area (and other companies as well—see related articles). However, these AI-generated designs can be so novel that they become hard to understand and, thus, unusable without complete interpretability.
- Emerging technologies, including the use of *optical components* on silicon chips (see related research).

While research is progressing rapidly, the question remains: What should be done with old hardware or outdated data centers? (For example, the French government's initiative to consolidate and shut down smaller data centers.) Another issue is the limited warranty period of machines, which often leads to their premature replacement despite sufficient remaining capacity.

5.2 *Different Tools for Contributing to Sustainability*

Enterprise Architecture frameworks such as TOGAF aim to align digital technologies with the overall goals of the organization. Environmental objectives are increasingly considered as important as goals related to profitability or market share. The use of digital twins also plays an essential role, as they make it possible to simulate and evaluate the impact of both business and technological decisions across the entire system.

Green IT practices are often closely connected to digital accessibility standards such as the French *RGAA* (Référentiel Général d'Amélioration de l'Accessibilité) and *RGESN* (Référentiel Général d'Écoconception des Services Numériques). These frameworks promote eco-design approaches that reduce environmental impact while maintaining accessibility. Examples include:

- Limiting the amount of data exchanged between servers and clients;
- Avoiding the automatic playback of videos or the automatic loading of interactive maps without user action, as required by the RGESN;
- Reducing the size of web pages through server-side processing that minimizes the data to be transferred. This must be balanced with the need to load several JavaScript libraries during the first visit, which are then kept in the cache for later use. Some JavaScript frameworks such as Svelte, Preact, or Alpine JS help reduce the total amount of code that needs to be loaded;
- Improving the semantic structure of applications in order to transmit higher-level, meaningful data rather than raw or redundant information. The DIJA project [64] explores this semantic approach for more efficient data exchange.

6 Conclusions

This chapter has presented a comprehensive overview of how digital architectures can contribute to sustainability. It has shown that sustainability in IT is no longer limited to energy efficiency or hardware optimization but extends across the entire digital ecosystem, from the physical infrastructure of data centers to the design of software systems and enterprise-wide strategies.

The evolution of digital architecture demonstrates a continual effort to manage complexity, increase efficiency, and align technology with societal and environmental objectives. Sustainable IT systems are built on key architectural principles such as flexibility, modularity, scalability, transparency, and standardization. These principles facilitate adaptability to shifts in technology while creating the least possible impact on the environment.

This chapter also examined the techniques and tools that illustrate how to facilitate sustainability using enterprise architecture frameworks, digital twins, and plans associated with Green IT in both the operational and strategic aspects of organizations. Additionally, developments in techniques, concepts, and tools, from energy aware hardware design to sophisticated data management and processing, illustrate that sustainability and performance can function together as complementary pursuits.

Creating sustainable digital architectures will require an important shift in thinking and working. Rather than a constraint, sustainability must be treated as a core design principle. By integrating environmental, social, and economic sustainability at all stages of the IT life cycle, organizations will be able to transform their digital ecosystems. The result will offer robustness, efficiency, and social responsibility in support of the broader aim of sustainable development.

Disclaimer: Note that the authors have assembled the information above to the best of their knowledge as available at the time when the present book was written. Any interpretation is the personal opinion of the authors. Also, the regulation framework can change and evolve any time. The only legally binding text is the one provided in the documents formally published in the European Union Official Journal.

References

1. O'Regan G (2018) Von Neumann architecture. In: The innovation in computing companion. Springer, Cham. https://doi.org/10.1007/978-3-030-02619-6_54
2. Chaudhry R (2024) Evolution of enterprise architecture (EA-Part-3)—enterprise architecture—medium. https://medium.com/razi-chaudhry/evolution-of-enterprise-architecture-ea-part-3-94beda636464
3. Ceruzzi PE (2003) A history of modern computing, 2nd edn. MIT Press, Cambridge, MA, USA
4. Dijkstra EW (1968) Go to statement considered harmful. Commun ACM 11(3):147–148
5. Wirth N (1971) Program development by stepwise refinement. Commun ACM 14(4):221–227
6. Royce WW (1970) Managing the development of large software systems: concepts and techniques. In: Proceedings of IEEE WESCON, pp 1–9
7. Shaw M, Clements P (2006) The golden age of software architecture. IEEE Softw 23(2):31–39. https://doi.org/10.1109/MS.2006.58
8. Wardle C (1984) The evolution of information systems architecture. In: ICIS 1984 proceedings, vol 4. https://aisel.aisnet.org/icis1984/4
9. Codd EF (1970) A relational model of data for large shared data banks. Commun ACM 13(6):377–387
10. Zachman JA (1987) A framework for information systems architecture. IBM Syst J 26(3):276–292
11. Berners-Lee T, Cailliau R, Groff JF, Pollermann B (1992) World-Wide Web: the information universe. Electron Netw: Res Appl Policy 2(1):52–58
12. Castells M (2001) The internet galaxy: reflections on the internet, business, and society. Oxford University Press
13. Booch G (1991) Object-oriented design with applications. Benjamin/Cummings Publishing Company
14. Gamma E, Helm R, Johnson R, Vlissides J (1994) Design patterns: elements of reusable object-oriented software. Addison-Wesley
15. Linthicum DS (2000) Enterprise application integration. Addison-Wesley
16. Ross JW, Weill P, Robertson DC (2006) Enterprise architecture as strategy: creating a foundation for business execution. Harvard Business School Press. The Open Group. (Ongoing editions). The TOGAF Standard. The Open Group. United States Office of Management and Budget (OMB). Federal Enterprise Architecture Framework (FEAF)
17. Mell P, Grance T (2011) The NIST definition of cloud computing. National Institute of Standards and Technology Special Publication, pp 800–145
18. Newman S (2015) Building microservices: designing fine-grained systems. O'Reilly Media
19. Fowler M, Lewis J (2014) Microservices: a definition of this new architectural style. Retrieved from martinfowler.com/articles/microservices.html
20. Forsgren N, Humble J, Kim G (2018) Accelerate: the science of lean software and DevOps: building and scaling high performing technology organizations. IT Revolution Press
21. Eckerson WW (2007) Extending the data warehouse with data governance. TDWI Best Practices Report, Q4

22. Dumitriu D, Popescu MAM (2020) Enterprise architecture framework design in IT management. Procedia Manuf 46:932–940, ISSN 2351-9789. https://doi.org/10.1016/j.promfg.2020.05.011
23. Judijanto L, Hindarto D, Wahjono SI, Djunarto (2023) Edge of enterprise architecture in addressing cyber security threats and business risks. Int J Softw Eng Comput Sci (IJSECS) 3(3):386–396. https://doi.org/10.35870/ijsecs.v3i3.1816
24. Lombardi F, Fanton A (2023) From DevOps to DevSecOps is not enough. CyberDevOps: an extreme shifting-left architecture to bring cybersecurity within software security lifecycle pipeline. Software Qual J 31:619–654. https://doi.org/10.1007/s11219-023-09619-3
25. Gaie C (2021) An API-intermediation system to facilitate data circulation for public services: the French case study. Int J Comput Syst Eng 6:(4):201–210. https://doi.org/10.1504/IJCSYSE.2021.120292
26. Wulf F, Lindner T, Westner M, Strahringer S (2021) IAAS, PAAS, or SAAS? The why of cloud computing delivery model selection—vignettes on the post-adoption of cloud computing. In: Proceedings of the annual Hawaii international conference on system sciences. https://doi.org/10.24251/hicss.2021.758
27. Semenov AB (2021) Data centers environmental impact assessment features. In: E3S web of conferences, vol 311, p 04007
28. Ragheb A, El-Shimy H, Ragheb G (2015) Green architecture: a concept of sustainability. Procedia Soc Behav Sci 216:778–787
29. Hustad E, Olsen DH (2021) Creating a sustainable digital infrastructure: the role of service-oriented architecture. Procedia Comput Sci 181:597–604. https://doi.org/10.1016/j.procs.2021.01.210
30. van de Wetering R, Mikalef P, Helms R (2017) Driving organizational sustainability-oriented innovation capabilities: a complex adaptive systems perspective. J Clean Prod 166:788–799
31. Murugesan S (2008) Harnessing green IT: principles and practices. IT Professional 10(1):24–33. An often-cited article from a practitioner-oriented but scholarly journal, summarizing early principles
32. Hilty LM, Herche S (2009) The ICT ecology—from product to infrastructure. In: ICT Innovations for sustainability. Springer, pp 3–20. Though this is a chapter from 2009, Hilty's work, along with others, was foundational in the preceding years and this chapter synthesizes it well
33. Harmon RR, Auseklis N (2009) The ten imperatives of green IT. Inf Syst Manag 26(1):17–23. A widely cited paper outlining core principles
34. Malmodin J, Bergmark P (2015) The power consumption of mobile and fixed networks. In ICT innovations for sustainability. Springer, pp 55–66. Again, a chapter, but the broader body of work by these authors and others often provides longitudinal data and historical context on energy consumption trends in ICT
35. Molla A (2009) Green IT: an historical perspective and a future research agenda. Australas J Inf Syst 16(2):5–21. A good example of an academic paper specifically looking back at the evolution of the field up to that point
36. Hilty LM, Aebischer B (eds) (2015) ICT innovations for sustainability: smart solutions for a green economy. Springer. While not solely a historical analysis, chapters within this and similar books often review the evolution of concepts like Green IT and Sustainable ICT. Hilty's long-standing work in the field provides a strong historical thread
37. Williams E (2004) Energy intensity of computer manufacturing: hybrid assessment combining process and economic input-output methods. Environ Sci Technol 38(22):6166–6174. https://doi.org/10.1021/es035152j
38. McLaren D, Rudenko I (2009) Sustainability and supply chain management in ICT manufacturing. J Clean Prod 17(14):1251–1260. https://doi.org/10.1016/j.jclepro.2009.03.004
39. Kuehr R, Williams E (2003) Computers and the environment: understanding and managing their impacts. Springer, United Nations University
40. Li J, Lopez N, Liu L (2013) Regional e-waste recycling: present practices and future perspectives. Waste Manage 33(3):660–670. https://doi.org/10.1016/j.wasman.2012.11.008

41. Brown HS, de Jong M, Levy DL (2009) Building institutions based on information disclosure: lessons from GRI's sustainability reporting. J Clean Prod 17(6):571–580. https://doi.org/10.1016/j.jclepro.2008.12.009
42. Jenkin T, Webster J, McShane L (2011) An agenda for 'green' information technology and systems research. Inf Organ 21(1):17–40
43. Dedrick J (2010) Green IS: concepts and issues for information systems research. Commun Assoc Inf Syst 27(1):11–18
44. Basel Convention (2022) Decisions adopted by the fifteenth meeting of the conference of the parties to amend annexes II, VIII and IX to the convention by adding or inserting entries on electrical and electronic waste. Amendments effective January 2025
45. Mertens W et al (2020) The role of digital technologies for sustainable development goals. J Clean Prod 276:124036
46. Henderson P et al (2020) Towards the sustainable development of artificial intelligence. In: Proceedings of the AAAI conference on artificial intelligence, vol 34, no (9), pp 13659–13666
47. Saberi S et al (2019) Blockchain technology and its applications in sustainable supply chain management. Int J Prod Econ 207:1–14
48. Monteleoni C et al (2013) Forecasting climate change using machine learning. In: Proceedings of the 2013 IEEE international conference on big data
49. Hustad E (2021) Sustainable digital infrastructures. J Clean Prod 316:128221
50. Correia P et al (2020) A review of current approaches for environmental impact assessment of ICT products. J Ind Ecol 24(6):1328–1342
51. Baldé CP, Kuehr R, Yamamoto T, McDonald R, D'Angelo E, Althaf S, BelG, Deubzer O, Fernandez-Cubillo O, Forti V, Gray V, Herat S, Honda S, Iattoni G, Khetriwal DS, di Cortemiglia VL, Lobuntsova Y, Nnorom I, Pralat N, Wagner M (2024) Global E-waste monitor. In: International telecommunication union (ITU) and United Nations Institute for Training and Research (UNITAR). Geneva/Bonn
52. Gaie C (2020) From secured legacy systems to interoperable services (the careful evolution of the French tax administration to provide new possibilities while ensuring the primary tax recovering objective). Int J Comput Syst Eng 6(2):76. https://doi.org/10.1504/ijcsyse.2020.111204
53. Gaie C (2021) An API-intermediation system to facilitate data circulation for public services: the French case study. Int J Comput Syst Eng 6(4):201. https://doi.org/10.1504/ijcsyse.2021.120292
54. Vitali M (2022) Towards greener applications: enabling sustainable-aware cloud native applications design. In: Franch X, Poels G, Gailly F, Snoeck M (eds) Advanced information systems engineering. CAiSE 2022. Lecture Notes in Computer Science, vol 13295. Springer, Cham. https://doi.org/10.1007/978-3-031-07472-1_6
55. Katal A, Dahiya S, Choudhury T (2023) Energy efficiency in cloud computing data centers: a survey on software technologies. Cluster Comput 26:1845–1875. https://doi.org/10.1007/s10586-022-03713-0
56. Venters CC, Capilla R, Nakagawa EY, Betz S, Penzenstadler B, Crick T, Brooks I (2023) Sustainable software engineering: reflections on advances in research and practice. Inf Softw Technol 164:107316. https://doi.org/10.1016/j.infsof.2023.107316
57. Atadoga NA, Umoga NUJ, Lottu NOA, Sodiya NEO (2024) Tools, techniques, and trends in sustainable software engineering: A critical review of current practices and future directions. World J Adv Eng Technol Sci 11(1):231–239. https://doi.org/10.30574/wjaets.2024.11.1.0051
58. Carrillo O, Barrios Hernandez CJ, Le Mouel F, Barrera H, Denneulin Y et al (2022) Scalable architectures to support sustainable advanced information technologies. In: CLUSTER 2022—2022 IEEE international conference on cluster computing. Heidelberg, Germany, pp 512–515. https://doi.org/10.1109/CLUSTER51413.2022.00065
59. Kumar S, Buyya R (2012) Green cloud computing and environmental sustainability. In: Murugesan S, Gangadharan GR (eds) Harnessing green IT: principles and practices. Wiley, pp 377–394. https://doi.org/10.1002/9781118305393.ch16

60. Tolegenov B, Rafif K, Rafi M, Herawati R, Abdullah A, Sari LK (2024) Digital transparency and consumer awareness: digital platforms for educating consumers about product origins and sustainability. J Agrimart 12(2):277–287. https://doi.org/10.29244/jam.12.2.277-287
61. Gupta A, Boas I, Oosterveer P (2020) Transparency in global sustainability governance: to what effect? Int Environ Agreements: Polit Law Econ 20(1):84–97. https://doi.org/10.1080/1523908X.2020.1709281
62. Santarius T, Dencik L, Diez T, Ferreboeuf H, Jankowski P, Hankey S, Hilbeck A, Hilty LM, Höjer M, Kleine D, Lange S, Pohl J, Reisch L, Ryghaug M, Schwanen T, Staab P (2023) Digitalization and sustainability: a call for a digital green deal. Environ Sci Policy 147:11–14. ISSN 1462-9011. https://doi.org/10.1016/j.envsci.2023.04.020
63. Gill AS, Germann S (2022) Conceptual and normative approaches to AI governance for a global digital ecosystem supportive of the UN sustainable development goals (SDGs). AI Ethics 2:293–301. https://doi.org/10.1007/s43681-021-00085-5
64. Denis L, Gardan Y, Perrin E (2004) A framework for a distributed CAD system. Comput-Aided Des 36(9):761–773, ISSN 0010-4485. https://doi.org/10.1016/j.cad.2003.09.004

Chapter 6
Defining Synergies Between Cybersecurity and Sustainability

Mirosław Karpiuk and Krzysztof Kaczmarek

Abstract With the development of civilization and advances in technology, almost every aspect of human activity is interrelated with modern digital technologies. However, these technologies represent not only opportunities but also threats. This implies the need for cybersecurity principles. This also applies to sustainable development, the implementation of which is based on modern digital tools. The present chapter focuses on identifying synergies between the two areas while highlighting the main cyber threats which sustainability systems and critical infrastructure elements are exposed to. The conclusions point to the need for a simultaneous and joint development of sustainability and cybersecurity elements, which at the same time has implications for the broader security of societies and states. Finally, it will be discussed how European Legislations build on the duality of sustainability and cybersecurity to shape access to the European Union Single Market.

Keywords Cybersecurity · Sustainability · Critical infrastructure · Artificial intelligence · Circular economy

1 Introduction

Analyses and consideration of the relationship between cybersecurity and sustainable development should commence by defining the concept of sustainability, which is most often defined as allowing present needs to be fulfilled and enabling the needs of future generations to be satisfied [1]. Therefore, it can be assumed that the idea refers to the use of available resources to the extent that they make it possible to meet

M. Karpiuk (✉)
University of Warmia and Mazury in Olsztyn, Olsztyn, Poland
e-mail: miroslaw.karpiuk@uwm.edu.pl

K. Kaczmarek
Koszalin University of Technology, Koszalin, Poland
e-mail: krzysztof.kaczmarek@tu.koszalin.pl

C. Gaie and M. Mueck (eds.), *Digital Sustainability: Legislations, Standards, Solutions and Related Case Studies*, Advances in Sustainability Science and Technology,
https://doi.org/10.1007/978-981-92-0458-8_6

the needs of modern man rather than an unlimited use of these just because they are available.

According to Wiesław Sztumski, on the other hand, social consensus on the most important issues in people's lives with a global dimension lies at the heart of sustainable development. Arriving at this consensus requires consistent adherence to the basic principles of human coexistence that environmentalism proclaims: the principle of tolerance and the principle of synergy. Adherence to the first principle stems from the will to preserve peace under conditions of rapidly increasing environmental condensation (primarily as a result of rapid population growth); adherence to the second one follows from the pursuit of the fundamental interest of humanity, that is, the preservation of life and the survival of the human species under conditions of progressive degradation of our natural and social environment [2].

The current civilizational and technological developments are changing almost all spheres of human activity, including social interaction, much of which has moved online. On the other hand, in the context of sustainable development, the use of information and communication technologies (ICT) offers a possibility of optimizing services and supplies, thus reducing the need for energy or raw materials. The use of the Internet of Things (IoT) in intelligent transport solutions may serve as an example here. However, it is important to remember that ICT does not only represent opportunities but also risks. This is particularly relevant in the context of a tense international situation and ongoing cyber warfare, the effects of which can directly threaten health, life, and property.

A possible scenario is that a cyberattack on infrastructure, including critical infrastructure, causes power outages without which a modern digital society cannot function properly. And, in such a situation, communications, product sourcing, transport control systems, or non-cash payment methods do not work. Thus, a society being cut off from network services is at the same time cut off from basic services. Services related to the implementation of sustainable development elements do not work then, either. A cyberattack may also result in a malfunctioning of systems designed to optimize the consumption of goods, services, and energy. Indeed, it is possible to imagine a situation in which traffic control systems operate in such a way as to make movement impossible and systems responsible for electricity distribution cause overloading of electricity grids and damage to them.

Thus, a properly secured digital infrastructure is necessary to ensure the continuity of ICT systems used in strategic sectors that guarantee sustainable development. Cybersecurity safeguards the development both of public and private spheres, and adequate protection against cyber threats is to ensure the security of users of information systems, which is linked to an introduction of sustainable solutions which, on the one hand, allow free, fast, and efficient operation in cyberspace and, on the other, protect against cyberattacks [3], including data leakage and damage to systems.

The link between cybersecurity and sustainable development is particularly important and evident in the case of critical infrastructure, which is responsible for sectors that are vital both to the state and its security but also to society or the economy.

The development of new technologies, including artificial intelligence, must be sustainable, and this balance is to be provided by cybersecurity. Modern digital tools must not only be seen unilaterally, and therefore not only as enabling faster, more efficient, and cheaper operations (fulfillment of needs, provision of services, performing tasks, and provision of entertainment). They must also be secure and resistant to cyberattacks. Figure 1 below illustrates the immediate synergy between digital sustainability and cybersecurity to provide user satisfaction and protect systems against cyber threats and attacks.

In view of the above, the objectives of this chapter are to identify and define the synergies of cybersecurity and sustainable development, and to evaluate regulations and strategies to combat digital threats in the implementation of sustainable development systems. The research hypothesis is that cybersecurity constitutes an indispensable element in the implementation of sustainable development strategies. In order to verify it, a methodology based on source literature analysis, expert threat analysis, and a scenario method describing possible consequences to sustainability of the lack of digital insecurity was used.

Fig. 1 Synergies between cybersecurity and sustainability

2 Sustainability and Cybersecurity

Pursuing socially accepted values at the core of a security culture should be compatible with the needs of sustainable development [4]. This also applies to cybersecurity as one of security domains.

Sustainable development is the subject of many policies, strategies, or programs, including both national and international ones. The resolution adopted by the United Nations General Assembly on September 25, 2015: The 2030 Agenda for Sustainable Development also addresses this issue. This agenda commits the global community to achieving sustainable development in three dimensions: economic, social, and environmental [5].

One of the objectives of the sustainable development agenda is to build stable infrastructure, promote sustainable industrialization, and foster innovation. This is to be achieved by building a reliable, sustainable, durable, and stable infrastructure of good quality, including regional and cross-border infrastructure that supports economic development and human well-being. In doing so, it is important to ensure equal access to such an infrastructure to all people at affordable prices. It is also important to promote sustainable and socially inclusive industrialization. Achieving this objective also requires a modernization of the infrastructure and a modernization of industry to ensure its sustainability, while increasing resource efficiency. The use of clean and environmentally friendly technologies and production processes is also necessary, but this must be of global dimension and not merely a national one. Building a stable infrastructure, promoting sustainable industrialization, and fostering innovation require intensifying research and raising the technological level of the industrial sector in all countries (and, in particular, in developing countries).

In order to achieve the aforementioned objective, the development of a sustainable and stable infrastructure must be facilitated through increased financial and technological support. It is also important to foster national technological development, research and innovation, also through policies favoring industrial diversification. Access to information and communication technologies should also be greatly increased, and efforts should be made to ensure inexpensive universal access to the Internet (especially in the least developed countries) [6]. In order to ensure that the objectives of the Sustainable Development Agenda and the related tasks do not turn out to be mere declarations and utopian ideas, it becomes necessary to adopt a multilevel mechanism for their implementation, also taking into account the fact that individual regions differ to a greater or lesser extent [7]. This also applies to building stable infrastructure, promoting sustainable industrialization, and fostering innovation.

Sustainable development goals can only be realized if civil society, businesses, local communities, and other stakeholders actively participate in the process and take responsibility for it. Participatory governance mechanisms must ensure that civil society is involved at all levels, from the local and national levels to the European and global levels. Achieving the Sustainable Development Goals requires cooperation in an integrated manner across sectors [8].

The aim of sustainable development is to build a stable infrastructure. Critical infrastructure, which also uses ICT systems for its operation, will be of particular importance here. The Polish legislator, in Article 3(2) of the Act on Crisis Management (i.e., Journal of Laws of 2023, item 122, as amended), defines critical infrastructure as systems, as well as their constituent and functionally related objects, including buildings, equipment, installations, and services that are key to the security of the state and its citizens, as well as those that serve to ensure the proper operation of public administration bodies, institutions, and entrepreneurs. Critical infrastructure includes the following systems: (1) supply of energy, energy raw materials, and fuels; (2) communications; (3) ICT networks; (4) financial; (5) food supply; (6) water supply; (7) health care; (8) transport; (9) emergency; (10) ensuring continuity of public administration; and (11) production, storage, and use of chemical and radioactive substances, including pipelines with dangerous substances. ICT systems that are responsible for the stability of the state and its economy must be duly protected [9], which is particularly true for systems used to ensure the proper functioning of critical infrastructure. It could even be argued that the critical infrastructure (including digital infrastructure) is essential for sustainable development; without it, it would even be impossible.

Fostering innovation as a sustainable development objective is also of a great importance. A trend that is conducive to the implementation of the postulates of such development is new technologies. On the one hand, through them, the postulates of sustainable development can be implemented faster and more effectively, but, on the other hand, they are exposed to cyber threats. The use of modern tools must be sustainable and responsible, with the right balance between implementing new technologies and protecting their users from threats.

Digital transformation requires an involvement of numerous actors, whether public or private, and appropriate policies on the part of the state, which bears responsibility for this transformation. It is also important to emphasize that digital transformation processes are also driven by international considerations; therefore, these cannot be developed in an isolation from international policies. New technologies are not the domain of one or a few countries; they are of a global dimension, and they affect the international economy [10]. Global digitalization, using new technologies, including artificial intelligence, may not be expansive in its nature; it must consider legally protected rights, including privacy or human rights, and must respond to the challenges of ensuring cybersecurity. It must be sustainable, based on law, and should take into account ethical principles.

Despite the great potential of artificial intelligence and the benefits that arise from its use, the balance between technological progress and the protection of privacy and human rights must not be forgotten. When implementing artificial intelligence tools, ethical as well as social aspects must also be considered in order for this technology to serve common interests [11]. On the one hand, artificial intelligence can be used as a tool against societies, posing a real threat, while on the other hand, by playing an important analytical role, it may enable rapid identification and combating of threats [12]. It is certainly a tool that can influence sustainable development and, if

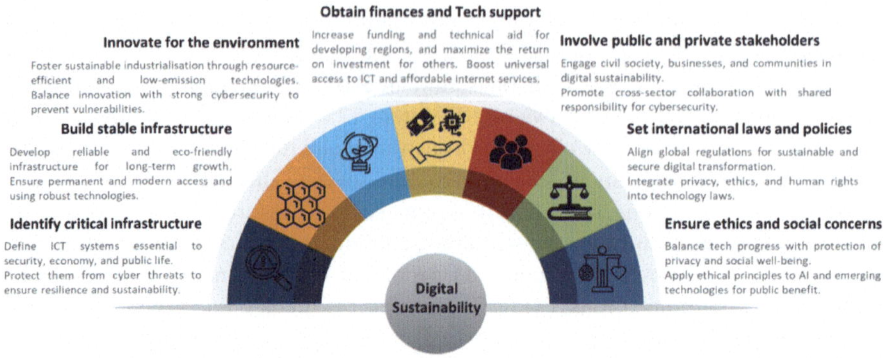

Fig. 2 Different dimensions to achieve digital sustainability

used properly and being cybersecure, can stimulate it. Figure 2 below illustrates the dimensions described in this section that contribute to achieving digital sustainability.

3 International Case Studies and Lessons Learned

3.1 International Case Studies

The understanding of synergies between cybersecurity and sustainability is strengthened by the experiences of various advanced countries. In this section, we analyze, through selected case studies, how cybersecurity is integrated into sustainable development agendas and how best practices can be adapted to diverse social, economic, and geopolitical contexts. These examples show that a strong cybersecurity posture is not only a defensive necessity but also a driver of sustainable digital transformation. Below, we consider the cases of Estonia, Singapore, Germany, and France.

The most prominent case of digitized public administration is Estonia. The country is globally recognized as a pioneer of e-governance, with nearly all public services digitized and accessible online. Its Cybersecurity Strategy 2024–2030 promotes national resilience against cyber threats while aligning with sustainable development goals through efficient resource use and reduced administrative overhead [13]. The success of Estonia's model is rooted in governance principles based on transparency, trust, and security, which foster entrepreneurship and public confidence. Advanced digital identity systems, blockchain integration, and nationwide cyber-awareness campaigns are central pillars of its approach.

By digitalizing and automating public administration, Estonia has reduced service delivery time and costs for both government and citizens, while mitigating risks of data misuse [14]. However, further improving public service delivery requires addressing persistent challenges such as the digital divide, emerging cyber threats, and the need for quantum-resistant security. Estonia's experience offers valuable lessons for other nations seeking to balance innovation, cybersecurity, and sustainable development.

Another notable example is Singapore's Smart Nation initiative, which demonstrates how advanced digitalization can be strategically integrated into sustainability objectives. The program promotes the use of Internet of Things networks, artificial intelligence, and data analytics across public services, urban planning, and environmental management [15]. Moreover, sustainability is addressed through the deployment of smart grids, intelligent transport systems, and optimized water resource management, resulting in more efficient energy use and reduced environmental footprints. These systems are safeguarded by a comprehensive cybersecurity framework that incorporates sector-specific codes of practice, continuous risk monitoring, and incident response capabilities to protect critical infrastructure.

The city-state's approach seeks to introduce new services with strong user value while ensuring the high reliability of its infrastructure. This model allows innovation to progress without compromising social equity, public trust, or environmental integrity. Additionally, Singapore's strategy emphasizes the importance of strengthening population skills through targeted investments in cybersecurity education, applied research, and public–private partnerships. These measures aim to ensure that the technical systems supporting sustainable urban living remain secure and resilient.

Technology is also embedded in the everyday experiences of citizens. For example, new Housing and Development Board estates integrate real-time monitoring via sensor networks to detect faults in services such as lifts or water pumps. Residents can report municipal issues via the OneService app or chatbot, where AI-based routing directs cases to the appropriate agencies for resolution. Advanced modeling tools, such as the Integrated Environmental Modeller, simulate environmental factors—wind flow and air temperature—to guide the placement of buildings and communal spaces, improve thermal comfort, and identify optimal locations for solar energy capture [16].

Singapore's experience highlights the interdependence of cybersecurity and sustainability in twenty-first-century cities. Securing digital infrastructure is not simply a defensive necessity but also a key enabler of sustainable development in highly connected, resource-efficient urban environments.

A third illustrative case is Germany's approach to integrating cybersecurity with sustainability. The country's Industry 4.0 strategy promotes the digital transformation of manufacturing through automation, artificial intelligence, and data-driven decision-making—positioning security as an enabler rather than a barrier to sustainability objectives. By embedding cybersecurity requirements from the outset, German industries ensure that interconnected production systems remain resilient against cyber threats while simultaneously achieving efficiency gains, reducing waste, and conserving resources.

This integration relies on long-standing efforts described by Schallbruch and Skierka [17] including the National Plan for Information Infrastructure Protection (NPSI) and subsequent cybersecurity strategies. These initiatives have strengthened critical infrastructure resilience through coordinated public–private implementation plans, the development of trusted German IT products (notably in encryption), and active participation in European and international standard-setting bodies. At the EU level, the NIS2 Directive further reinforces these efforts by mandating robust security measures for critical and digital infrastructure, safeguarding the integrity of industrial and environmental data flows essential for innovation.

The cybersecurity framework directly supports the climate-neutrality objectives of the European Green Deal, which rely on secure, digitally optimized processes for energy management, logistics, and supply chains. Scharte [18] stresses in his analysis of the German National Security Strategy that resilience should be understood as adaptive capacity rather than mere resistance. This perspective highlights the need to address potential tensions between efficiency and resilience and to give greater weight to the social dimensions of security.

Germany's policy landscape demonstrates that cyber resilience is a prerequisite for sustainable industrial transformation. By aligning technological innovation, security, and environmental goals, Germany has developed an integrated model that both maintains global competitiveness and advances sustainability.

3.2 Lessons Learned

From these examples, several conclusions emerge. Cybersecurity is integral to national sustainability agendas, with countries that align their cyber strategies to environmental and social objectives achieving both resilience and innovation. Secure digital identity systems and trust frameworks play a key role in encouraging public adoption of sustainable e-services, while investment in human capital proves as important as technological advancement. This investment concerns the development of cyber skills, public awareness, and cross-sector coordination.

Furthermore, international cooperation on cyber norms and incident response strengthens the sustainability of global digital systems, reflecting the borderless nature of cyber threats. Collectively, the case studies confirm that sustainable development in the digital era is unattainable without a robust cybersecurity foundation and that the integration of secure infrastructure, public trust, and inclusive innovation stands out as a shared success factor across all the examined national models (Table 1).

4 Technological Synergies and Challenges

The association of cybersecurity and sustainability offers both opportunities and challenges through the use and deployment of multiple technologies. This section outlines the relevance of implementing solutions based on artificial intelligence, blockchain and the Internet of Things, smart grids combined with edge computing, and connected infrastructures.

Table 1 Learnings from the different case studies

Country	Initiative(s)	Cybersecurity—sustainability link	Key takeaways
Estonia	Cybersecurity strategy 2024–2030: cyber-conscious Estonia	Strong integration of national digital governance with cybersecurity Ensures resilient e-services and trust in state platforms	Demonstrates that small nations can be leaders of the domain Facilitated by cybersecurity into national identity and service delivery
Singapore	Smart nation initiative	Uses secure IoT and AI-driven urban systems Provides efficient energy use, intelligent transport, and citizen engagement	Highlights that urban digitalization and sustainability require adaptive cybersecurity measures
Germany	Industry 4.0, NIS2 directive, European green deal	Aligns industrial digitalization with climate goals and cybersecurity regulations across member states	Shows how supranational policy can harmonize tech innovation, cyber resilience, and environmental commitments

Each of these technologies can drive progress toward sustainability by improving efficiency, transparency, and resilience. At the same time, however, they expand the digital attack surface and introduce new vulnerabilities. Understanding these synergies and challenges is essential to ensure that technological innovation strengthens, rather than undermines, sustainable development.

4.1 *Artificial Intelligence: Opportunities and Risks*

Artificial intelligence is increasingly embedded in systems designed to advance sustainability. Its capacity to process large volumes of data enables significant efficiency gains, reducing waste and supporting the transition to low-carbon economies. At the same time, AI already contributes to solving pressing global challenges. In the water sector, smart management systems powered by AI replicate human learning to optimize decisions and guide investment in water infrastructure [19].

In agriculture, projects such as PlantVillage at Penn State University use AI to support smallholder farmers—who produce much of the developing world's food—by diagnosing crop diseases and improving yields [20]. AI applications also extend to sanitation and health, as demonstrated by Peter Ma's innovative work on monitoring the purity of water systems to detect waterborne diseases, with potential benefits for municipalities and individuals [21].

Despite these opportunities, AI systems are not without risks. Algorithms trained on biased data may reinforce inequality or inefficiency, while adversarial attacks can manipulate images or data outputs in many domains [22]. Ensuring transparency, explainability, and resilience in AI deployment is therefore essential for aligning technological progress with both cybersecurity requirements and sustainability objectives.

4.2 Blockchain and IoT in Sustainability Systems

Blockchain, artificial intelligence (AI), and the Internet of Things (IoT) represent transformative technologies with the potential to advance sustainability and support circular economy principles. Blockchain is a decentralized, peer-to-peer protocol for sharing and updating information securely, transparently, and irreversibly. Blockchain provides traceability in supply chains, facilitates fair-trade compliance, enables carbon credit accounting, and ensures data integrity across diverse sectors, including healthcare and resource management [23]. IoT sensors enable real-time monitoring of environmental indicators, allowing cities and industries to optimize energy, water, and traffic management [24].

Despite their potential, these technologies introduce challenges that must be addressed. Blockchain faces issues such as scalability, high energy consumption, and cybersecurity vulnerabilities, whereas IoT devices often suffer from weak security, raising concerns about privacy and data breaches. Moreover, these technologies involve high implementation costs, complex technical requirements, and demand effective governance. The absence of lightweight cryptography and the lack of standardized protocols are additional barriers to adoption.

To maximize the benefits of these technologies while mitigating risks, it is essential to promote multidisciplinary research, fund R&D initiatives, encourage collaboration among academia, industry, and government, and develop sustainable implementation strategies. Ethical considerations, data governance, and secure information exchange must be prioritized to ensure responsible innovation.

4.3 Smart Grids, Edge Computing, and Critical Vulnerabilities

Smart grids represent a transformative evolution from conventional centralized electrical networks to decentralized, participatory systems that integrate renewable energy and distributed demand management. Edge computing enhances this paradigm by processing data near its source, enabling real-time monitoring, reducing latency, improving energy efficiency, and supporting adaptive control strategies [25].

The deployment of smart grids involves a complex integration of sensors, communication protocols, and advanced control algorithms. These components facilitate efficient energy distribution, enhance reliability, and allow dynamic adaptation to fluctuations in both supply and demand. However, the interconnection of multiple technologies introduces challenges in interoperability, scalability, and communication infrastructure.

Cybersecurity is a critical concern in this ecosystem. Smart grids depend on continuous collection and transmission of sensitive data from millions of sensors and smart meters, exposing them to a range of cyber threats. Vulnerabilities exist across three levels: the communication network, the users (customers), and the operators managing the system. Passive attacks, such as eavesdropping and traffic analysis, compromise data confidentiality, while active attacks, including false data injection, denial-of-service (DoS), and replay attacks, threaten data integrity and availability [26].

Historical cyber incidents, including the Ukraine 2015 power grid attack, and major North American blackouts highlight the potential impact of cyberattacks on grid stability and the critical importance of robust security frameworks. Standards such as those proposed by NIST provide guidelines for authentication, authorization, confidentiality, integrity, and availability to safeguard smart grid operations.

As a consequence, the current challenge lies in balancing the benefits of these new technologies with the need to ensure robust security. Future developments require resilient and adaptive cybersecurity strategies, updated protocols, and comprehensive stakeholder engagement to maintain a secure, reliable, and sustainable grid operation.

5 Disinformation and Cybersecurity vs. Sustainable Development

When analyzing the synergy of sustainability and cybersecurity, digital threats should not be considered as ones that are independent of other factors. In societal terms, cyberspace is not an independent entity but forms a part of the environment where humans function. At the same time, ICTs often provide new tools to achieve political and economic goals. One of the means used to achieve these goals is disinformation, through which the behavior of individuals, societies, and, indirectly, the policies of states, can be influenced. At the same time, disinformation operations, being part of hybrid operations, are combined with such cyberattacks as DDoS, phishing, and attacks on critical infrastructure. At this point, it is worth to recall that sustainable development is, among other things, about the social system being optimally ordered within each of its structures individually [27]. By contrast, the aim of disinformation is to undermine social structures and the state, which may occur in a manner so discreet as to be almost imperceptible. It does, however, pose a real threat.

By influencing societal behavior, disinformation can exploit emotions aroused by sustainability issues. It can already be seen that societies are becoming increasingly differentiated in their attitudes and behavior toward sustainability threads [28]. However, it should be emphasized that this issue does not only concern disinformation related to hybrid activities but can also be used by internal actors to achieve political goals, e.g., seizing power as a result of winning elections.

In contrast, what is a particular type of risk associated with the use of ICT to carry out sustainability-related changes is errors in artificial intelligence training material or discrete cyberattacks on its algorithms and an introduction of changes that may result in future malfunctions. It should be noted that errors in training data can also result from intentional human actions. However, the most important point is that erroneous operations of artificial intelligence algorithms can significantly limit, or even prevent, sustainability. Indeed, it can be assumed that it is likely that systems designed with the help of artificial intelligence will not be working properly, which may be the result of even minor errors in the training materials. Furthermore, one may not rule out that these materials contain previously undetected false data and information.

One must not forget that it is human beings that constitute the weakest element in any security system, including digital security. Indeed, there are no technical means to protect against the mistakes of those responsible for cybersecurity, and nowadays, everyone using the network is responsible for it. However, in the context of sustainability, it is also important how everyone using network services takes care of every aspect of their own digital security. Indeed, it should be emphasized that a malware-infected device consumes more electricity and is less durable, which in turn increases the demand for the raw materials needed to produce it.

Therefore, it is paramount that when using ICT for sustainability planning, no aspect of cybersecurity is overlooked. Both technical and human factors should be taken into account. It is also important to keep in mind that a significant proportion of cyber threats do not yet exist but will certainly emerge in future. Therefore, cybersecurity is also about anticipating possible, even unlikely, scenarios.

6 European Legislations Related to Sustainability and Cybersecurity

While there is a natural motivation to address sustainability and cybersecurity objectives during a product development phase, there are a number of European Legislations which furthermore enforce related requirements through appropriate regulation of market access.

A prime example is **Regulation (EU) 2024/1735 of the European Parliament and of the Council of 13 June 2024 on establishing a framework of measures for strengthening Europe's net-zero technology manufacturing ecosystem** [29]. This Regulation is *focusing on the manufacturing of net-zero technologies in terms* *of final products, specific components, and specific machinery that is primarily used to produce them.* In this context, *"net-zero technologies" means the technologies listed in Article 4* of [29], such as solar technologies, onshore wind and offshore renewable technologies, and battery and energy storage technologies, *where they are final products, specific components, or specific machinery primarily used for the production of those products.* While cybersecurity is not the prime focus of the regulation, it is repeatedly listed as a key requirements and thus demonstrates how sustainability and cybersecurity objectives are closely intertwined.

To give some examples, the following parts of Regulation **(EU) 2024/1735** specifically address cybersecurity requirements:

- Article 25 of Regulation (EU) 2024/1735 relates to **sustainability and resilience contribution in public procurement procedures**. It explicitly introduces cybersecurity requirements by stating that *contracting authorities and contracting entities shall apply at least one of the following conditions … (b) a requirement to demonstrate* ***compliance with applicable cybersecurity requirements*** *provided for in a cyber resilience regulation, including, where appropriate and where available, through a relevant European cybersecurity certification scheme; …*
- The following Recital clarifies the desire to meet cybersecurity and data security objectives: *(63) In order to promote the responsible business conduct of bidders, the cybersecurity and data security of the technologies used, projects and related infrastructures and to promote the delivery of projects fully and on time, public authorities designing auctions for the deployment of energy from renewable sources should include pre-qualification criteria related to responsible business conduct, cybersecurity and data security and the ability to deliver the project fully and on time.*

Regulation (EU) 2024/1735 is thus a suitable example to understand how the European Legislator is effectively combining sustainability and cybersecurity requirements in a single legislation.

As a further example, the Cyber Resilience Act (CRA) [30] is a key component in Europe's Cybersecurity Strategy. It broadly applies to any ***products with digital elements***, i.e., any *software or hardware product and its remote data processing solutions, including software or hardware components being placed on the market separately.* Through its *essential requirements*, the CRA requires products with digital elements to be secure by design and to remain so throughout their life cycle. This approach fosters long-term trust, reduces the need for costly, frequent overhauls due to security breaches, and encourages companies to build more durable and resilient products. Sustainable digital products are those that are not only secure from the start but also maintained with updates and vulnerability management, creating a stable and secure environment for consumers and businesses. In consequence, although the CRA is not relating to sustainability objectives explicitly, it is design to strengthen Europe's sustainability agenda by linking it to cybersecurity objectives.

There are many more European Legislations which efficiently link sustainability and cybersecurity objectives, either explicitly or implicitly. It is interesting that the legislator has understood this important duality and is injecting it into legislations shaping future access to the European Union Single Market.

7 Conclusions

What is rational is multidimensional and multifaceted, systematic, and intensified action for sustainable and balanced development, treated as a certain model (pattern). In all spheres of life and activities undertaken by people, governments, or businesses, in the sphere of science and technology, economy, consumption, education, media, politics, ideology, religion, social development, culture, entertainment, and sport, it is not only possible but also necessary to think and act bearing this model in mind, to seek agreement and think rationally, to avoid conflicts, to solve problems peacefully, to develop cooperation, and to have a sense of responsibility for the future. It is also necessary to accept an innovative approach in relation to technology, education, and organizational or political undertakings. There seems to be no other way than coordinated and systematic action for sustainable development by as many actors (local, national, and international) as possible [31]. The issue of sustainable development should not only remain in the public consciousness but should also be translated into actions both of national and global scopes [32].

In the case of sustainable development, appropriate regulations are necessary at the international level. Only in this way can the implementation of this principle be universal and done according to universal and verifiable standards [33].

In addition to the interpretative key of the idea of sustainable development, according to which it is seen as a solution to contemporary and future problems of mankind, there is also another one which tends to perceive ideology in an extended concept of sustainable development in its current form [34]. The principle of sustainable development leads to a formulation of complex strategies of action, relating to diverse planes of human activity. These strategies do not always succeed in being implemented, and some of those introduced fail to function properly [35].

Cybersecurity is an indispensable element of sustainable development, and a simultaneous and joint development of these two areas allows for more effective implementation of sustainability-based systems. Thus, the hypothesis that cybersecurity constitutes an indispensable element in the implementation of sustainability strategies has been verified positively.

However, both cybersecurity and sustainability strategies depend heavily on the policies pursued by countries and on the public awareness both of risks and benefits, including future ones. One element of sustainability is circular economy, in which products and materials are designed to be reused, repaired, and recycled at the end of their life cycle, creating a closed loop of materials and resources. This approach reduces waste, saves resources, and minimizes environmental impacts such as greenhouse gas emissions and pollution [36].

In this context, a closed loop of raw materials can increase the level of digital security. This is due to the fact that most of electronic components needed to build a digital society are concentrated in a small number of countries, where both production and acquisition of raw materials are cheaper than, for example, in the European Union. This also applies to the production of renewable energy sources or their components. Thus, it cannot be ruled out that a lack of oversight of the production of such components may result in them being vulnerable to digital threats in ways unknown to their users. Nor can it be ruled out that they may contain deliberately introduced elements that facilitate unauthorized access to the systems that rely on them. However, for such items to be produced in Europe, a change is needed not only in economic policy but, above all, in policy philosophy. Those countries that do not introduce appropriate fiscal incentives and whose economies are primarily consumption-based are not able to develop in a sustainable manner. Otherwise, sustainable development will remain only a theoretical solution. Increasing cybersecurity, on the other hand, requires increased public awareness that, first and foremost, everyone is responsible for their own digital security.

8 Disclaimer

Note that the authors have assembled the information above to the best of their knowledge as available at the time when the present book was written. Any interpretation is the personal opinion of the authors. Also, the regulation framework can change and evolve any time. The only legally binding text is the one provided in the documents formally published in the European Union Official Journal.

Acknowledgements The authors would like to sincerely thank Dr. Christophe Gaie and Dr. Markus Mueck for their invaluable support and insightful contributions throughout the development of this book chapter. Their expertise, guidance, and constructive feedback greatly enhanced the quality and clarity of this chapter.

References

1. European Union (2024) Sustainable development. https://eur-lex.europa.eu/EN/legal-content/glossary/sustainable-development.html
2. Sztumski W (2006) Idea zrównoważonego rozwoju a możliwości jej urzeczywistnienia. Problemy ekorozwoju 1(2):73–76
3. Langlois-Berthelot J, Gaie C, Lebraty JF (2024) Epidemiology inspired cybersecurity threats forecasting models applied to e-government. Transforming public services—combining data and algorithms to fulfil citizen's expectations. Springer, Cham, pp 151–174
4. Filipek A, Wąsowska-Wołkow I (2016) Kultura bezpieczeństwa ekonomicznego a zrównoważony rozwój. Kultura Bezpieczeństwa 6:75–96
5. Karwowski M, Paszkiewicz A, Zubrzycka E (2022) Ujawnianie informacji dotyczących celów zrównoważonego rozwoju na przykładzie działalności leśnej i pocztowej. Kwartalnik Nauk o Przedsiębiorstwie 2:87–102

6. UN (2015) Resolution adopted by the general assembly on 25 September 2015. In: Transforming our world: the 2030 agenda for sustainable development. https://odpowiedzialnybiznes.pl/wp-content/uploads/2017/10/Agenda-na-rzecz-zr%C3%B3wnowa%C5%BConego-rozwoju-2030.pdf
7. Perkowski M, Kosicki A, Chrzanowski S (2023) Realizacja Celów Zrównoważonego Rozwoju Agendy 2030. Perspektywa regionalna Prawo i Więź 1:32–48
8. European Economic and Social Committee (2017) Item 4.6.1 of the opinion of the European economic and social committee. In: Communication from the commission to the European parliament, the council, the European economic and social committee and the committee of the regions—next steps towards a sustainable future for Europe. European actions for sustainable development. Official Journal EU C 345, pp 91–96
9. Bencsik A, Karpiuk M, Kelemen M, Włodyka E (2023) Cybersecurity in the visegrad group countries. Lex Localis Press
10. Gaie C, Karpiuk M, Spaziani A (2024) New technologies in public administration. Ius et Securitas 2:49–70
11. Kaczmarek K, Karpiuk M, Soler U (2024) The potential use of artificial intelligence in crisis management. Sicurezza, Terrorismo e Società 2:141–151
12. Gergelewicz T (2024) Bipolarity of artificial intelligence: chances and threats. Ius et Securitas 2:71–94
13. Republic of Estonia (2023) Cybersecurity strategy 2024–2030: cyber-conscious Estonia. Ministry of Economic Affairs and Communications. https://www.mkm.ee/en/cybersecurity-strategy
14. Espinosa VI, Pino A (2024) E-Government as a development strategy: the case of Estonia. Int J Public Adm 48(2):86–99. https://doi.org/10.1080/01900692.2024.2316128
15. Aigner I, Garai-Fodor M, Szemere TP (2023) Singapore's journey as a digitalized and innovative smart nation toward sustainability. Keleti Faculty of Business and Management, Óbuda University
16. Government of Singapore (2024) Smart nation 2.0 report. https://file.go.gov.sg/smartnation2-report.pdf
17. Schallbruch M, Skierka I (2018) The evolution of german cybersecurity strategy. In: Cybersecurity in Germany. SpringerBriefs in cybersecurity. Springer, Cham. https://doi.org/10.1007/978-3-319-90014-8_3
18. Scharte B (2023) Resilience misunderstood? Commenting on Germany's national security strategy. Eur J Sec Res 8:63–71. https://doi.org/10.1007/s41125-023-00095-1
19. Hill T (2018) How artificial intelligence is reshaping the water sector. Water Finan Manag 12:752
20. Lohr S (2018) From agriculture to art—the AI wave sweeps in. New York Times, Business Day. https://www.nytimes.com/2018/10/21/business/from-agriculture-to-art-the-ai-wave-sweeps-in.html
21. Goralski MA, Tan TK (2020) Artificial intelligence and sustainable development. Int J Manag Educ 18(1):100330. https://doi.org/10.1016/j.ijme.2019.100330
22. Qiu S, Liu Q, Zhou S, Wu C (2019) Review of artificial intelligence adversarial attack and defense technologies. Appl Sci 9(5):909. https://doi.org/10.3390/app9050909
23. Upadhyay A, Mukhuty S, Kumar V, Kazancoglu Y (2021) Blockchain technology and the circular economy: implications for sustainability and social responsibility. J Clean Prod 293:126130. https://doi.org/10.1016/j.jclepro.2021.126130
24. Chauhan M, Sahoo DR (2024) Towards a greener tomorrow: exploring the potential of AI, blockchain, and IoT in sustainable development. Nat Environ Pollut Technol 23(2):1105–1113
25. Moreno-Escobar JJ, Morales-Matamoros O, Tejeida-Padilla R, Lina-Reyes I, Quintana-Espinosa H (2021) A comprehensive review on smart grids: challenges and opportunities. Sensors 21(21):6978. https://doi.org/10.3390/s21216978
26. Tufail S, Parvez I, Batool S, Sarwat A (2021) A survey on cybersecurity challenges, detection, and mitigation techniques for the smart grid. Energies 14(18):5894. https://doi.org/10.3390/en14185894

27. Niewielska M (2021) Wyzwania zrównoważonego rozwoju w dobie społeczeństwa informacyjnego: podkreślenie problemów w zakresie więzi i relacji społecznych w dobie Internetu. Eunomia 1(100):79–90
28. Barlett R, Melnick R, Kay B (2025) Sustainability in the face of political polarization. ICMA. https://icma.org/articles/pm-magazine/sustainability-face-political-polarization
29. Regulation (EU) 2024/1735 of the European Parliament and of the Council of 13 June 2024 on establishing a framework of measures for strengthening Europe's net-zero technology manufacturing ecosystem and amending Regulation (EU) 2018/1724 (Text with EEA relevance). https://eur-lex.europa.eu/eli/reg/2024/1735/oj/eng
30. Regulation (EU) 2024/2847 of the European Parliament and of the Council of 23 October 2024 on horizontal cybersecurity requirements for products with digital elements and amending Regulations (EU) No 168/2013 and (EU) 2019/1020 and Directive (EU) 2020/1828 (Cyber Resilience Act) (Text with EEA relevance). https://eur-lex.europa.eu/eli/reg/2024/2847/oj/eng
31. Zacher L (2008) Trwały rozwój—utopia czy realna możliwość? Problemy Ekorozwoju 2:63–68
32. Raczkowska M, Mikuła A, Utzig M (2021) Zrównoważony rozwój w obszarze społecznym Unii Europejskiej, Wydawnictwo SGGW
33. Ziemblicki B (2016) Zrównoważony rozwój z perspektywy prawa międzynarodowego i Europejskiego. Prace Naukowe Uniwersytetu Ekonomicznego we Wrocławiu 452:149–161
34. Kosiek T (2015) Zrównoważony rozwój—rozwiązanie czy ideologia? Zeszyty Naukowe Politechniki Śląskiej. Organizacja i Zarządzanie 85:233–244
35. Pawłowski A (2006) Wielowymiarowość rozwoju zrównoważonego. Problemy Ekorozwoju 1:23–32
36. Szczech-Pietkiewicz E, Czerniak A (2024) Gospodarka obiegu zamkniętego jako potencjał zrównoważonego rozwoju polskich przedsiębiorstw. Studia i Prace Kolegium Zarządzania i Finansów 197:85–96

Chapter 7
Ethics and Sustainability

Dimitris C. Gkikas, Christophe Gaie, and Markus Mueck

Abstract The goals of ethics and sustainability are interdependent and interconnected where the basic principle refers to that people should live in a way that considers and protects the needs of both present and future generations. This chapter addresses the requirements as they emerge in digital systems, including design guidelines, strategy, ethical issues in sourcing, and general concerns regarding the convergence of the latter in the digital systems sectors, where technological development affects critical ethical and environmental issues. As the sustainable development moves beyond the ecological impacts, this chapter describes the key requirements for the ethical development of digital systems. The topics which are analyzed range from ethics, material sourcing, life cycle assessment, digital technologies, human-centered design, regulatory frameworks, and governance initiatives. Drawing on a wide range of expert sources, including policy frameworks, academic research, technical reports from relevant bodies, and industry practice, this chapter demonstrates how governments, industries, and educational institutions are promoting technological progress in accordance with sustainability goals as established by relevant global bodies to ensure quality of life, ethics, well-being, equity, and justice. Through an interdisciplinary lens, it offers design principles, best practices, and approaches for integrating ethics and sustainability into the life cycle of digital technologies. Finally, the chapter provides insights into how ethical awareness, legislation, and sustainable thinking can lead to a socially responsible and ecologically sustainable future for the entire ecosystem.

Keywords Digital systems · Environmental impact · Life cycle assessment · Ethical sourcing · Regulatory frameworks · AI governance · Sustainable development · Digital responsibility

---

D. C. Gkikas (✉) · C. Gaie · M. Mueck
University of Patras, Patras, Greece
e-mail: dgkikas@upatras.gr

C. Gaie and M. Mueck (eds.), *Digital Sustainability: Legislations, Standards, Solutions and Related Case Studies*, Advances in Sustainability Science and Technology,
https://doi.org/10.1007/978-981-92-0458-8_7

1 Introduction

Ethics and sustainability are deeply intertwined concepts that affect both human and technological progress. Ethics helps to define what is right and responsible in human actions. However, charting a course based on sustainability ensures that these actions support both the larger ecological and social systems. When combined, the two fields create the conditions for increasing the quality of the development of new technologies such as artificial intelligence (AI), data centers, and digital services that affect social equity, environmental protection, awareness, and sustainability, ensuring justice from generation to generation [1–3]. The relationship between innovation and sustainability must be guided by purposeful, value-based, and ethically oriented frameworks that guide technology toward human awareness for environmental protection, conservation and development. In Fig. 1, a funnel path beginning with the relationship between ethics and sustainability is shown [3–5].

The integration of ethical principles and sustainability rules in the design of digital systems is essential for the development of cultural infrastructure, the advancement of sciences, digital governance, and social progress and prosperity. The role of these systems is vital as they are systems that can either cause environmental damage, exploitation of labor, and biased information and communication or contribute as mechanisms to ensure justice, improve medical technologies and educational tools, and promote equality and equal rights. Figure 2 illustrates the technological domain where sustainability is deemed necessary [6].

This chapter provides a concise overview of the integration of ethics and sustainability into digital systems. It addresses theories, legal frameworks, and

Fig. 1 From ethics to sustainability

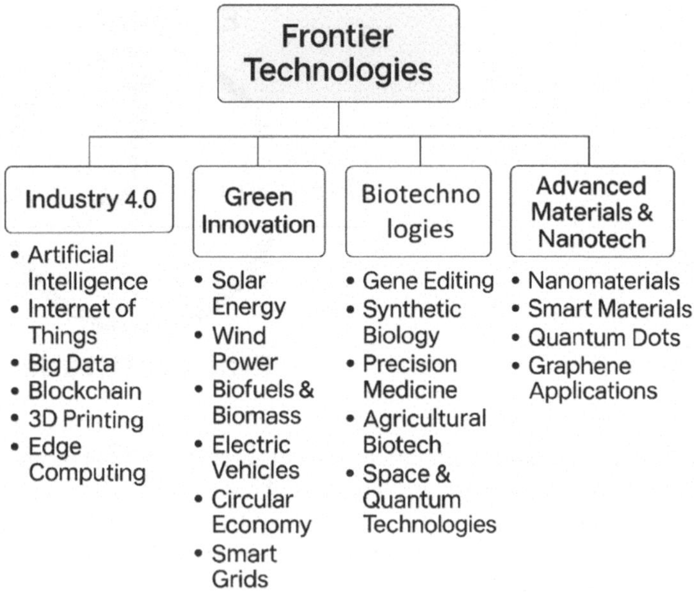

Fig. 2 Frontier technologies

emerging practices in philosophical principles, ethical issues, and sustainable practices, sourcing of raw materials and materials, life cycle approaches, and regulatory frameworks and governance practices, affecting both public and private sectors.

2 Philosophical Principles

Ethics and sustainability are often viewed as separate domains, but they are interdependent and interdependent. The concept of ethics incorporates moral values and issues of right and wrong, while sustainability is about how systems can operate within an eco-centric framework. When combined, these two concepts guide the proper human-centered development of technologies that contribute to human well-being by reducing the risk of destruction of life as we know it on the planet and ensuring social justice (Fig. 3) [4, 6].

2.1 Intergenerational Justice and Environmental Ethics

The fundamental ethical principle of sustainability is the assurance of intergenerational justice, which implies a duty and obligation on those involved to ensure that

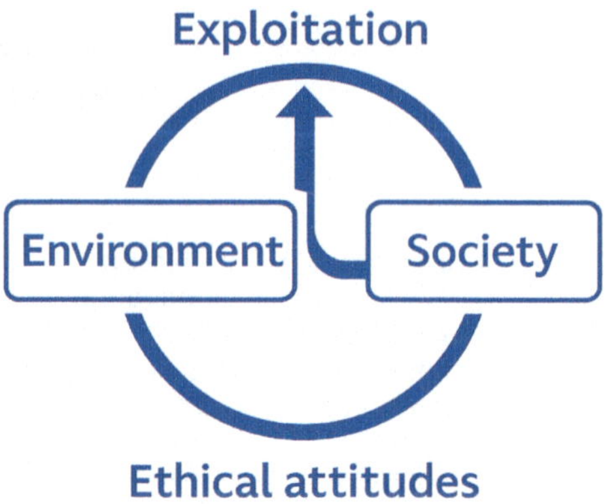

Fig. 3 Relationship between the environment and society

the needs and efforts of future generations are not sacrificed for opportunistic short-sighted gains [7]. Natural ecosystems deserve to be morally examined even though the question of whether they directly or not affect humans is answered. The environmental and ethical perspective broadens this view by understanding the value of life, ecosystem, and nature (Fig. 4) [8, 9].

Regulatory frameworks including ethics, ecology, and sustainability advocate humility in technological goals, paying greater attention to boundaries, self-restraint, responsibility, and self-awareness. These actions offer a moral load to the person who will take responsibility to offer technical solutions to the problem. In practice, regulatory frameworks push forward and promote digital systems assessment processes not only based on their performance and effectiveness but also based on their ecological effect and the way they contribute to optimizing the relationship between human and nature [10, 11].

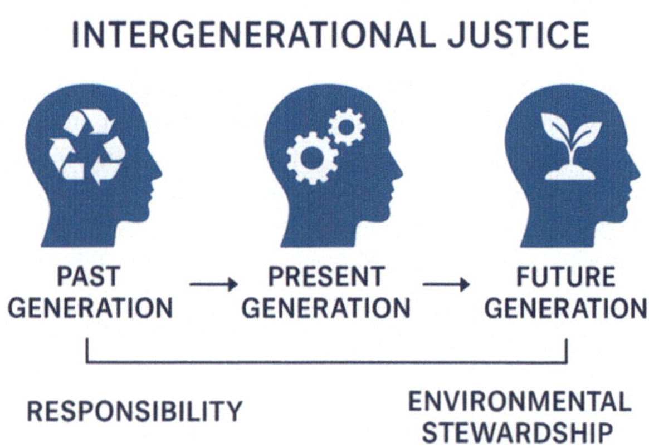

Fig. 4 Intergenerational justice

2.2 *Human Rights and Data Ethics*

The digital transformation highlights ethical questions about privacy, surveillance, independence, freedom, autonomy, justice, and digital inclusion. The human rights chart offers the basis for these discussions, promoting dignity and the right to active participation in altering the technological future of humanity. When these factors are combined with the ethics of sustainability, human rights-based theories suggest an obligation where, on the one hand, technologies must protect human freedoms and, on the other, reinforce the necessary conditions that will build the foundations for long-standing collective fairness and well-being [5, 12].

2.3 *Virtue Ethics and Global Responsibility*

The value of ethics moves the focus from performance and results to the character and intentions of the real actuators, whether they are programmers, designers, analysts, engineers, lawmakers, or policymakers. The kind of society that has become through the technological choices is since it values characteristics such as accountability, transparency, collaboration, and responsibility for generations to come [13].

Global ethical responsibility strongly suggests that sustainability is not limited to the area of a city or country but expands beyond the local politics and borders. It is a moral obligation to distant places, isolated communities, indigenous systems, and populations vulnerable to the impacts of climate change.

3 Sustainable and Ethical Design of Digital Systems

Ethical design offers values in technical projects and various types of developed systems. Sustainable practices in the design of these systems ensure that they operate within specific ecological boundaries, promoting long-term well-being. Together, they guide digital technologies toward an inclusive, meritocratic, just, and regenerative future.

3.1 *Principles of Justice and Sustainability*

According to Sasha Costanza-Chock, design justice challenges dominant scenarios of efficiency and profit, raising questions about who benefits and who is excluded from design processes. It promotes participatory, inclusive, and culturally sensitive design that confronts structural inequalities [1, 14, 15]. These principles are consistent with environmental design ethics, encouraging life cycle thinking, emission reduction,

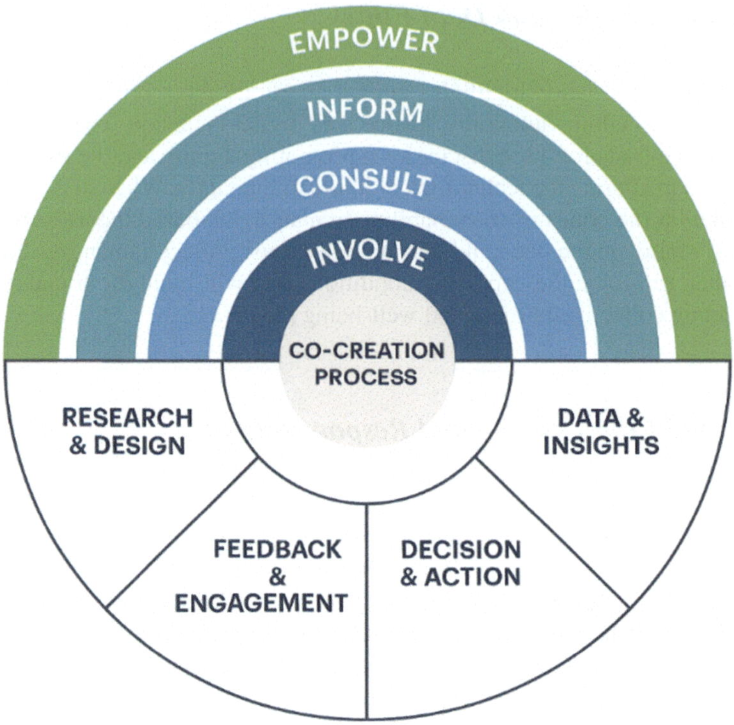

Fig. 5 Community engagement and participatory design processes

and innovation in biomimicry. Organizations such as IEEE and RIBA recommend design regulatory frameworks that introduce ethics and sustainability throughout the product life cycle [16].

These include:

- Low-carbon architectural design in hardware and data infrastructure;
- Renewable, modular, and repairable components;
- Human-centered interfaces that prevent exclusion, discrimination, and bias; and
- Eco-friendly and energy-efficient software that decrease resource consumption (Fig. 5).

3.2 Tools and Metrics for Responsible Design

Specific indicators for ethical and sustainable design are increasingly being defined by international standard setters and organizations. Some examples include:

- Responsible consumption and production is highlighted by Sustainable Development Goal 12 (SDG 12).

- Creation of criteria in the EU for Green Public Procurement Criteria.
- Integrating eco-design into management systems based on the ISO 14006 standard.

New environmental policy design tools such as Life Cycle Assessment (LCA), Environmental Product Declarations (EPD), and AI fairness checks help identify blind spots and improve accountability. According to DeepMind's 2024 report "The Era of Experience," experiential learning in AI is linked to energy efficiency and context awareness [17].

3.3 *The Carbon Cost of Digital Systems*

Digital systems have a significant and growing environmental footprint. It is noted that the emissions of digital systems such as large language models (LLMs) are comparable to the emissions of more than five cars over their entire operating life. LLMs' training can emit over 300 tons of CO_2, while large data centers consume 1–3% of the world's electricity, a large proportion of which is produced from fossil fuels. Ethical design should include energy-efficient programming, green cloud computing, active consumption, and low-impact algorithms as principles in the design process [17, 18].

4 Responsible Material Sourcing and Life Cycle Strategies

Addressing the entire life cycle of materials from mining to end of life is a key factor in achieving ethical sustainability in digital systems. The involvement of global value chains that produce and dispose of digital devices compounds the problem as it entails significant environmental and social costs [18, 19].

4.1 *Extraction and Supply Chain Ethics*

Cobalt, lithium, tungsten, and gold are rare earth elements that are essential for electronic hardware. They are mined in areas with high environmental risks and weak protection of human resources and working conditions. It should be noted that research has revealed exploitative inhumane labor practices, including child labor and precarious work in artisanal mining. These phenomena are mainly found in the Democratic Republic of Congo (Fig. 6) [6, 19, 20].

Ethical sourcing is a matter of corporate and political responsibility, not simply a matter of risk avoidance. Tools for ethical sourcing, including chain of custody audits, supplier certifications, and conflict mineral disclosures, are offered by organizations

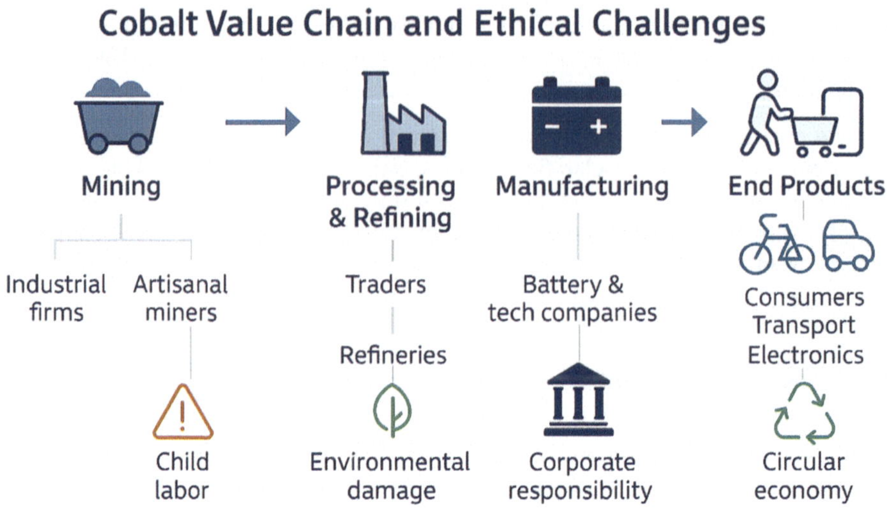

Fig. 6 True cost of cobalt mining

such as the Responsible Minerals Initiative (RMI) and the OECD Due Diligence Guidelines [21, 22].

4.2 Circular Economy and Sustainable Life Cycle Design

A circular approach to digital products is needed to promote the reuse, remanufacturing, and recycling of products, rejecting traditional linear product models, which involve taking, making, and discarding, which are harmful to the environment. Typical examples of ethical innovation in devices should include design for repairability, modular construction, and material traceability [1, 23].

Procurement strategies are informed by life cycle assessment (LCA) functions. Green Public Procurement (GPP) standards promote incentives for suppliers who manufacture products with low life cycle emissions, use recyclable and environmentally friendly materials, and develop energy-efficient designs. Life cycle assessment (LCA) is one of those tools that provides quantitative indicators of environmental impacts at all stages of production, distribution, use, and end of life. Governments and academic institutions are increasingly adopting it (Fig. 7) [24, 25].

4.3 E-waste, Disposal, and Ethics

Exceeding 50 million tons annually, of which less than 20% is officially recycled, e-waste is becoming one of the fastest growing causes of waste generation and disposal

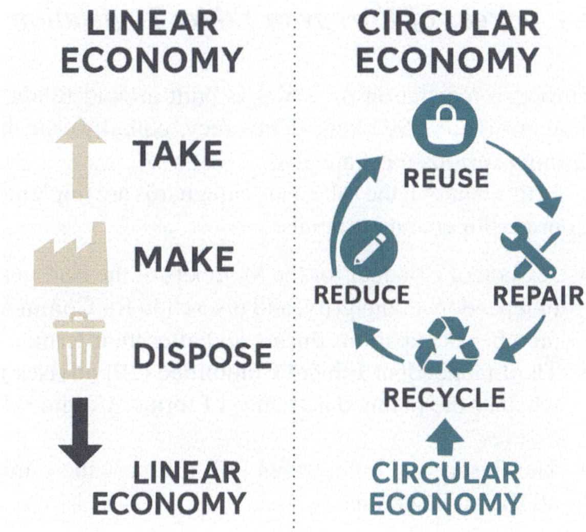

Fig. 7 Circular life cycle models for digital products

worldwide. Global South deals with most of the discarded devices which are sent there to be processed informally, causing a series of significant environmental and health threats [4].

Ethical end-of-life strategies include:

- Responsibility laws adopted by the companies and producers (e.g., Extended Producer Responsibility—EPR); and
- Take-back programs and eco-labels.

Right-to-repair legislation highlights environmental justice by reducing toxic exposures and helping sustainable jobs in recycling and refurbishment areas [26, 27].

5 European Union: AI, Procurement, and Green Innovation

The institutional commitment of local governments to global regulatory frameworks determines the effective integration of ethics and sustainability. In the last five years, digital ethics related to sustainability have begun to be reshaped by the rise of public sector policies.

5.1 Areas of European Ethics Legislation

European legislation on ethics is built around fundamental values like respect for human dignity, freedom, democracy, equality, and human rights and is enforced through various mechanisms.

In this context, the following safeguards are implemented with respect to European Commissioners and officials:

- A Code of Conduct for the Members of the European Commission [28] mandates independence, integrity, and discretion for Commissioners, including restrictions on other occupations during and after their term;
- The Independent Ethical Committee [29] advises the European Commission on whether the planned activities of former Commissioners are compatible with the treaties;
- The European Ombudsman [30] oversees the implementation of ethics rules for all EU civil servants.

For example, the European Ombudsman has recently intervened to ensure *transparency and accountability in the adoption of harmonized standards related to artificial intelligence in the context of the AI Act* [31].

The following safeguards apply:

- In terms of science and new technologies, the European Group on Ethics in Science and New Technologies (EGE) [32] provides independent recommendations on major ethical, societal, and fundamental rights issues related to science and new technologies. There are topics addressed which include AI (e.g., genome editing and planetary ethics).
- In terms of general research and funding:
  - Research projects must respond and align to the highest ethical standards and respect basic EU values like human dignity and equality [33];
  - For projects involving personal data and individual properties, compliance with the GDPR is mandatory and essential.
- In terms of fundamental rights, the EU Treaties and the Charter of Fundamental Rights of the EU [34] prohibit discrimination and require respect for human rights and linguistic diversity.
- In terms of enforcement and monitoring:
  - The Article 7 procedures of the Treaty on European Union (TEU) [35] are designed to deal with serious breaches of its founding values, including democracy and fundamental rights. They can be used as a last resort to suspend the rights of an EU member state that seriously breaches the founding principles.
  - The European Ombudsman monitors the implementation of ethics rules and pushes for improvements [30].

5.2 *European Union Digital Ecosystems Legislation*

A new risk classification system for AI systems sets out a series of legislative provisions to assess their environmental impact and human rights implications. This new system is part of the new EU law on artificial intelligence, the first comprehensive effort to regulate artificial intelligence, which, together with the European Green Deal, the Digital Services Act, and the Innovation Procurement First Finds report, establishes a comprehensive framework for responsible and sustainable digital ecosystems in the European Union [24, 36].

In public procurement, contracting authorities should select ethically and environmentally advanced offers, even if these come with higher initial costs, according to new EU directives that now impose life cycle costing criteria rather than price criteria (Fig. 8) [37].

Fig. 8 EU AI act trilogues: status of fundamental rights recommendations

RECOMMENDATION	STATUS
Remove blanket exemption for AI for national security	⚠
Limitation on export of prohibited systems outside EU	✗
Clear risk classification process (without loopholes)	✓
Full prohibition of remote biometric identification in public	✓
(Partial) prohibition on predictive policing	⚠
Biometric identification as high risk	?
Accessibility requirements for high-risk systems	✓
Use of high-risk systems registered in public database	✓
Fundamental rights impact assessment (for high-risk systems)	✗
Rights of affected people: complaint + explanation	✓

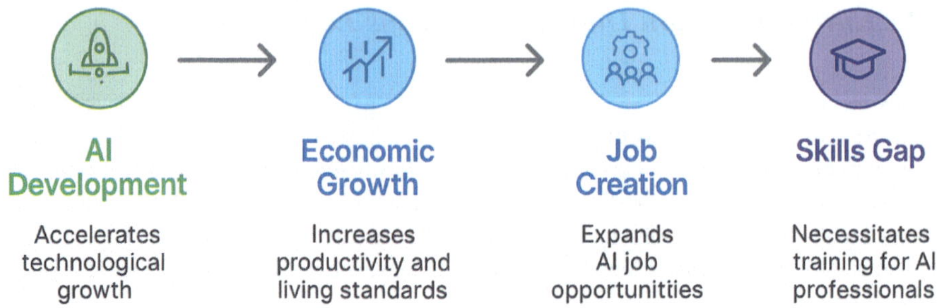

Fig. 9 UK Government's AI opportunities action plan

5.3 United Kingdom AI Opportunities and Public Ethics

The UK government's AI Opportunities Action Plan (2024) emphasizes "AI for the public good," highlighting the need for transparency in public procurement, the role of open consultation, and the importance of sustainability. It proposes a type of public benefit audit for AI in public services, which considers and evaluates ethical and environmental metrics alongside economic feasibility (Fig. 9) [38–40].

5.4 OECD and UNCTAD Global Ethical Governance

The OECD's "Anticipatory Governance" model highlights the importance of policy foresight, multi-actuators participation, and repetitive decision-making to reduce digital risks before they generate further problems. In the meantime, UNCTAD's Technology and Innovation Report 2025 introduces global collaboration and coordination on digital ethics, inclusive development, digital equality, justice, public access, and the development of sustainable infrastructure [6, 10] (Fig. 10).

6 Corporate Sector

This chapter offers a systematic approach to the procedures of incorporating ethics and sustainability into digital ecosystems. It manages to provide a combination of theories, legal frameworks, and best practices at the level of (1) philosophical principles, (2) ethical and sustainable strategic design norms, (3) material sourcing, (4) life cycle strategies, and (5) governance regulations, addressing both public and private sector actuators. Private sector innovation is the primary driver of digital progress and carries ethical and ecological responsibilities. Companies are increasingly facing pressure from regulators, investors, and civil society to adopt an ethical

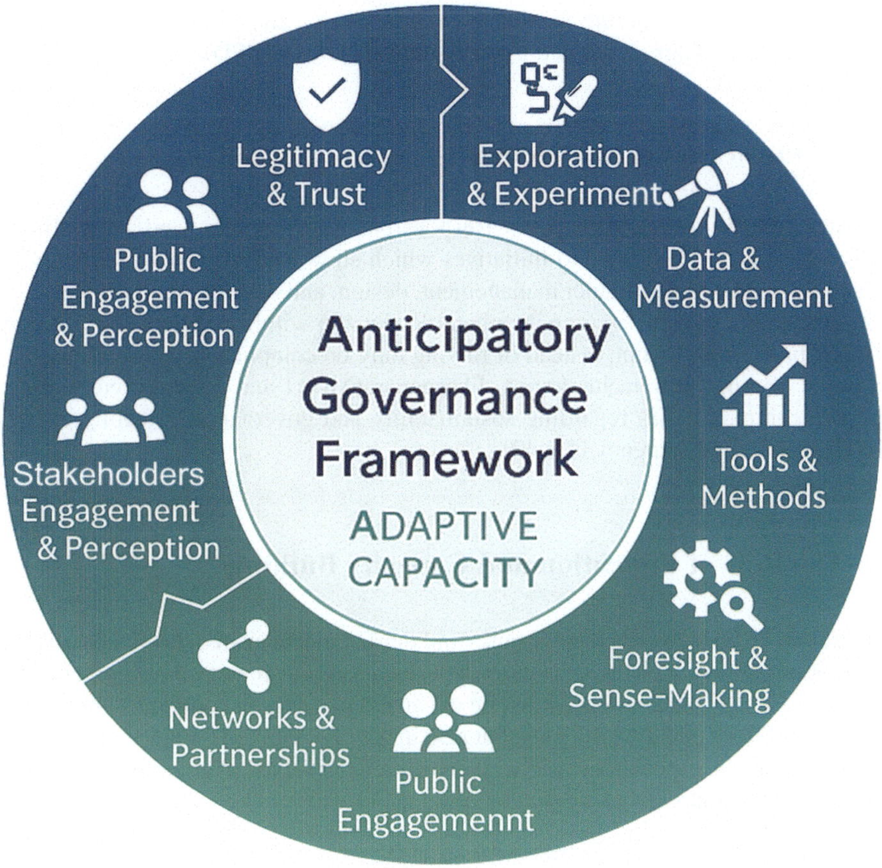

Fig. 10 Anticipatory governance framework

stance and approach to important issues, often based on the Environmental, Social, and Governance (ESG) standards, goals, and expectations.

6.1 ESG Reporting and Accountability

Referring to these developments, companies like IBM and SAP are publishing detailed ESG reports that incorporate analytical information on data protection, carbon footprint, and ethical supply chains [25]. International standards including ISO 26000 and third-party audits are essential to confirm real compliance and authenticity [14, 22]. The publication of these reports is supported by frameworks such as:

- Global Reporting Initiative (GRI);

- Sustainability Accounting Standards Board (SASB); and
- Task Force on Climate-related Financial Disclosures (TCFD).

6.2 Ethical Innovation Models

Mozilla's "Public AI," Microsoft's Responsible AI Principles, and DeepMind's Ethics & Society team refer to initiatives which show internal progress and try to incorporate ethics into product management, design, and development. McKinseys' "Superagency" framework propels ethical leadership within companies to support sustainability from within, instead of relying only on compliance. These efforts are essential because new technologies, like generative AI and quantum computing, provide complex issues regarding sustainability and governance, affecting energy use and model transparency [27, 40].

7 The Role of Education and Capacity Building

In a top-down change driven by policies and market mechanisms, sustainable digital transformation also relates to bottom-up education and human development and progress. Ethical sustainability must be expanded not only in companies, organizations, institutions and governments, but also in the civil daily life, habits, values, and skills of humans, especially those who are directly involved in creating, designing, building, regulating, and using digital systems [41] (Fig. 11).

7.1 Formal Education and Curriculum Integration

Educational institutions across the globe are incorporating ethical and sustainability principles into their curricula. Some examples include:

- HEC Paris provides a dedicated module on "Ethics and Sustainability" as part of its Postgraduate program in Management program.
- EAIE supports the best practices regarding the ethical business education adoption across European institutions and universities [14].
- WFEO implements international curriculum goals and criteria for engineering ethical standards and sustainable design [42].

Universities like the University of Amsterdam and SUNY include such principles through laboratories and learning environments, where students assess and reduce real case studies of sustainability problems globally occur using digital tools.

In computer science and data science teaching programs, schools include courses on AI ethics, environmental effect of computers, and inclusiveness. This not only

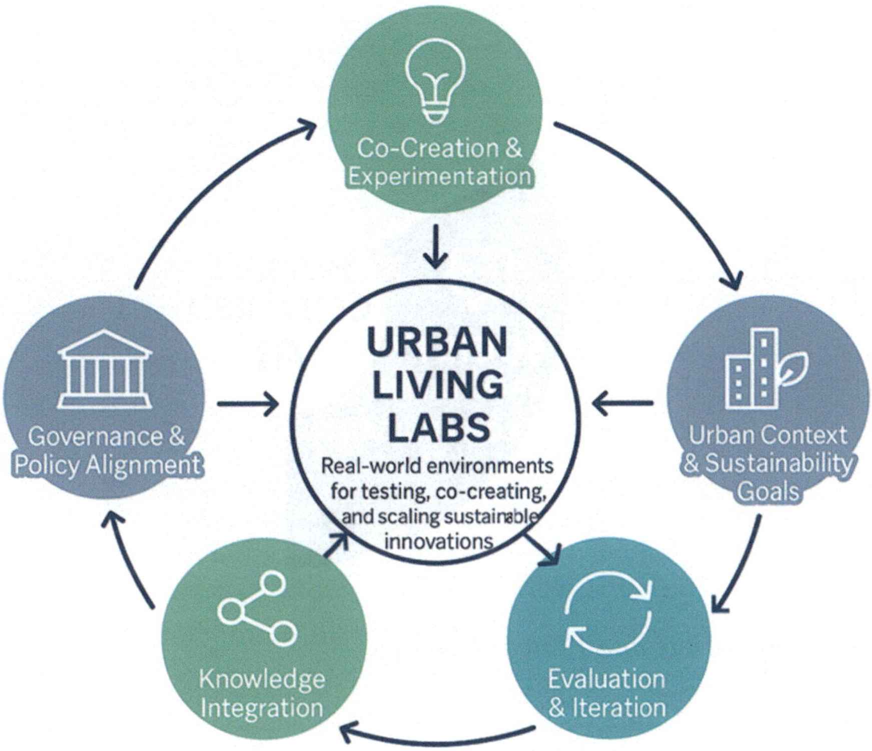

Fig. 11 Urban living labs infographic

sets the foundations for students to comprehend and shape future regulations but also communicates moral imagination and system thinking progress.

7.2 Lifelong Learning, MOOCs, and Certification

Web platforms for online learning, like Coursera, have democratized access to ethical values, regulatory frameworks, and sustainability education. Popular courses include:

- Sustainability Science by the University of Tokyo;
- AI for Good by Stanford University and the United Nations (UN); and
- Environmental and Climate Change Ethics by various universities [43].

Professional organizations like ACCA offer sustainability certifications for data scientists, architects, and procurement managers. These certifications are important in public procurement and ESG frameworks (Fig. 12) [25, 44].

Fig. 12 Human-centered AI

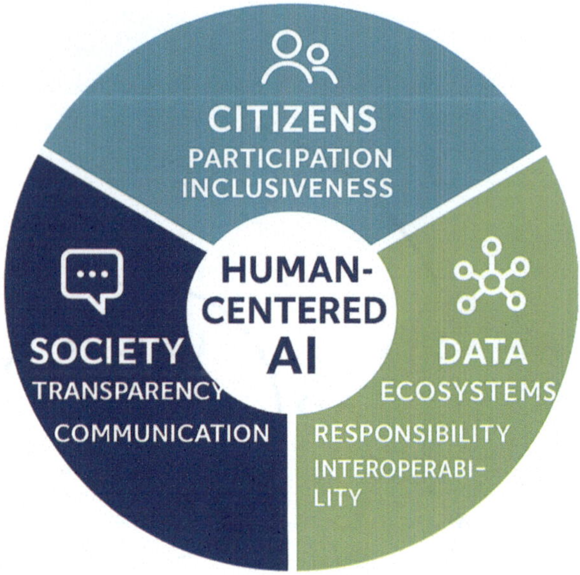

7.3 Civil Society and Ethical Literacy

Globethics.net and UNESCO support for wider ethical literacy, incorporating ethics into civilian participation, journalism, education, and digital rights campaigns. Ethical literacy responds to the ability to identify, perform, and act on value-laden decisions in digital contexts [5, 44].

Mozilla's "Public AI" program and the UK's open AI consultations operate as models for incorporating non-experts in changing the landscape of ethics in technology. This advocates that digital governance is not only shaped by experts but it is also democratically legitimate [27, 39].

8 Anticipatory, Participatory, and Pluralistic Governance

Referring to the core principle of anticipatory governance, governance must act, react, and control crises as a proactive strategy that helps societies to drive digital innovation toward sustainability before dangerous situations occur [10].

8.1 Anticipatory Governance

Anticipatory governance uses tools like:

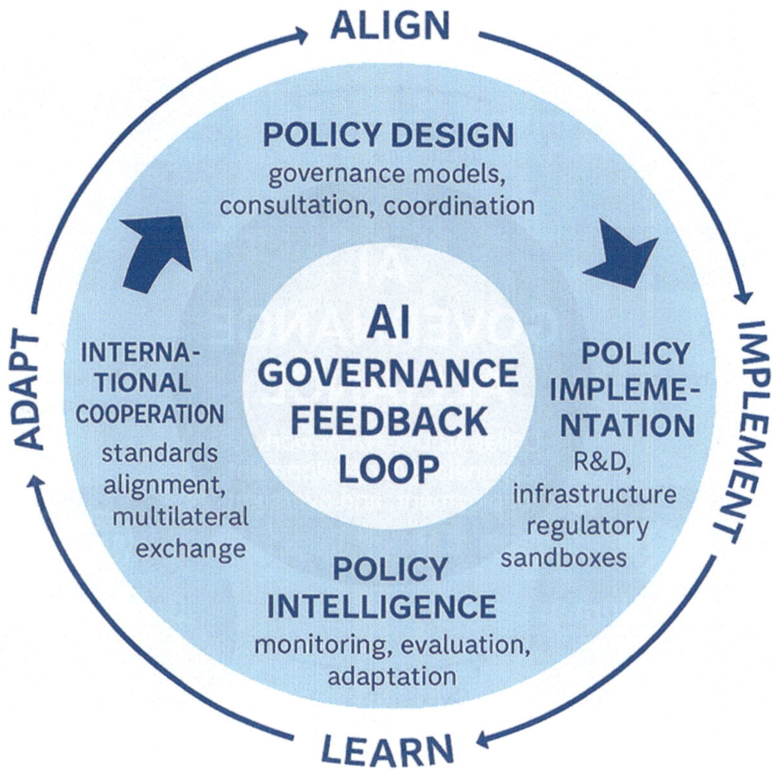

Fig. 13 AI policy cycle

- Foresight exercises;
- Ethical impact performance assessments;
- Dynamic regulatory exploration and experimentation with new products; and
- Diverse groups collaborative events.

OECD's investing on AI foresight and the UK Centre for Data Ethics and Innovation enables methods to guide adaptive policymaking. The goal is not to control every technological possibility but to adapt systems with changing principles such as societal values and long-term objectives (Fig. 13) [10, 45].

8.2 Participatory Governance and Public Trust

Participation helps create the legitimacy that ethics and sustainability need. Some examples include:

Fig. 14 AI governance alliance

- Mozilla's Public AI Toolkit which helps communities assess risks and values in AI systems [27];
- UK's public consultations impact white papers through inclusive dialogue [39]; and
- Seoul AI Summit pinpoints international collaboration on common AI ethics principles [46].

Participatory governance is especially essential for citizenship and sustainability reasons, mostly because the problems caused by unsustainable development like pollution, exclusion, and vulnerability, are unevenly shared. Thus, those impacted need to participate in the processes (Fig. 14) [10, 17, 48].

8.3 Pluralism and Ethical Diversity

Global digital governance must be wide and stand above cultures avoiding ethical monoculture. The meaning of pluralism enables recognition of different and various cultural traditions of eco-sensitivity and sustainability worldwide from Buen Vivir in Latin America to Ubuntu in Africa and indigenous ethics worldwide. UNESCO and the European Parliament are supporting intercultural dialogue and opinion exchange during the formation of sustainability policymaking. Instead of exporting one single size or form that fits to all cultural backgrounds and frameworks, ethical governance should allow the generation of context-sensitive, culturally grounded solutions and principles that respect diversity while keeping shared values and principles [5, 27, 48].

9 Tensions, Contradictions, and Trade-Offs

Ethics and sustainability do not always align seamlessly, and various tensions are created which must be recognized, comprehended, evaluated and addressed.

9.1 Innovation Versus Precaution

Technological development usually marginalizes ethical reflection and regulation. Extreme caution is highly advised when developing robust sophisticated systems, such as image detection, facial, or voice recognition without a thorough comprehension of the dangers. Nevertheless, excessive legislation could prevent beneficial innovation. Thus, a balance must be established where governance should be flexible, proactive, iterative, and responsive, according to ethical and democratic accountability [49].

9.2 Equity and Efficiency

Optimizing systems performance can exclude vulnerable and sensitive groups. Ethical sustainability demands that designers aim to build inclusion-sensitive systems, even if it means sacrificing some effectiveness.

9.3 Local Adaptation and Global Standards

The SDGs, the Paris Agreement, and the EU AI Act Global known as legal or ethical regulatory frameworks provide consistency, but they may not agree with local practices, traditions, customs, capabilities, or infrastructures. Ethical governance should allow contextual adaptation where values and core principles, like justice, trustworthiness, transparency, and environmental responsibility, should be respected and promoted.

10 Conclusions

New technologies like generative AI brought up a series of questions about ethics, justice, inclusion, sustainability, and governance, including a range of serious issues starting from energy consumption to civil rights and ecosystem protection.

The ethical and sustainable transformation and adaptation of digital systems is important, direct, comprehensible, urgent, and achievable. It demands reevaluation, reassessing and reconstructing material sourcing, e-governance, and educational models that advocate new-age technology infrastructures. Ethics and moral principles work as the moral compass, when sustainability, conservation, and restoration work as the long-term goal. Incorporating all notions is imperative for building a sustainable, fair, inclusive, and environmentally responsible digital future.

10.1 Key Findings

During this literature research, a series of key findings emerged, and they are essential to be gathered and briefly mentioned.

- As mentioned, ethics and sustainability are conceptually different notions, but they are practically intertwined and related.
- Justice through generations, equality, access to services, environmental ethics, and human rights protection offer the foundations for human-centered responsible digital innovation.
- Design processes must change course from efficacy-driven approaches to incorporate sustainability, equity, fairness, accessibility, and repairability, driven by justified mandatory design protocols and frameworks such as ISO 14006.
- ESG benchmarks, EPR laws, and life cycle assessments standards should affect the supply chains and sourcing materials and ways of sourcing which must deal with important ethical implications and responsible ethical sourcing protocols.
- EU AI Act, UK procurement policy, and OECD foresight government regulations and corporate frameworks including IBM's AI ethics, McKinsey's agency, and Mozilla's Public AI are aligning digital innovation and transparency with sustainability and societal values.
- Education and capacity building are important for altering institutional cultures and context to be able to face the challenges and the upcoming adversities.
- Participatory and anticipatory governance are imperative for keeping legitimacy and resilience. This incorporates pluralization, ethical frameworks, open access, and public participation.

10.2 Policy and Recommendations

A series of recommendations may assist to incorporate morality, ethics, equal rights, and sustainability in digital systems across sectors and regions:

1. Institutionalize independent audits on ethics and sustainability for all major digital technology projects involving living beings and tighten based on the magnitude of environmental impacts.
2. Develop life cycle procurement where governments and organizations should adopt practices that prioritize circularity, recyclability, and inclusive design instead of focusing only on the lowest cost.
3. Expand education and request mandatory certification where universities should enable ethics and sustainability modules into all tech-oriented degrees.
4. Expand professional certifications requesting mandatory ESG training.
5. Institutionalize public participation to serious matters that affect society. Public trust is essential for ethical digital systems to perform. Procedures like citizens' assemblies, open consultations, and stakeholder panels should be established locally, nationally, and internationally.
6. Global ethics and sustainability should pursuit another approach instead of a top-down enforcement. Cross-regional learning is the key factor that will enable regions like Latin American and African sustainability models to improve equity and contextual relevance.
7. Governments, think tanks, and research groups should invest and provide funds to foresight laboratories and anticipatory ethics to eliminate the upcoming risks of generative AI, climate modeling, and the convergence of biotech and data.
8. Invest in collaboration and promote Open Ethics Data and Benchmarks where sustainability in digital systems would benefit from open ethical benchmarks, valuable indicators, case studies, and performance metrics.

In an age of global insecurity, great uncertainty, and fast digital growth, incorporating ethics and sustainability principles would benefit humanity. Such options are no longer optional but more essential for both moral and practical survival. The legislation, regulatory, anticipatory frameworks define a certain level of readiness since society possesses many of the necessary tools to manage this digital transformation.

What is needed now is commitment, coherence, and courage—from institutions, industries, and individuals. Only by aligning our technological abilities with our moral and ethical responsibilities can we create digital systems that promote and support life, democracy, and human dignity for future generations.

Acknowledgements The author would like to sincerely thank Dr. Christophe Gaie and Dr Markus Mueck for their invaluable support and insightful contributions throughout the development of this book chapter. Their expertise, guidance, and constructive feedback greatly enhanced the quality and clarity of this chapter.

Disclaimer Note that the authors have assembled the information above to the best of their knowledge as available at the time when the present book was written. Any interpretation is the personal opinion of the authors. Also, the regulation framework can change and evolve any time. The only legally binding text is the one provided in the documents formally published in the European Union Official Journal.

References

1. Russ J (2023) Sustainability and design ethics. Routledge, London
2. Meinhold R (2023) Business ethics and sustainability. Routledge, London. https://doi.org/10.4324/9781003127659
3. Torelli R (2021) Sustainability, responsibility and ethics: different concepts for a single path. Soc Respons J 17(5):719–739. https://doi.org/10.1108/SRJ-03-2020-0081
4. United Nations (2012) The future we want: outcome document of Rio+20 conference. UN, NewYork.https://wedocs.unep.org/bitstream/handle/20.500.11822/13662/N1238164.pdf?sequence=1&amp%3BisAllowed=
5. UNESCO (2021) Recommendation on the ethics of artificial intelligence. UNESCO, Paris. https://unesdoc.unesco.org/ark:/48223/pf0000380455
6. UNCTAD (2025) Technology and innovation report 2025: shaping the future—driving equitable, inclusive and sustainable transformation. United Nations, Geneva. https://unctad.org/system/files/official-document/tir2025_en.pdf
7. World Commission on Environment and Development (WCED) (1987) Our common future. Oxford University Press. https://sustainabledevelopment.un.org/content/documents/5987our-common-future.pdf
8. Nelson MP, Vucetich JA (2012) Sustainability science: Ethical foundations and emerging challenges. Nat Educ Knowl 3(10):12
9. KMCO (2024) Applying environmental ethics for sustainability. https://kmco.co.ke/wp-content/uploads/2024/04/Applying-Environmental-Ethics-for-Sustainability.pdf
10. OECD (2024) Anticipatory governance for sustainable innovation. OECD.AI Policy Observatory, Paris. https://www.oecd.org/content/dam/oecd/en/publications/reports/2025/02/steering-ai-s-future_70e4a856/5480ff0a-en.pdf
11. Nature Scitable (2024) Sustainability science: ethical foundations and emerging challenges. https://www.researchgate.net/publication/291830206_Sustainability_Science_Ethical_Foundations_and_Emerging_Challenges
12. Montreal AI Ethics Institute (2023) The AI ethics brief series. Available at: https://montrealethics.ai
13. (2022) Science must respect the dignity and rights of all humans. Nat Hum Behav 6:1029–1031. https://doi.org/10.1038/s41562-022-01443-2
14. European Association for International Education (EAIE) (n.d.) Our commitments: Social responsibility. https://www.eaie.org/about-us/our-commitments/social-responsibility.html
15. ACCA Global (2023) Ethical sustainability reporting. ACCA Professional Insights, London. https://www.accaglobal.com/content/dam/ACCA_Global/professional-insights/ethical-dilemmas/PI-ETHICAL-SUSTAINABILITY-REPORTING%20v4.pdf
16. RIBA Ethics and Sustainable Development Commission (2019) Final report of the RIBA ethics and sustainable development commission. Royal Institute of British Architects. https://www.architecture.com/knowledge-and-resources/resources-landing-page/ribas-ethics-and-sustainable-development-commission-final-report
17. DeepMind (2024) The era of experience. DeepMind Technologies, London. https://storage.googleapis.com/deepmind-media/Era-of-Experience%20/The%20Era%20of%20Experience%20Paper.pdf
18. Re-thinking The Future. Community engagement and participatory design processes. https://www.re-thinkingthefuture.com/architectural-community/a10233-community-engagement-and-participatory-design-processes/

19. Parry I, Black S, Roaf J (2022) More countries are pricing carbon, but emissions are still too cheap. IMF Blog. International Monetary Fund. Available at: https://www.imf.org/en/Blogs/Articles/2022/07/21/blog-more-countries-are-pricing-carbon-but-emissions-are-still-too-cheap
20. Human Trafficking Search (2022) Modern slavery: the true cost of cobalt mining. Available at: https://humantraffickingsearch.org/resource/modern-slavery-the-true-cost-of-cobalt-mining/
21. OECD (2018) OECD due diligence guidance for responsible business conduct. Organisation for Economic Co-operation and Development. Available at: https://mneguidelines.oecd.org/due-diligence-guidance-for-responsible-business-conduct.htm
22. Responsible Minerals Initiative (RMI) (n.d.) Responsible minerals initiative: advancing responsible sourcing worldwide. Available at: https://www.responsiblemineralsinitiative.org/
23. HAW Hamburg (2023) Assessing ethics and sustainability standards in corporate practices. Soc Responsibil J. https://www.haw-hamburg.de/fileadmin/LS/FTZ-NK/PDF/Publications/2023-Assessing_ethics_and_sustainability_standards_in_corporate_practices.Social_Responsibility_Journal..pdf
24. European Commission (2024) Public procurement directives evaluation. Brussels. https://www.ecec.net/news/detail/news/ecec-feedback-to-the-evaluation-of-the-public-procurement-directives/?tx_news_pi1%5Bcontroller%5D=News&tx_news_pi1%5Baction%5D=detail&cHash=158672a25dcf830e428c766ac913be39
25. Ellen MacArthur Foundation (2019) Circular design guide: methods for designing circular products and services. Available at: https://www.circulardesignguide.com/
26. IBM (2024) For the planet and people: IBM's focus on AI ethics in sustainability. https://www.ibm.com/think/insights/for-the-planet-and-people-ibms-focus-on-ai-ethics-in-sustainability
27. Mozilla Foundation (2024) Public AI consultation report. https://www.mozillafoundation.org/en/research/library/public-ai/
28. Code of conduct for the members of the European Commission, European Commission, available at https://commission.europa.eu/about/service-standards-and-principles/ethics-and-good-administration/commissioners-and-ethics/code-conduct-members-european-commission_en
29. The independent ethical committee, European Commission, available at https://commission.europa.eu/about/service-standards-and-principles/ethics-and-good-administration/commissioners-and-ethics/independent-ethical-committee_en
30. The European Ombudsman, European Commission, available at https://www.ombudsman.europa.eu/en/home
31. How the European Commission ensures transparency and accountability in the adoption of harmonised standards related to artificial intelligence in the context of the AI Act, European Commission, available at https://www.ombudsman.europa.eu/en/opening-summary/en/212090
32. European Group on Ethics in Science and New Technologies (EGE), European Commission, available at https://research-and-innovation.ec.europa.eu/strategy/support-policy-making/scientific-support-eu-policies/european-group-ethics_en
33. Horizon Europe, European Commission, available at https://research-and-innovation.ec.europa.eu/system/files/2022-09/ec_rtd_ethics-aspects-cluster6-coordinators-day.pdf
34. Charter of fundamental rights of the European union, European Parliament, available at https://www.europarl.europa.eu/charter/pdf/text_en.pdf
35. Article 7 procedures, European Council, available at https://www.consilium.europa.eu/en/policies/article-7-procedures/
36. European Parliament (2025) EU AI act: regulation on artificial intelligence. Brussels. https://eur-lex.europa.eu/legal-content/EN/TXT/?uri=CELEX%3A32024R1689
37. European Commission (2024) Better Regulation: Law-making process. Retrieved from https://commission.europa.eu/law/law-making-process/better-regulation_en
38. European Digital Rights (EDRi) (2024) EU AI Act Trilogues: status of fundamental rights recommendations. Retrieved from https://edri.org/our-work/eu-ai-act-trilogues-status-of-fundamental-rights-recommendations/
39. UK Government (2024) AI opportunities: action plan. Department for Science, Innovation and Technology. Available at: https://www.gov.uk/government/publications/ai-opportunities-action-plan/ai-opportunities-action-plan

40. McKinsey & Company (2024) Superagency in the workplace. https://www.mckinsey.com/capabilities/mckinsey-digital/our-insights/superagency-in-the-workplace-empowering-people-to-unlock-ais-full-potential-at-work
41. Brown University (2023) Campus as living lab: sustainability & resiliency initiative [Infographic]. Retrieved from https://sustainability.brown.edu/education/campus-living-lab
42. WFEO (2023) Engineering ethics and sustainability. World Federation of Engineering Organizations. https://www.wfeo.org/code-of-ethics/
43. Stanford Encyclopedia of Philosophy (2024) Environmental ethics. https://plato.stanford.edu/entries/ethics-environmental/
44. Globethics.net (2024) Global ethics network: resources for ethics in higher education. Retrieved from https://globethics.net/
45. OECD (2023) AI Policy Observatory: AI governance policy cycle infographic [Infographic]. Organisation for economic co-operation and development. Retrieved from https://oecd.ai/en/
46. Republic of Korea (2024) AI seoul summit 2024: building inclusive and trusted AI governance [Conference website]. Retrieved from https://aiseoulsummit.kr/
47. Frontiers in Artificial Intelligence (2023) Human-centricity in AI governance: a systemic approach [Figure]. Frontiers Media. Retrieved from https://www.frontiersin.org/articles/https://doi.org/10.3389/frai.2023.1174652/full
48. Hitachi Research Institute (2023) AI governance in a rapidly evolving world: from development to regulation [infographic]. Retrieved from https://www.hitachi-hri.com/english/research/contribution/vol19_01_3en.html
49. Jankovic SD, Curovic DM (2023) Strategic integration of artificial intelligence for sustainable businesses. Sustainability 15(6):15208. https://doi.org/10.3390/su152115208

Chapter 8
From Concept–Knowledge (C-K) and Theory of Inventive Problem Solving (TRIZ) to Net Technological Assessment (NTA): Structured Decision-Making for the European Green Deal

Christophe Gaie and Jean Langlois-Berthelot

Abstract This chapter examines how governments can align digital innovation with the European Green Deal. Technical strategies—such as shutting down unused components, adapting performance to demand, or reducing communication overhead—illustrate sustainability at the system level, but they do not solve the challenge of prioritizing technologies under policy, budget, and Environmental, Social, and Governance (ESG) constraints. Formal innovation methods like Concept–Knowledge (C-K) theory and Theory of Inventive Problem Solving (TRIZ)support structured exploration but remain insufficient for governmental decision-making. The chapter, therefore, introduces **Net Technological Assessment (NTA)** as its core contribution: a novel framework that integrates maturity, investment potential, and sustainability weighting to translate innovation options into accountable, Green Deal–aligned policy choices.

Keywords Net technological assessment · European green deal · Sustainability policy · Innovation methodologies · C–K theory · TRIZ · ESG weighting · Technology assessment · Digital transformation · Governance

---

C. Gaie (✉)
French Prime Minister Office, Laboratoire de Recherche Magellan, Université Jean Moulin, Lyon, France
e-mail: christophe.gaie@gmail.com

J. Langlois-Berthelot
A.S.CENT Centre of Advanced Studies (UNIPA), Palermo, Italy

C. Gaie and M. Mueck (eds.), *Digital Sustainability: Legislations, Standards, Solutions and Related Case Studies*, Advances in Sustainability Science and Technology,
https://doi.org/10.1007/978-981-92-0458-8_8

1 Introduction

The European Green Deal sets the most ambitious framework for sustainability ever attempted within the European Union (EU) [1]. It has notably the following major objectives: achieving climate neutrality by 2050, decoupling economic growth from resource use, and ensuring a just transition. The success of this strategic orientation requires not only regulatory action but also the systematic mobilization of innovation. Government administrations are at the center of this transformation: they must translate European objectives into national strategies, guide investment, and anticipate the technological shifts that will shape industry, security, and society, as illustrated by Romania's transposition of the policy [2].

However, innovation cannot rely solely on spontaneous creativity or on policies initiated by private or public organizations. Fortunately, in recent decades, formal innovation methodologies have been developed to structure the design process and enhance the generation of novel solutions. Among these, the authors selected Concept–Knowledge (C–K) theory [3], TRIZ (Theory of Inventive Problem Solving) [4], and Bayesian approaches [5]. This choice aims to retain rigorous frameworks that transform creativity into a structured and often mathematical exploration of possibilities. Moreover, these methodologies prove to be very effective for expanding concept spaces, resolving contradictions, and managing uncertainty.

In 2019, the publication of the Green Deal established targets that government organizations must achieve. Meeting the goals set by this regulation requires generating innovative ideas and then selecting the most effective solutions. More precisely, public services must determine where to invest and how to ensure compliance with sustainability criteria. In this context, formal innovation methodologies are useful for exploration, but they do not fully meet the requirements of public decision-making.

In this chapter, the authors aim to fill this gap and propose a structured framework dedicated to technology assessment. The Net Technical Assessment (NTA) makes it possible to evaluate the maturity, investment potential, and operational impact of technologies. Unlike classic exploratory methods as illustrated below, it also provides a means for prioritizing potential solutions. Additionally, when complemented with ESG (Environmental, Social, and Governance) and sustainability weighting [6], NTA offers governments a decision-support framework aligned with the strategic imperatives of the European Green Deal (Fig. 1).

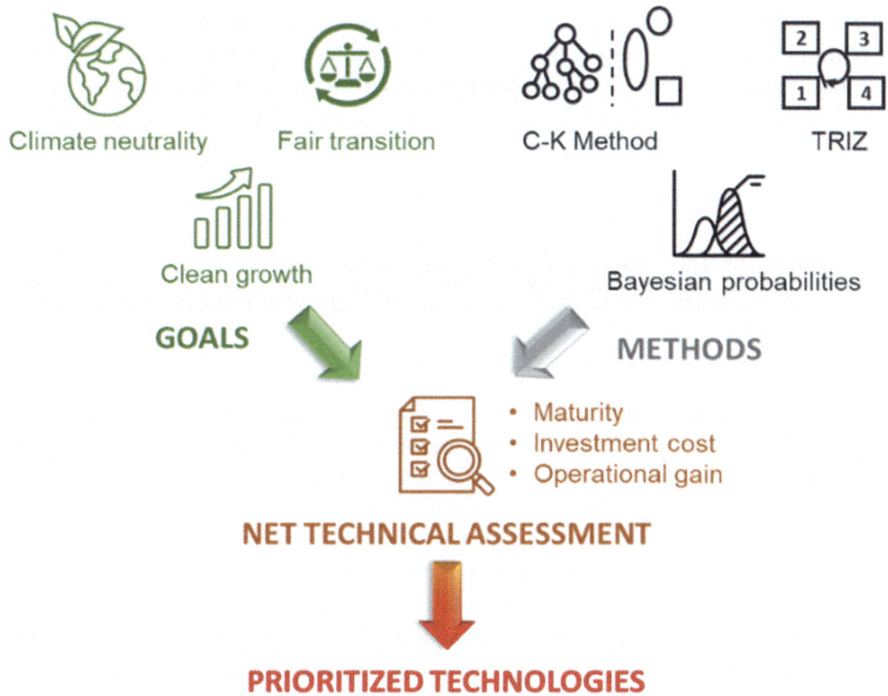

Fig. 1 A method to prioritize technologies and reach sustainability goals

2 The European Green Deal and Governmental Responsibilities

2.1 Overview of the Green Deal's Systemic Ambitions

The European Green Deal is not a single environmental program but a comprehensive strategy designed to reshape the Union's economic, industrial, and social architecture. Its systemic ambitions are threefold: to achieve climate neutrality by 2050, to decouple growth from resource use, and to secure a just and inclusive transition for all regions and communities.

What distinguishes the Green Deal from earlier European initiatives is its scope and integration. Climate policy is no longer confined to the energy sector but is linked to agriculture, mobility, digitalization, biodiversity, and industrial competitiveness [7]. Every major policy field is integrated into a common framework that seeks to align growth with ecological limits. In this respect, the Green Deal provides a roadmap for transforming Europe's development model.

Thus, sustainability is established as one of the organizing principles of European integration, and the Green Deal redefines the relationship between policy, economy, and society. It seeks not only to mitigate environmental risks but also to reorient investment, innovation, and social welfare systems toward resilience and long-term prosperity. This ambition underscores that climate neutrality is not a narrow environmental target but a lever for transforming the European Union in a resource-constrained world [8].

2.2 *The Specific Role of Public Administrations in Translating EU Objectives into National and Sectoral Policies*

The ambitions of the Green Deal need to be translated by governments into national laws. This offers the opportunity to adjust implementation to national contexts, which often vary considerably in terms of economic structure, energy dependence, and social priorities. For example, there is a fundamental difference between Germany, a European leader in industry with the largest population, and Malta, the smallest country, mainly focused on tourism. In the digital domain, Estonia is recognized as a pioneer and is expected to define ambitious measures for digital sustainability.

While this chapter focuses on the digital domain, it is essential to consider energy as a key parameter of sustainability. The rise of large data centers to support cloud infrastructure, data storage, and computation capacity for artificial intelligence has made this even more pressing [9]. National laws should therefore define transformation pathways that integrate both national constraints and strengths. For example, Germany produces much of its energy using coal, while France relies on nuclear power. As a result, their strategies for developing sustainable energy should be different. Germany may need to expand renewable energy to reduce coal usage, while France should modernize its low-carbon system.

Government administrations should not simply translate European objectives. Indeed, they also need to define investment strategies that prioritize technologies in a context of high budget deficits. This choice should prioritize technologies based on their expected benefits, as well as their potential for rapid and cost-effective implementation.

In this regard, several national and EU initiatives illustrate how administrations can dedicate resources toward sustainable digital and scientific innovation. At the European level, the Digital Europe Programme provides targeted funding for supercomputing, artificial intelligence, cybersecurity, and advanced digital skills. These findings are essential to ensure that the Green Deal's digital transition is both competitive and sustainable [10].

France is investing significantly in innovation projects, research, and development to link ecological transition with digital transformation. To this end, the Centre National de la Recherche Scientifique (CNRS) and other national agencies are operationally driving the programme entitled "France 2030 Strategy." It provides funding for clean energy, advanced materials, quantum technologies, and sustainable mobility, thereby concretely aligning industrial policy with the ambitions of the Green Deal [11]. Similarly, the Agence Nationale de la Recherche (ANR) plays a key role in managing public investment in sustainable development research, supporting numerous projects that integrate environmental, technological, and social dimensions [12].

Public administrations also act as facilitators of international collaboration. The creation of the "Choose France for Science" platform illustrates how national governments can attract international researchers and innovative projects, thereby reinforcing Europe's position as a global leader in sustainable and digital research [13]. This outlines that the role of administrations goes beyond the mere legal translation: they must also orchestrate strategic investment, innovation governance, and international cooperation to ensure that the objectives of the Green Deal are realized in practice across both national and sectoral contexts.

Another key role of government administration is to coordinate across multiple sectors, as the systemic nature of the Green Deal means that progress in one area cannot come at the expense of another. Artificial intelligence provides a clear example: while it can enhance public service delivery and efficiency, it also requires vast amounts of data and significant energy to train models and respond to user requests [14]. To avoid situations where choices must be made between adopting new technologies and protecting the environment, governments must simultaneously advance AI development, optimize data center energy consumption, and reduce the energy footprint of their own digital services.

In this regard, administrations are uniquely positioned to facilitate synergies between the public and private sectors. As the deployment of AI is expected to accelerate and intensify its environmental impact in the coming years, public institutions should not only regulate and set sustainability standards but also act as role models by integrating energy-efficient solutions within their operations [15]. By doing so, governments can ensure that adopting AI contributes positively to the objectives of the Green Deal.

To ensure effective monitoring and compliance, administrations must translate sustainability goals into measurable indicators. Tools such as the European Green Deal Index (EGDI) [16] provide a composite measure that tracks Member States' progress in implementing the Green Deal. By capturing both performance levels and trajectories (e.g., "Outperforming" countries such as Estonia, Sweden, and Finland versus "Diver" countries like Germany, France, and Spain), the EGDI allows policymakers to identify convergence and divergence trends across the EU. This form of benchmarking highlights the need for harmonized standards, enabling comparability, transparency, and early detection of lagging sectors or regions.

Complementing such composite indices, Monitoring and Evaluation (M&E) frameworks [17] offer continuous feedback loops to assess the effectiveness, efficiency, and long-term impact of Green Deal policies. M&E not only validates whether objectives are met but also informs adaptive management by highlighting which instruments are delivering tangible outcomes.

Digital tools may also be applied to enhance governmental frameworks. For instance, blockchain can ensure the transparency of available resources within government services. Another example is AI tools, which may be used to track energy consumption and emissions of public and private organizations. By using these technologies, governments are taking measures that contribute to the concrete implementation of the Green Deal. This helps ensure that EU Member States will reach this objective through transparent governance and effective monitoring.

2.3 Tensions Between Short-Term Operational Requirements and Long-Term Sustainability

The implementation of the Green Deal in government services presents a major challenge. Indeed, there is a need to balance short-term incremental improvements with long-term, higher-value goals. On the one hand, government services must secure many public policies (such as energy supply and protecting employment), and simultaneously maintain economic stability (by limiting inflation and protecting vital services). These requirements tend to favour incremental improvements as well as the deployment of effective technologies. On the other hand, transforming government organizations requires developing and deploying high-value innovations. For example, the use of AI for decarbonization or biodiversity restoration could bring significant benefits but represents a difficult challenge in the context of government services. Indeed, these high-impact objectives require expert skills as well as sustained investment over decades.

The opposition between long-term and short-term goals is known as the "tragedy of the time horizon" [18]. Indeed, decision-makers often prioritize short-term outcomes because these provide immediate effects and political rewards. Similarly, projects within government services also tend to favour small-scale gains, as these are more accessible in a context of limited resources. This makes it difficult to modernize systems in-depth and can lead institutions to become locked in suboptimal situations [19]. To overcome these limitations, recent research emphasizes the need for strategic approaches that explicitly manage this duality. Such approaches aim to achieve a balance between short-term strategies that offer small improvements at low risk and long-term strategies that are essential for achieving in-depth sustainability [20].

To accomplish the goals of the Green Deal, environmental needs must be balanced with social stability, political feasibility, and public support. The authors emphasize that this requires decision-making frameworks that combine pragmatic judgment with long-term ambition. In this context, Net Technological Assessment (NTA) is presented as a particularly valuable method, as it helps administrations evaluate technologies, understand trade-offs, and define clear directions for deep systemic transformation. In addition, approaches inspired by sustainability research can reduce the risk of short-termism and support better alignment between immediate priorities and long-term structural change.

Another dimension of this tension lies in the budgetary cycle and political time horizon. National administrations typically operate within annual or multi-year budget frameworks and under electoral mandates of four to five years, whereas Green Deal objectives extend across decades. This structural misalignment incentivizes policies that deliver visible, short-term results (such as subsidies for energy price relief) over investments in systemic transformation. The latter, including next-generation grids or low-carbon industrial clusters, generate more significant benefits but only materialize after one or two decades. Electoral cycles, in fact, encourage governments to adopt policies with immediate visibility, often at the expense of measures that deliver deeper structural change [21]. Similarly, short-term political priorities frequently undermine long-term adaptation and public well-being, reinforcing the risks of postponing transformational investments [22].

To achieve long-term objectives, governments need systems that protect sustainability projects from short-term fluctuations. A very relevant example is the European Climate Law [23], which states net zero emissions by 2050 and a 55 percent greenhouse gas emissions reduction by 2030, providing a mechanism for periodic review and for establishing targets, including that of 2040, under the supervision of an independent scientific advisory body. The law includes mechanisms for periodic review, target setting, including for 2040, and a supervision system managed by an independent scientific advisory entity. By embedding these commitments into law, it ensures that the goals persist over time and provides stability and continuity for Europe's long-term climate strategy.

Other ways to enhance resilience are independent Green Investment Funds with long-term mandates and cross-party agreements for continued investment regardless of the outcome of an election. Establishing a long-term commitment through the establishment of independent funds will help to reduce the impact of political cycles on government strategic sustainability planning and limit the potential for short-term crises to impact sustainability objectives. The Green Deal will redefine the role of public sector organizations from passive regulatory agents to system-builders, charged with coordinating Europe's largest-ever transition, balancing competing demands between current and future needs.

3 Formal Innovation Methodologies as Foundations of Structured Exploration

Innovation is not merely the product of spontaneous creativity. Especially in the context of the European Green Deal, innovation must be structured, reproducible, and accountable to sustainability criteria. Over the past decades, several methodologies have emerged to systematize the design and invention process. Among the most prominent are Concept–Knowledge (C–K) theory, TRIZ (Theory of Inventive Problem Solving), and Bayesian approaches to design under uncertainty.

These frameworks serve as an additional exploration tool; they are not intended to supplant the existing political framework, but rather create a more comprehensive environment regarding design possibilities, alternative solutions, and managing uncertainty. In this section, the authors examine the guiding principles of these frameworks, demonstrate how these frameworks may be applied to problems relating to sustainability, and discuss the drawbacks of applying them in government contexts.

In this section, we structure the discussion into three subsections: Sect. 8.3.1 introduces Concept–Knowledge (C–K) theory as a framework for systematically expanding the space of possible innovations; Sect. 8.3.2 turns to TRIZ as a method for resolving design contradictions through recurring inventive principles; and Sect. 8.3.3 examines Bayesian approaches as probabilistic tools for managing uncertainty in technological development. Together, these methodologies illustrate how structured exploration can support administrations in addressing sustainability challenges while highlighting the limits of purely exploratory methods in governmental contexts.

3.1 Concept–Knowledge (C–K) Theory

C–K theory was developed in engineering design research to formalize the process of generating novel concepts [3]. It relies on two interdependent spaces:

- **C-space (Concept space):** the set of ideas that are not yet fully defined or proven. For example, "a data center that consumes negative net energy."
- **K-space (Knowledge space):** the set of validated knowledge, scientific laws, or technological facts. For example, "photovoltaic conversion efficiency is bounded by the Shockley–Queisser limit."

Innovation occurs when concepts in C-space are expanded using elements of K-space, and when new knowledge is generated in K-space to validate or refute concepts. Formally, we can express the operation as:

$$\begin{aligned} C_{t+1} &= f(C_t, K_t) \\ K_{t+1} &= g(K_t, C_t) \end{aligned} \tag{1}$$

Here:

- f represents the **expansion of concepts** through available knowledge.
- g represents the **enrichment of knowledge** through conceptual exploration.

Applied to sustainability, C–K theory helps administrations generate new ideas beyond incremental improvements. For example:

- **Concept**: urban logistics with zero-emission fleets.
- **Knowledge expansion**: available battery technologies, hydrogen storage, vehicle-to-grid systems.

- **Concept expansion**: zero-emission logistics based on hydrogen drones.
- **Knowledge requirement**: research into lightweight hydrogen tanks.

Thus, C–K creates a **tree of possibilities**, guiding exploration without prematurely committing to one solution.

3.2 TRIZ (Theory of Inventive Problem Solving)

TRIZ was developed by Genrich Altshuller in the Soviet Union [2]. It is based on the empirical observation that most innovations follow recurring patterns of problem solving. The core idea: Many engineering problems involve contradictions, meaning that improving one parameter often worsens another. For example, "increasing computational capacity" versus "reducing energy consumption."

TRIZ formalizes contradictions in a matrix of 39 parameters and proposes 40 inventive principles for resolution, such as:

- Segmentation (divide the system into parts),
- Dynamics (make the system adjustable),
- Local quality (different properties in different parts).

Applied to sustainability in digital systems, TRIZ might resolve the contradiction:

- Requirement: high availability of data services.
- Conflict: energy consumption grows with capacity.
- TRIZ principle: "segmentation" → distribute workload dynamically across smaller modular servers that can be shut down individually.

TRIZ thus offers administrations and industries a toolbox for systematic creativity, avoiding the trap of incrementalism and reusing patterns from previous technological breakthroughs.

3.3 Bayesian Approaches

While C–K and TRIZ focus on expanding possibilities, Bayesian approaches provide a **probabilistic framework** for managing uncertainty.

In Bayesian terms, innovation is a process of updating beliefs about the feasibility and potential of technologies:

$$P(H|D) = \frac{P(D|H)P(H)}{P(D)} \tag{2}$$

where H is a hypothesis (e.g., "photonic processors reduce energy per computation by 80%"), D is observed data, $P(H)$ is prior belief, and $P(H|D)$ is the updated belief after evidence.

Governments face severe uncertainty: many emerging technologies (quantum computing, hydrogen storage, AI chips) have incomplete data and unclear adoption curves. Bayesian models allow:

- Integration of expert judgment as priors,
- Dynamic updating, such as pilot projects or experiments, provides new data,
- Decision-making under uncertainty by comparing expected utilities of competing technology paths.

For example, an administration evaluating carbon capture technologies might begin with a neutral prior. As demonstration projects succeed, Bayesian updating increases the posterior probability of maturity, justifying further investment.

3.4 Exploring Innovation: C–K, TRIZ, and Bayesian Methods

Despite differences, C–K, TRIZ, and Bayesian methods share essential features:

- They structure exploration, avoiding unbounded brainstorming.
- They are repeatable and transparent, enabling accountability in innovation processes.
- They generate a broad portfolio of possibilities rather than a single solution.

They provide the first step in a pipeline: enlarging the set of feasible and novel ideas that can contribute to Green Deal objectives.

Application to the Green Deal

In the Green Deal context, administrations need to explore radical innovations across multiple domains: energy systems, transport, agriculture, and digital technologies.

- C–K theory helps envision entirely new systems (e.g., "climate-neutral smart cities" or "self-healing materials for construction").
- TRIZ provides structured solutions to recurring contradictions, such as high-performance vs. low-energy, or speed vs. environmental footprint.
- Bayesian methods allow decision-makers to rationally manage uncertainty about the maturity and risk of these innovations.

Limits in the Government Context

However, these methodologies stop at exploration. They do not answer:

- Which technology should be funded first?
- How to rank options according to sustainability, maturity, and feasibility?

- How to allocate scarce resources under political and budgetary constraints?

For governments, exploration is only the beginning. The Green Deal requires prioritization and sequencing. Innovation methodologies provide the "what could be," but administrations must answer "what should we do now."

This is the gap that leads to Net Technological Assessment (NTA).

Moving towards NTA?

Formal methodologies for innovation (C–K, TRIZ, and Bayesian) are essential to increase the design space of governments and industries by allowing consideration of a complete portfolio of possible solutions. Formal innovation methodologies enable systematic creativity, solution contradiction resolution, and uncertainty incorporation.

However, formal innovation methodologies do not provide governments with the tools needed to prioritize technologies for investment and systematically incorporate sustainability constraints.

Therefore, formal methodologies provide the foundation for structured exploration but need to be partnered with evaluation tools, NTA, for example, in order to transition from a list of possible innovations to a prioritized, sustainable strategy for implementation.

4 Why Formal Innovation Methodologies Are not Enough for Administrations

5 Introduction

The European Green Deal is as much a governance challenge as a technical one. Governments must allocate limited resources, design long-term rules, and manage trade-offs. Formal innovation methods—C–K, TRIZ, Bayesian exploration—help generate ideas but do not fully meet administrative decision needs. Three gaps exist: they focus on generation rather than prioritization, lack integration with resource and political constraints, and weakly embed sustainability metrics. A complementary framework, such as Net Technological Assessment (NTA), is required.

Exploration versus Decision-Making

Exploration methodologies expand options:

- **C–K theory:** conceptual expansions
- **TRIZ:** contradiction resolution
- **Bayesian:** probability updates

However, administrations must choose, justify, and fund options. Formal optimization is required:

Option set:

- I set of all potential innovation options generated by exploration methods.
- $i \in I$ a single innovation option (e.g., a new technology, process, or policy idea).

Utility function:

- $U(i)$ measure of the "value" of option i. Can include multiple dimensions such as expected economic impact, environmental benefit, or social acceptance.

Constraints:

- Budget: available financial resources for investment or implementation.
- Maturity: technological readiness level (TRL) or feasibility.
- Sustainability: compliance with environmental, social, and governance (ESG) criteria.

$$\max_{i \in I} U(i)$$
$$\text{With subject to} \begin{cases} B_i \leq B_{max} \\ M_i \geq M_{min} \\ S_i \geq S_{min} \end{cases} \tag{3}$$

Formal optimization:

Exploration generates ideas *{i}*, but administrations must **rank and select** based on utility $U(i)$ under real-world constraints (B_i, M_i, S_i). Indeed, exploration methods enlarge the option set but do not optimize or rank alternatives.

Resource, Policy, and Governance Constraints

Exploration tools enlarge the option set **I,** but do not solve the problem of **selection under constraints**. Unlike firms, governments operate with strict budget ceilings, accountability requirements, and sustainability commitments. They cannot freely speculate; every chosen option must be defensible, fundable, and legitimate. Formally, administrations face a **multi-criteria, constrained optimization problem**. When options must be chosen collectively, governments face a **portfolio problem**. Let $y_k \in \{0,1\}$ indicates whether option k is selected. The optimization becomes:

$$\begin{aligned} &\max_{y \in \{0,1\}^K} \sum_{k=1}^{K} w_k U(i_k) y_k \\ &\textit{Subject to}: \sum_{k=1}^{K} B_{i_k} . y_k \leq B_{\max} \\ &\sum_{k=1}^{K} M_{i_k} . y_k \geq M_{\min} \\ &\sum_{k=1}^{K} S_{i_k} . y_k \geq S_{\min} \end{aligned} \tag{4}$$

where w_k are policy weights reflecting priorities (e.g., climate urgency, industrial sovereignty). This formulation highlights three constraints absent from pure exploration methods:

1. **Resource constraints** indicate that public funds are limited, and every choice implies opportunity costs.
2. **Policy alignment** means that governments must select options consistent with strategic goals (e.g., emission reduction, industrial sovereignty). C–K, TRIZ, and Bayesian methods are domain-neutral and do not embed such filters.
3. **Governance and legitimacy** require that, in democratic contexts, decisions can be explained and accountable. They may also be negotiated across parliaments, ministries, NGOs, and citizens. Additionally, methodologies developed for engineering purposes (such as contradiction matrices or probability-based approaches) do not account for situations in which the legitimacy of decisions is contested.

Therefore, using exploratory methodologies to create options for the public good will only allow for the use of government when the methodologies provide a solution to an opportunity-based approach to optimizing current resources, policy, and governance. Until that time, it will be impossible to transition innovation outcome decisions from idea generation to legitimate, sustainable, and fundable public choices.

Illustrative Example: Digital Infrastructure under the Green Deal Exploration outcomes might include:

- **C–K:** carbon-negative data centers
- **TRIZ:** modular activation for efficiency
- **Bayesian:** cooling technology uncertainty

The government's task is to decide funding, align with the energy strategy, and meet 2030 milestones. This requires prioritization, not mere exploration, as depicted in the detailed example of Annex 1 of the present chapter.

Bridging the Gap: From Exploration to Assessment

Exploration methodologies expand possibilities, but governments need prioritization, budgeting, sustainability integration, and legitimacy. The missing link is a formal assessment framework bridging creativity with decision-making. NTA fills this role.

C–K, TRIZ, and Bayesian exploration remain valuable for structured creativity but are insufficient for administrations. They do not prioritize, budget, align with policy, or ensure legitimacy. Net Technological Assessment complements them, enabling evaluation, prioritization, and sustainability integration, turning exploration into actionable policy decisions.

6 The Net Technological Assessment (NTA)

6.1 Introduction and Context

While formal innovation methodologies such as C–K theory and TRIZ provide powerful mechanisms for generating ideas, governments require structured tools to evaluate, prioritize, and adopt technologies in alignment with strategic objectives. This need has been particularly acute in the defense and security domain, where NATO in recent years has sought systematic ways to monitor technological trajectories and their strategic implications.

In 2024, Langlois-Berthelot introduced the concept of Net Technological Assessment (NTA) within French Army and NATO circles as a formal methodology not only for decision-support but also for the structured **prioritization** of emerging technologies. NTA extends the long-standing practice of Net Assessment—traditionally focused on balances of power across military, economic, and organizational domains—toward a quantitative analysis of technology maturity, investment potential, and sustainability impact, enabling administrations to both decide and prioritize under constraint [24].

This section of the paper gives an overview of the principles of NTA, including most of the mathematics behind NTA, and the relevance of NTA to European state administrations in their efforts to fulfil the Green Deal. The 2024 environment for Defence was shaped both by advances in technology at an astounding rate (e.g., quantum computing, AI, hypersonics, energy tech) as well as the fact that military strategic policies and resource constraints are in competition with commercial objectives.

Therefore, the traditional definition of the Technology Readiness Level (TRL) is overly simplistic, since they only determine whether the technology is technically viable, without assessing the degree of adoption barriers, sustainability, and the systemic effects of those technologies once adopted.

6.2 Net Technological Assessment (NTA) - Formal Framework

State Space

We consider n technologies, indexed $i = 1..n$.

Each technology is evaluated according to m dimensions, indexed $j = 1..d$.

The dimensions notably represent technological maturity, economic viability, social acceptance, environmental, social, and governance performance.

Each technology has a **state vector expressed as:**

$$x_i(t) = \{x_{i,j}(t)\}_{j=1..d} \tag{5}$$

The **global system state** is simply the collection of all technology states:

$$X(t) = \{x_i(t)\}_{i=1..n} \in \mathbb{R}^n \tag{6}$$

This means we describe the technological ecosystem at time t by gathering all relevant dimensions for all technologies into one large state vector.

Dynamics

The state of each technology evolves over time according to:

$$\frac{dx_i(t)}{dt} = f_i[x_i(t), \gamma_i(t), \xi_i(t)] + \sum_{j \neq i} \alpha_{i,j}(t) h(x_i(t), x_j(t)) \tag{7}$$

where:

- $\gamma_i(t)$ is a **control** (e.g., policy, investment, or regulation)
- $\xi_i(t)$ is a **random disturbance** (e.g., unexpected shocks, crises, breakthroughs)
- $\alpha_{i,j}(t)$ is a **coupling coefficient** that measures how strongly technology j affects technology i
- $h(x_i(t), x_j(t))$ is a **coupling function** quantifying the effect of technology j on technology i

If each technology is described by a single scalar maturity variable $m_i(t)$ rather than a vector $v_i(t)$, the coupling function $h(x_i, x_j)$ is often assumed to depend on the difference in maturity:

$$h(x_i, x_j) \sim m_j - m_i \tag{8}$$

This is based on diffusive interactions: technologies tend to "pull" each other toward similar maturity levels. Thus, the scalar dynamics become:

$$\frac{\partial m_i}{\partial t} = \sum_{j \neq i} \alpha_{i,j}(m_j - m_i) + F_i(m_i, u_i) \tag{9}$$

where $F_i(m_i, u_i)$ represents endogenous dynamics and control inputs.

- If $m_i < m_j$ the gradient term $(m_j - m_i)$ is positive, so m_i increases.
- If $m_i > m_j$ the gradient term $(m_j - m_i)$ is negative, so m_i decreases.

This is the diffusion-like effect, which is why the gradient appears naturally from $h(x_i, x_j)$.

Matrix Form and the Graph Laplacian

The coupling term can be rewritten as:

$$\sum_{j \neq i} \alpha_{i,j}(m_j - m_i) = -\left(\sum_{j \neq i} \alpha_{i,j}\right) m_i + \sum_{j \neq i} \alpha_{i,j} m_j \tag{10}$$

This corresponds to a graph Laplacian, L, encoding the network of technological interactions. Finally, the dynamics in matrix form are:

$$\frac{\partial m}{\partial t} = DLm + F(m, u) \tag{11}$$

where:

- m is the vector of all technology maturities
- L is the network Laplacian derived from the gradient/difference terms
- D is the diffusion coefficient (strength of interaction)
- $F(m,u)$ encodes endogenous dynamics and policy controls

In words: each technology has its own natural trajectory, shaped by deliberate choices and random shocks, but also influenced by what happens to other technologies around it.

Viability Constraints

Constraints are represented by a function

$$g : \mathbb{R}^{nd} \to \mathbb{R}^k \tag{12}$$

i.e., g maps the full system state $X \in \mathbb{R}^{nd}$ to k scalar constraint values.

The **viability kernel** is defined as:

$$V : \{X(\cdot), g(X(t)) \leq 0, \forall t \geq 0\} \tag{13}$$

This is the set of all system trajectories that always respect admissible limits (for example, total budget limits, emission caps, or minimum social thresholds).

Decision Problem

A utility functional is defined as:

$$U : \mathbb{R}^{nd} x\mathcal{U} \to \mathbb{R}^k \tag{14}$$

It assigns a real-valued measure of performance or desirability to each system state $X(t)$ under a chosen control $u(t) \in$. The state vector $X(t)$ represents the key variables of the system (e.g., budget usage, emission levels, or user activity), while the control u(t) represents the admissible actions or policies (e.g., resource allocation, technology investment, or policy interventions) that influence system evolution. The output $U(X(t),u(t))$ quantifies the instantaneous value of being in state $X(t)$ under control *u(t)*, balancing multiple objectives such as efficiency, sustainability, or social welfare.

The decision problem is to choose a control policy *u(t)* that maximizes the expected cumulative utility over a time horizon *[0,T]*, while ensuring that the system remains within the viability kernel *V*, i.e., respecting all admissible constraints:

$$\max_{u(\cdot)} \mathbb{E}\left[\int_0^T U(X(t), u(t))dt\right] \text{subject to} X(t) \in V, \forall t \geq 0 \tag{15}$$

with [·] represents the expected value of the utility over all possible realizations of the stochastic system.

In other words, the goal is to select the policies u(t) that maximize overall system value while ensuring that the trajectory of the system respects all physical, social, and environmental limits at all times.

6.3 Existence, Uniqueness, and Stability

Theorem (Existence, Uniqueness, and Uniqueness)

If the functions $f_i()$ and *h()* of Eqs. (7) and (8) are Lipschitz continuous, controls are bounded and measurable, and constraint sets are convex and compact, then:

- Trajectories exist and are unique for all initial states $X(0) \in V$,
- At least one viable control exists,
- If the utility is convex or concave in controls, an optimal policy $u^*(\cdot)$ exists.

This ensures the system is mathematically well-posed: the dynamics behave predictably, and viable (even optimal) policies can be determined.

Proposition (Stability of Equilibria)

For a fixed-point X*, local stability is determined by the Jacobian J(X*):

$$\mathrm{J}(\mathrm{X}^*) = \left[\frac{\partial f_i}{\partial \mathrm{x}_j}(\mathrm{X}^*)\right]_{i,j} \tag{16}$$

- If all eigenvalues of J(X*) have negative real parts, X* is **asymptotically, stable**.
- If at least one eigenvalue has a positive real part, X* is **unstable.**

Explanation: The Jacobian captures how small perturbations in one dimension affect the others. by studying eigenvalues, we measure whether perturbations shrink (negative real part) or grow (positive real part) over time.

Corollary (Bifurcation and Tipping Points)

If X^* depends on a parameter λ, and $J(X^*(\lambda_c))$ has an eigenvalue crossing the imaginary axis at λ_c :

- For $\lambda < \lambda_c$, the equilibrium may be stable
- For $\lambda > \lambda_c$, stability is lost, leading to major shifts in technology adoption or sustainability

Explanation: This describes points where small changes in parameters trigger large shifts in outcomes (e.g., sudden adoption of a new technology, collapse of outdated systems, or transition to unsustainable pathways).

6.4 Summary

The NTA framework consists of:

1. Controlled stochastic dynamical systems on $\mathbb{R}^{nd}$.
2. Network couplings modelled via Laplacians.
3. Viability kernels encoding admissible constraints.
4. Stochastic control for optimization under constraints.
5. Existence and uniqueness guarantee for trajectories and controls.
6. Local stability is characterized by Jacobian eigenvalues.
7. Bifurcation analysis identifying tipping points.

Altogether, NTA provides a mathematically rigorous way to assess technologies as an evolving networked system under control, with explicit treatment of constraints, stability, and critical transitions.

7 Integrating ESG Weighting into Net Technological Assessment

7.1 Revisiting the NTA Framework

The NTA framework was originally developed to support technology choices in security and defense contexts. As it matured, its use expanded to a broader range of public governance situations. NTA provides a structured approach to assessing technological readiness and the opportunity to invest in emerging technologies. It also helps determine whether a technology can be deployed effectively.

Nonetheless, despite its expanding usage in various government sectors, NTA continues to use technical and economic analysis. To date, there still isn't a structured method for incorporating environmental and societal effects surrounding a technology. Because of the pressing demands associated with addressing and safeguarding the environment, there clearly needs to be an examination of technologies from a sustainability perspective. Because of the rising need for ecologically sound production and, more so, due consideration for social equity, it becomes necessary for NTA to be modified based on Environmental, Social, and Governance factors.

7.2 Sustainability Weighting as Corrective Mechanism

The Green Deal reframes what it means for a technology to be "fit for purpose." Energy efficiency, material circularity, labour fairness—none of these can be sidelined as secondary. If sustainability remains an external add-on rather than an internal criterion, decision-making will continue to reward technologies that meet short-term benchmarks while undermining long-term goals. It's not a hypothetical risk.

A highly efficient processor architecture may rely on rare earth minerals extracted under potentially exploitative conditions. Conversely, a more conventional device may use abundant materials and support local job creation, even if both systems deliver similar levels of performance. Consequently, it becomes imperative to identify these situations and provide explainable and structured information to help decision-makers. To this end, weighing ESG factors will enable a balanced perspective on competing interests, but decision-makers will retain the final decision.

7.3 Anchoring ESG Dimensions in Evaluation

A structured integration of ESG requires a departure from checklist thinking. Indicators need to be specific enough to compare, but broad enough to account for variation across technologies and sectors. The following three central dimensions are considered:

- **Environmental:** it covers a larger scope than reducing greenhouse gas emissions. Indeed, it also includes lifecycle impacts, resource inputs, recyclability, land disturbance, and energy intensity during operations. Moreover, biodiversity effects, while difficult to quantify, must at least be acknowledged.
- **Social:** This area is often the most neglected. Metrics should include job creation and quality, skill development, equitable access, and regional distribution. Although they are difficult to trace, labour standards throughout the supply chain remain essential.
- **Governance:** Beyond regulatory compliance, this includes data protection, transparency, input traceability, and alignment with EU sustainability taxonomies. It's less about internal corporate governance than about systemic accountability.

These three dimensions are essential but not exhaustive. They give administrators a clear starting point for considering sustainability and allow them to improve its integration over time.

7.4 Making the Integration Operational

For ESG weighting to work within NTA, it must follow a method that is both repeatable and justifiable—but not necessarily elaborate. One possible approach follows six general steps:

1. **Translate Policy into Thresholds.** Begin with what's already mandated. If, for example, emissions reduction is a priority, then technologies unable to demonstrate meaningful cuts—say, 40% or more—can be deprioritized early.
2. **Choose Indicators Carefully.** The temptation to include too many must be resisted. Fewer indicators, drawn from reliable sources (Eurostat, EEA, LCAs), provide more clarity than an expansive but inconsistent list.
3. **Scale and Score.** Each metric is normalized—0 to 100 is common—relative to clearly stated benchmarks. This isn't about precision; it's about comparability.
4. **Apply Weightings Reflectively.** These should reflect political priorities and sectoral context. In some domains, environmental concerns will dominate. Elsewhere, social cohesion or governance may deserve greater emphasis.
5. **Integrate into Composite Assessment.** ESG scores aren't meant to replace NTA's existing criteria. They extend them. Maturity, investment logic, and operational potential remain important—but now sit alongside broader impact measures.
6. **Test for Fragility.** Sensitivity analysis reveals whether slight shifts in assumptions produce wildly different rankings. If they do, the method—or the data—requires revision.

This process isn't about producing perfect outputs. It's about ensuring transparency and proportionality in how technologies are judged.

7.5 A Comparative Illustration

Imagine two technologies under evaluation. One is a high-efficiency AI processor—technically sophisticated, market-ready, and energy-saving. But its material inputs are scarce, difficult to recycle, and ethically sourced only in part. The second is a cooling system retrofit. Technologically modest, yes—but built from standard components, labour-intensive to install, and fully compatible with regional workforce capacity. On paper, using only conventional NTA, the processor likely scores higher. Once ESG is factored in, however, the balance may shift—depending on weightings, perhaps decisively. That shift doesn't represent bias; it reflects a more complete picture.

7.6 *Institutional Architecture and Oversight*

The credibility of ESG weighting rests not only on method, but on governance. Multiple roles must be clearly defined:

- A **coordinating authority**—possibly a cross-sectoral agency—tasked with maintaining and updating the framework, ensuring consistency, and overseeing implementation.
- A network of **external reviewers** drawn from academia, civil society, and audit bodies to interrogate assumptions and verify data integrity.
- **Stakeholder panels**, bringing together industry representatives, unions, and NGOs to advise on thresholds and weightings.
- **Legislative oversight** to periodically revisit the framework and keep it aligned with democratic processes.

Without this institutional layering, the framework risks either politicization or technocratic insularity—neither of which serves the public interest.

7.7 *Recognizing the Limits*

No model escapes limitations. ESG data, especially social metrics, remain incomplete in many sectors. Lifecycle assessments are expensive and often incomparable across technologies. And weighting decisions—while unavoidable—can be politically charged. This doesn't imply failure. Rather, it signals the need for what might be termed *adaptive sufficiency*: use what's reliable, disclose what's not, and design for revision. Overshooting complexity helps no one.

7.8 *Strategic Fit with the Green Deal and Closing Thoughts*

Despite its limitations, incorporating ESG considerations into the NTA framework has one significant advantage; it makes policy judgement criteria usable for decision-making purposes by converting abstract ideas such as 'social cohesion' and 'climate neutrality' to usable criteria that businesses and individuals can use to make decisions. Furthermore, by enabling the identification of potential trade-offs prior to making decisions on a project; businesses and individuals will face additional difficulty justifying their decision-making processes when diverging from established sustainability goals. Thus, in addition to being used as a selection tool, it functions as a mechanism that connects public investments with public commitments.

The inclusion of ESG considerations within the NTA framework does not merely add to the complexity of the NTA, rather it provides a response to the ongoing evolution of what is required from innovation policy. The future success of European

Technology governance will be reliant on the ability of the governing bodies to find a balance between short-term technical performance and long-term ecological/social responsibilities. While integrating ESG considerations into the NTA framework alone will not ensure that a balance is reached; without it, there is virtually no chance that such a balance could be achieved.

8 From Exploration to Decision: Synthesizing Methods for Sustainable Innovation

Innovation, particularly when guided by public objectives such as those embedded in the European Green Deal, requires more than ideation. It demands a structured progression—from the generation of novel concepts to their evaluation and, ultimately, to selection under real-world constraints. Within this continuum, formal innovation methodologies and assessment frameworks do not function in isolation. Rather, they serve complementary roles that together enable coherent, transparent, and accountable decision-making.

At the exploratory stage, structured methodologies such as Concept–Knowledge (C-K) theory, TRIZ (the Theory of Inventive Problem Solving), and Bayesian design logic provide essential tools. These approaches support creative problem framing while grounding invention in logical inference or probabilistic reasoning. C-K Theory, for instance, enables the controlled expansion of design space without falling into incoherence. TRIZ, rooted in engineering heuristics, reveals patterns in past innovations that may inform future directions. Bayesian methods, meanwhile, allow for incremental updating of expectations based on emerging evidence—particularly useful in contexts of uncertainty or partial knowledge.

However, while these techniques facilitate innovation, they are not a means for prioritizing options. Evaluative frameworks will fulfil this need. The Net Technological Assessment (NTA) model provides a means for systematically comparing technologies based on their readiness, investment desirability, and operational viability. When combined with sustainability weighting, as mentioned in previous sections, NTA becomes more than a means of ranking technologies. Instead, it is a filtering mechanism to connect creative exploration with coherent policy development.

This creates a continuum of function. On one side is structured creativity. On the opposite side are the decision-making frameworks that align with societal mandates. In between is a progressively narrowing funnel where decision-makers can analytically and judiciously progress from an infinite number of innovative options to a manageable and defensibly selected priority list of strategic options. Without the complete journey from idea generation to a structured review process, through to selecting sustainable options, there is considerable risk that governments will select technologies that are either technologically unviable, socially misaligned, and/or ecologically unsustainable.

9 Conclusion

Today, there is a strong need for European institutions to accelerate the transition to renewable energy, digital infrastructure, and public services. As a result, there is a growing need for an easily understood and thoroughly justified innovation strategy. Structured methodologies such as C-K Theory, TRIZ, and Bayesian exploration provide vital information. They help government agencies develop structured creativity and provide a means of applying discipline when working with complex problem spaces instead of searching for solutions based on random associations.

However, a creative idea cannot support an investment decision unless it is evaluated. To do this, the extended NTA framework provides a valuable tool. The purpose of the NTA framework is to provide a process by which an organization can evaluate technical, economic, and sustainability aspects of new technology development on one platform. When you add ESG weighting to the NTA framework, the structure will shift from a simple comparative analysis to an ethically aligned and strategically aligned structure that supports the principles of the Green Deal. This structured approach will ensure that the technology chosen for support by a public entity will not only support energy efficiency but also provide long-term support for resilience and policy coherence.

The synthesis of formal innovation tools with ESG-augmented NTA provides governments with a pathway—from exploration to selection, from novelty to norm. In a policy landscape increasingly defined by constraint and urgency, this integration is not a luxury. It is a necessity. Only by aligning creativity with structured evaluation can European administrations pursue innovation that is not only ambitious but sustainable.

Annexes

Annex 1: Illustrative Example with Real Computations

Suppose a government must choose among **three innovation options** generated by exploration (C–K, TRIZ, Bayesian).

Option	Description	Cost Bi (M€)	Maturity Mi (TRL 1–9)	Sustainability Si (0–100)	Utility U(i) (0–100)
i_1	Carbon-negative data center (C–K)	60	7	90	85
i_2	Modular server activation (TRIZ)	40	6	75	70
i_3	Advanced cooling technology (Bayesian)	502	5	60	65

(continued)

(continued)

Option	Description	Cost Bi (M€)	Maturity Mi (TRL 1–9)	Sustainability Si (0–100)	Utility U(i) (0–100)

The policy thresholds (constraints) are defined as follows:

- **Budget**: $B_{max} = 100$ M€
- **Minimum maturity**: $M_{min} = 6$
- **Minimum sustainability**: $S_{min} = 70$

Step 1: Apply feasibility constraints

- i_3 fails the sustainability filter ($S_3 = 60 < S_{min} = 70$)
- i_3 fails the maturity filter ($M3 = 5 < M_{min} = 6$)

→ **Excluded** from the feasible set.
Feasible options: i_1,i_2.

Step 2: Check the budget constraint

- $B1 + B2 = 60 + 40 = 100 \leq B_{max}$

→ Both can be funded together.

Step 3: Portfolio optimization

We want to maximize:

$$\max_{y \in \{0,1\}^K} \sum_{k=1}^{K} w_k U(i_k) y_k \tag{17}$$

Suppose equal policy weights: $w_1 = w_2$.

- If only i_1 is chosen: total utility = 85.
- If only i_2 is chosen: total utility = 70.
- If both are chosen: total utility = 85 + 70 = 155within budget.

Optimal solution: fund both i_1 and i_2.
Step 4: Interpret governance implications

- **Resource constraint**: Only possible because the budget allowed exactly both; otherwise, trade-offs would arise.
- **Policy alignment**: i_3 was promising technologically, but was rejected because it failed sustainability and maturity thresholds.
- **Legitimacy**: Decision can be justified publicly—funding options i_1 and i_2 deliver maximum value while respecting climate and feasibility goals.

Conclusion

This small calculation illustrates why exploration methods alone are insufficient. They generated three creative options, but only formal optimization under budget, policy, and governance constraints produced a defensible portfolio. NTA (Net Technological Assessment) provides exactly this missing layer: linking creative exploration with transparent, accountable decision-making.

Annex 2: Illustrative Example with Real Computations

Example: Net Technological Assessment for Energy Transition Technologies.

We consider three technologies:

1. Solar PV (T_1)
2. Offshore Wind (T_2)
3. Green Hydrogen (T_3)

Each is evaluated on four dimensions (d = 4):

- Technological Maturity (TRL-scale, 0–9 normalized to [0,1])
- Economic Viability (normalized LCOE or cost-competitiveness index [0,1])
- Social Acceptance (survey-based, normalized [0,1])
- Environmental, Social & Governance (ESG) performance (normalized [0,1])

Step 1: State space

Each technology has a state vector:

$$x_i(t) = \{x_{i,j}(t)\}_{j=1..4} \tag{18}$$

Example initial conditions (t = 0):

- Solar PV: $x_1(0) = (0.9,0.7,0.8,0.85)$
- Offshore Wind: $x_2(0) = (0.8,0.6,0.7,0.9)$
- Green Hydrogen: $x_3(0) = (0.5,0.4,0.5,0.6)$x_3(0)

Global system state:

$$X(t) = \{x_1(t), x_{12}(t), x_1(t)\} \in \mathbb{R}^{3x4} \tag{19}$$

Step 2: Dynamics

For demonstration, use scalar **maturity variables** $m_i(t)$.

- Initial maturities:
  - Solar PV: $m_1(0) = 0.9$
  - Offshore Wind: $m_2(0) = 0.8$
  - Hydrogen: $m_3(0) = 0.5$

The diffusive coupling rule gives:

$$\frac{\partial m_i}{\partial t} = \sum_{j \neq i} \alpha_{i,j}(m_j - m_i) + F_i(m_i, u_i) \tag{20}$$

where $F_i(m_i, u_i)$ represents endogenous dynamics and control inputs.

Step 3: Coupling Structure (Network Laplacian)

Suppose the coupling matrix α is:

$$\alpha = \begin{bmatrix} 0 & 0.2 & 0.1 \\ 0.2 & 0 & 0.3 \\ 0.1 & 0.3 & 0 \end{bmatrix} \tag{21}$$

with.

- Solar PV (T_1) influences hydrogen moderately ($\alpha_{13} = 0.1$, cost synergies).
- Offshore wind (T_2) has strong influence on hydrogen ($\alpha_{23} = 0.3$).
- Hydrogen feeds back weakly (small α_{31}, α_{32}).

The Laplacian L is:

$$L = \begin{bmatrix} -0.3 & 0.2 & 0.1 \\ 0.2 & -0.5 & 0.3 \\ 0.1 & 0.3 & -0.4 \end{bmatrix} \tag{22}$$

So in compact form:

$$\frac{\partial m}{\partial t} = DLm + F(m, u) \tag{23}$$

Step 4: Viability Constraints

Constraints g(X):

1. **Budget Constraint**: Total Subsidies ≤ 5% GDP

$$g_1(X) = \sum_i u_i - 0.05GDP \leq 0 \tag{24}$$

2. **Carbon Cap**: Emissions ≤ 1 Gt/year

$$g_2(X) = E(X) - 1 \leq 0 \tag{25}$$

3. **Social acceptance floor**: Average Social Score ≥ 0.6

$$g_2(X) = 0.6 - \frac{1}{n}\sum_i x_{i,3} \leq 0 \tag{26}$$

Viability kernel = all trajectories satisfying g(X) ≤ 0.

Step 5: Utility functional

Utility function balances sustainability, cost, and social benefit:

$$\text{U(X, u)} = w_1 \cdot \text{avg maturity(X)} + w_2 \cdot \text{social acceptance(X)} - w_3 \cdot \text{policy cost(u)} \tag{27}$$

Example weights:

$$w = (0.5, 0.3, 0.2) \tag{28}$$

Decision problem:
$\max\limits_{u(\bullet)} \mathbb{E}\left[\int_0^T U(X(t), u(t))dt\right]$ subject to $X(t) \in V, \forall t \geq 0$
Step 6: Stability Analysis
Suppose an equilibrium maturity vector is:

$$m^* = (0.95, 0.9, 0.8) \tag{29}$$

Jacobian of endogenous dynamics F:

$$J(m^*) = \begin{bmatrix} -0.2 & 0.05 & 0.02 \\ 0.05 & -0.25 & 0.08 \\ 0.02 & 0.08 & -0.15 \end{bmatrix} \tag{30}$$

Eigenvalues: (−0.22, − 0.19, − 0.11) → all negative → **stable equilibrium**.
If policy support drops and one eigenvalue turns positive (say, -0.02 → + 0.05), hydrogen may destabilize, leading to tipping point (collapse in adoption).

Step 7: Interpretation

Coupled Diffusion: Solar PV and wind pull hydrogen upward, accelerating its maturity.

- **Controls:** Subsidies targeted at hydrogen amplify its endogenous $F_3(m_3,u_3)$.
- **Constraints:** Ensure budget/emission/social acceptance limits are never violated.

- **Utility:** Captures trade-off between accelerating adoption and policy costs.
- **Stability:** Indicates resilience of the system against perturbations.

References

1. COMMUNICATION FROM THE COMMISSION: The European Green Deal, COM(2019) 640 final, 11th December 2019, available at https://eur-lex.europa.eu/legal-content/EN/TXT/?uri=celex:52019DC0640
2. Ciot M-G (2021) On European green deal and sustainable development policy (the case of Romania). Sustainability 13(21):12233. https://doi.org/10.3390/su132112233
3. Hatchuel A, Weil B (2009) C-K design theory: an advanced formulation. Res Eng Design 19(4):181–192. https://doi.org/10.1007/s00163-008-0043-4
4. Al'tshuller GS (1999) The innovation algorithm: TRIZ, systematic innovation and technical creativity. Technical Innovation Center, Inc.
5. Pearl J (2009) Causality: models, reasoning, and inference (2nd ed.). Cambridge University Press
6. Friede G, Busch T, Bassen A (2015) ESG and financial performance: aggregated evidence from more than 2000 empirical studies. J Sustain Finan Inves 5(4):210–233. https://doi.org/10.1080/20430795.2015.1118917
7. Almeida DV, Kolinjivadi V, Ferrando T, Roy B, Herrera H, Gonçalves MV, Van Hecken G (2023) The "Greening" of empire: the European green deal as the EU first agenda. Polit Geogr 105:102925. https://doi.org/10.1016/j.polgeo.2023.102925
8. Schunz S (2022) The 'European Green Deal'–a paradigm shift? Transformations in the European Union's sustainability meta-discourse. Politic Res Exchange 4(1). https://doi.org/10.1080/2474736X.2022.2085121
9. Mokhtari A, Jonglez B, Ledoux T (2024) Towards digital sustainability: involving cloud users as key players. In: 2024 IEEE international conference on cloud engineering (IC2E), pp 126–132. https://doi.org/10.1109/IC2E61754.2024.00021
10. European Commission (2025) Funding digital—Digital Europe Programme. https://digital-strategy.ec.europa.eu/en/activities/funding-digital
11. CNRS (2025) France 2030. https://www.cnrs.fr/en/our-research/france-2030
12. Agence Nationale de la Recherche (ANR) (2025a) Supporting research for sustainable development. https://anr.fr/en/anrs-role-in-research/commitments/supporting-research-for-sustainable-development/
13. Agence Nationale de la Recherche (ANR) (2025b) Choose France for Science: Launch of the dedicated platform for applications to host international researchers. https://anr.fr/en/latest-news/read/news/choose-france-for-science-launch-of-the-dedicated-platform-for-applications-to-host-international-r/
14. Al-Besher A, Kumar K (2022) Use of artificial intelligence to enhance e-government services. Measure Sens 24:100484. https://doi.org/10.1016/j.measen.2022.100484
15. Mhlanga D (2023) Artificial intelligence and machine learning for energy consumption and production in emerging markets: a review. Energies 16(2):745. https://doi.org/10.3390/en16020745
16. Olczyk M, Kuc-Czarnecka M (2025) European green deal index: a new composite tool for monitoring European union's green deal strategy. J Clean Prod 495:145077. https://doi.org/10.1016/j.jclepro.2025.145077
17. Puaschunder, Julia M., Monitoring and Evaluation (M&E) of the Green New Deal (GND) and European Green Deal (EGD) (2021) Puaschunder JM (2021) Monitoring and Evaluation (M&E) of the Green New Deal (GND) and European Green Deal (EGD). In: 21st Research Association for Interdisciplinary Studies (RAIS) conference Proceedings, pp. 202–206, March

1, 2021., Available at SSRN: https://ssrn.com/abstract=3817224 or https://doi.org/10.2139/ssrn.3817224
18. Böhme K (2023) The tragedy of the time horizon: Navigating short-termism for long-term sustainability. Trans Assoc Europ Schools Plann 7(1):1–5. https://doi.org/10.24306/TrAESOP.2023.01.001
19. Francesco B, Sabin R (2004) Balancing long-term and short-term strategies in a sustainability game. iScience 27(6):110020, ISSN 2589–0042. https://doi.org/10.1016/j.isci.2024.110020
20. Lin W, Nachiappan S, Muhammad DA, Chang L, Kulwant SP (2017) Short-term versus long-term benefits: balanced sustainability framework and research propositions. Sustain Product Consump 11:18–30, ISSN 2352–5509. https://doi.org/10.1016/j.spc.2016.09.003.
21. Kai S (2021) Policy characteristics, electoral cycles, and the partisan politics of climate change. Global Environ Polit 21(2):44–72. https://doi.org/10.1162/glep_a_00593
22. Sarah K, Jeff Birchall S (2023) Climate change adaptation: how short-term political priorities trample public well-being. Environ Sci Policy 146:144–150, ISSN 1462–9011. https://doi.org/10.1016/j.envsci.2023.05.012
23. .Regulation (EU) 2021/1119 of the European Parliament and of the Council of 30 June 2021 establishing the framework for achieving climate neutrality and amending Regulations (EC) No 401/2009 and (EU) 2018/1999 ('European Climate Law') http://data.europa.eu/eli/reg/2021/1119/oj
24. Langlois-Berthelot J (2024) Net technological assessment: a framework for technology investment, Reserved Note, French Army

Chapter 9
Emerging Tools in Security and Defense for Sustainable Systems: Stochastic Optimization, Quantum Decision Models, and Technological–Human Convergence

Jean Langlois-Berthelot and Christophe Gaie

Abstract This work originates from joint efforts launched in 2018 between an artificial intelligence and cognitive science lead at the French Defense Innovation Agency and the first author of this article. These collaborations progressively linked approaches from applied mathematics, algorithmics, and cognitive science to issues of sustainability in defense and security. The present text is a non-confidential synthesis, focused solely on methods and formalisation, without sensitive elements or specific operational contexts.

Keywords Security and defense · Tools · Sustainable systems · Stochastic optimization · Quantum decision models · Technological–human convergence

1 Introduction

Sustainability in security and defense can no longer be treated as peripheral. It has become a condition of legitimacy, effectiveness, and resilience for organisations operating in increasingly constrained environments [1]. While the immediate success of a mission remains the primary objective, it cannot be assessed in isolation. It must be related to resource consumption, environmental impact, robustness of logistical and technological infrastructures, and the level of human readiness required for operations [2]. These dimensions form a set of interdependent and sometimes conflicting criteria that must be integrated within a coherent analytical framework.

J. Langlois-Berthelot (✉)
A.S.CENT Centre of Advanced Studies (UNIPA), Palermo, Italy
e-mail: jeanlanglois4@gmail.com

C. Gaie
French Prime Minister Office, Laboratoire de Recherche Magellan, Université Jean Moulin, Lyon, France

C. Gaie and M. Mueck (eds.), *Digital Sustainability: Legislations, Standards, Solutions and Related Case Studies*, Advances in Sustainability Science and Technology,
https://doi.org/10.1007/978-981-92-0458-8_9

This calls for a formalisation in terms of multi-criteria decision-making under uncertainty. Decision-makers must compare options across heterogeneous dimensions that cannot be reduced to a single metric. Furthermore, implementation is marked by uncertainty: variable environments, adversarial actions, organisational constraints, and fluctuations in the human factor [3]. The goal is not only to identify the average optimal strategy, but to ensure that a decision remains viable and robust even in unfavourable scenarios, which requires specific approaches for managing uncertainty in Multi-Criteria Decision Analysis (MCDA) [4].

The need for integrated approaches is not new. Since the 1970s, Net Assessment has aimed to provide a long-term, systemic view of power balances. Developed initially within the US Department of Defense, it compares not only immediate military capabilities, but also broader economic, technological, organisational, and cognitive factors. Net Assessment does not reduce to an inventory of weapons; it seeks to reveal structural asymmetries and enduring trends. Historically, it has relied heavily on qualitative expertise and analysts' judgment. The first direction explored in this article is to show that it can be reformulated as a Multi-Criteria Decision Analysis (MCDA) problem, thereby enhancing transparency, reproducibility, and robustness [5].

A second dimension is the technological–human convergence (THC). Operational performance depends not solely on technologies or on human skills considered in isolation, but on their interaction. Advanced technology can multiply effectiveness if assimilated by an operator at adequate readiness. Conversely, it can introduce fragility if its demands exceed current human capacities. Similarly, a highly trained team may compensate for technological limits, but at the cost of increased fatigue or reduced resilience. Integrating sustainability, therefore, requires explicit modelling of this convergence as a hybrid system combining continuous variables (human readiness) and discrete modes (technological configurations) [6]. This resonates with hybrid intelligence models developed for critical infrastructure protection [7].

The third dimension is optimisation under uncertainty. Defense and security operations unfold in contexts where outcomes are uncertain. Constrained Markov decision processes (CMDPs) provide a rigorous framework for representing sequential decision-making under constraints. Monte Carlo methods yield empirical estimates of performance and risk through simulation. Distributionally robust optimisation (DRO), based on Wasserstein distances, ensures that decisions remain valid even when probability distributions of uncertainties are imprecisely known [8, 9]. These tools strengthen the ability to embed risk into trade-offs.

Beyond classical approaches, the article opens toward quantum decision models. Quantum mechanics, taken here purely as a mathematical framework rather than a physical description, introduces new possibilities. Superposition represents states in which several options coexist before a choice is made. Amplitude interferences explain preference paradoxes observed in decision psychology [10]. Quantum walks explore state spaces faster than traditional Markov chains. Some combinatorial optimisation problems can be formulated as the search for ground states of Hamiltonians, paving the way for future use of variational quantum algorithms [5].

The article adopts a methodological stance. It does not present operational tools or institutional arrangements, but general methods with minimal examples and academic pseudocode. A Python prototype is provided as an illustration, showing computational feasibility and supporting reproducibility. The aim is to enrich strategic reflection by providing a formal, transparent language that complements qualitative expertise. This work, originating from the 2018 collaborations, thus connects applied mathematics, algorithmics, and cognitive science with sustainability challenges in security and defense. The text presented here synthesises these efforts in non-confidential form: methods and formalisation only, without sensitive elements or operational contexts.

A comparative summary of the various methods is illustrated in Fig. 1.

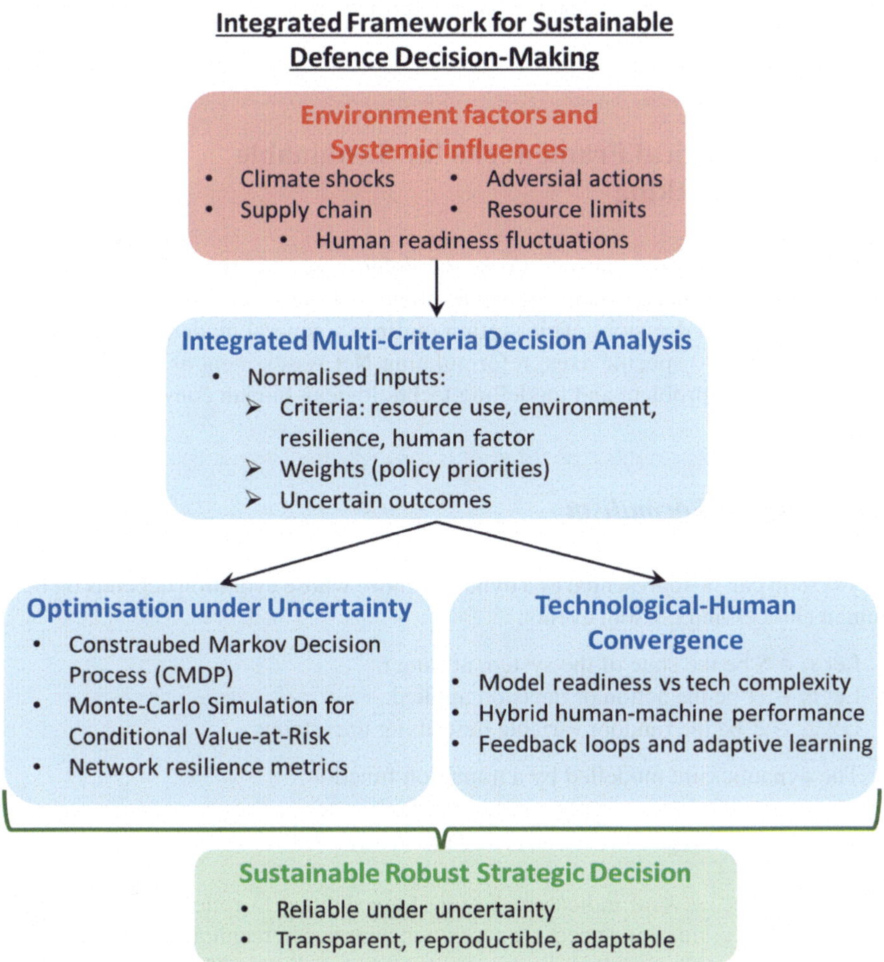

Fig. 1 Integrated framework for sustainable defence decision making

The article is structured into four main sections after this introduction. The first presents methodological frameworks for sustainable decision-making, including the general formalism, the reformulation of Net Assessment, and the technological–human convergence model. The second focuses on optimisation under uncertainty and resilience analysis, through CMDPs, Monte Carlo simulations, network studies, and portfolio problems. The third discusses theoretical guarantees, algorithms, and their complexity, along with a modular Python prototype. The fourth introduces quantum decision models. A conclusion then summarises contributions and outlines perspectives.

This approach shows that sustainability in security and defense can be addressed through rigorous analytical frameworks combining stochastic optimisation, multi-criteria modelling, hybrid dynamics, and quantum perspectives. It provides a transparent, reproducible, and adaptable basis to enrich strategic reflection in uncertain and complex environments.

2 Methodological Frameworks for Sustainable Decision-Making

Integrating sustainability into security and defense requires methodological tools that can address a dual challenge: the multiplicity of criteria and the structural uncertainty affecting all decisions. This section outlines a general modelling framework, then examines two specific axes: reformulating Net Assessment as a multi-criteria decision-support problem and modelling technological–human convergence.

2.1 General Formalism

Any system can be represented as a dynamic entity whose evolution depends on both human choices and random events.

- Let $s_t \in S$ be the state of the system at time t.
- Let $u_t \in U$ be the action or decision applied.
- Let $\xi_t \in \Xi$ be the random variable describing uncertainty.

The dynamics are modelled by a transition function:

$$s_{t+1} = f(s_t, u_t, \xi_t) \tag{1}$$

In this space, several indicators $X_i(s,u,\xi)$ are defined to reflect the dimensions of sustainability: mission success, energy consumption, greenhouse gas emissions, logistical or organisational resilience, and human readiness. These heterogeneous indicators are normalised by functions v_i mapping them into the interval [0,1].

The overall score associated with a strategy u is then given by:

$$S(u) = \sum_{i=1}^{m} w_i v_i(X_i(s, u, \xi)) \tag{2}$$

where the weights w_i express the relative importance of each criterion and satisfy

$$\sum_{i=1}^{m} w_i = 1 \tag{3}$$

The general decision problem can then be written as:

$$\begin{aligned} &\max_{u \in U} S(u) \\ &\text{Subject to} \forall j = 1..n,\ g_j(s, u, \xi)) \leq 0 \end{aligned} \tag{4}$$

The constraints g_j represent carbon budgets, minimum readiness thresholds, or structural viability conditions.

Illustrative example

Two strategies, A and B, are evaluated on three criteria: mission, consumption, and emissions. After normalisation:

- For A: (0.8, 0.6, 0.4)
- For B: (0.7, 0.9, 0.5)

With uniform weights $w_i = 1/3$, the scores are such that $S(A) = 0.6 > S(B) = 0.7$.

So strategy B is preferred. If the weights are adjusted to give more importance to the mission criterion, the gap reduces or may even reverse. This shows that results depend not only on data but also on methodological choices regarding weighting.

2.1.1 Net Assessment Reformulated as Multi-criteria Decision Support

Net Assessment is a strategic analysis method developed in the 1970s within the US Department of Defense. Its aim is to evaluate long-term balances of power, comparing not only immediate military capabilities but also broader factors: economic, technological, organisational, and cognitive. Net Assessment is not a simple accounting of armaments; it seeks to identify structural advantages, vulnerabilities, and long-term trends. Historically, it has relied heavily on qualitative expertise and analysts' judgment.

For sustainability, this method can be reformulated as a Multi-Criteria Decision Analysis (MCDA) problem, where the criteria explicitly capture sustainable outcomes such as mission success, energy consumption, emissions, and resilience [9].

Several aggregation models are possible:

- Additive model: weighted average of criteria. Transparent and easy to interpret, but assumes independence and perfect substitutability.
- Multiplicative model: imposes stronger penalties on extreme deficits, thus preventing strategies that sacrifice sustainability in one dimension from being considered optimal.
- CES functions (Constant Elasticity of Substitution): adjust the degree of substitutability between criteria.
- Choquet integral: captures synergies or redundancies between criteria using a capacity defined on coalitions [10].

In parallel, outranking methods such as ELECTRE [11, 12] or PROMETHEE [13] do not seek a single score but instead establish dominance relations between options, incorporating preference or indifference thresholds. These methods are particularly suited to defense contexts where qualitative expertise must be combined with quantitative sustainability indicators.

A central element of the framework is the choice of weights, which can be made adaptive to reflect evolving priorities. Instead of relying on fixed values, scenario-based weights allow decision-makers to adjust the importance of each dimension depending on the objective (e.g., promoting green energy, reducing carbon emissions). Weights can be determined by:

- Analytic Hierarchy Process [14]
- Entropy methods (weighting by data variability) [15]
- Simos method (enriched ordinal ranking) [16]
- Swing weighting (based on marginal variations) [17]

To enhance transparency, sensitivity analysis on weights and aggregation models should always be performed, showing decision-makers where rankings are robust or fragile. This is critical for building legitimacy in sustainability-sensitive domains [18].

Uncertainty is integrated by stochastic extensions:

- Conditional Value-at-Risk (CVaR) evaluates performance in worst-case scenarios instead of average [19];
- Distributionally Robust Optimisation (DRO), based on Wasserstein distances, ensures solutions remain valid even when the probability distribution of uncertainties is not precisely known [20].

These models can be complemented by viability constraints (e.g., carbon budgets, readiness thresholds) to ensure that recommended strategies are not only preferable but also feasible in the long term [21].

Example: Three options are evaluated on two criteria (mission, emissions).

- In an additive model, a balanced option may dominate.
- In a multiplicative model, an option with a strong deficit is heavily penalised.
- In a CVaR model, the riskiest option in terms of emissions is eliminated, even if its average performance is acceptable.

Thus, final rankings depend on the chosen model, underscoring the need for methodological transparency, explainability, and integration of sustainability-focused constraints.

2.2 Technological–Human Convergence

Performance

Operational performance cannot be reduced to technology or the human factor in isolation. It results from their interaction: the principle of technological–human convergence (THC).

- **Human readiness** is modelled as a continuous variable

$$h_t \in [0, 1]$$

- **Technological configuration** is modelled as a discrete mode variable

$$m_t \in \{1, 2 \ldots M\}$$

Performance is represented by:

$$P(h_t, m_t) = \exp\left(-\kappa h_t - \mathrm{h}^*(m_t)^2\right) \tag{5}$$

where $\mathrm{h}^*(m_t)$ is the ideal readiness for the mode considered, $\|\cdot\|$ is the norm operator, and $\kappa > 0$ measures sensitivity to deviations.

This model formalises a simple intuition: advanced technology is optimal only if operator readiness aligns with its requirements. A degraded, less demanding technology may outperform if it better matches current human capacity.

Viability kernel

We can define a viability kernel: the set of trajectories (h_t,m_t) that keep P above a minimum threshold. The problem is to choose control strategies (rest, training, technological reconfiguration) that maintain the system in this kernel K_α despite randomness.

$$K_\alpha = \{(h_t, m_t) : P(h_t, m_t) \leq \alpha\} \tag{6}$$

Optimisation Formulations

If readiness evolves smoothly over time, for example, through linear recovery after rest, the problem can be expressed as a convex optimisation task, where the objective is to maintain h_t close to its target.

In the case of multiple technological modes, the challenge becomes a mixed-integer linear program (MILP), since one must balance continuous readiness dynamics with discrete technological reconfiguration.

Finally, decomposition methods such as the Alternating Direction Method of Multipliers (ADMM) allow the problem to be separated into two subproblems:

- the human dimension, concerning continuous processes like recovery and training scheduling,
- the technological dimension, involving discrete mode switching.

Numerical example

To illustrate the model, consider a simple case in which operator readiness is compared against two possible technological modes. By computing performance under each mode, we can see how alignment between human capacity and technological demands determines the effective outcome.

- Operator readiness: $h = 0.6$
- Two modes are defined:
  - Normal mode with $h^* = 0.8$ h
  - Degraded mode with $h^* = 0.5$
- For $\kappa = 10$, the performance is such that
  - $P = 0.13$ in normal mode
  - $P = 0.61$ in degraded mode

Thus, the degraded technology is more effective in this case. This shows that sustainability depends on dynamic alignment between readiness and technology, not on always seeking the most advanced option.

Sustainability can therefore be modelled as a multi-criteria problem, integrating multiple dimensions and constraints, enriched by both the reformulation of Net Assessment and the modelling of technological–human convergence. These frameworks form the basis for the next step: optimisation under uncertainty and resilience analysis.

3 Optimisation Under Uncertainty and Resilience

Integrating sustainability into planning and operations cannot stop at a static comparison of options. Decisions unfold dynamically over time under unpredictable randomness. Strategies must both maximise effectiveness under strict constraints and remain robust in uncertain environments. Four main approaches are central: constrained Markov decision processes, Monte Carlo simulation, network resilience analysis, and intervention portfolio formulation.

3.1 Constrained Markov Decision Processes (CMDPs)

A Markov Decision Process (MDP) is defined by:

- state space S,
- action space U,
- transition law $P(s' \mid s, u)$,
- reward function $r(s, u)$.

A policy π maps each state to an action. The standard objective is to maximise expected cumulative reward:

$$\max_{\pi} \mathbb{E}\left[\sum_{t=0}^{\infty} \gamma^t r(s_t, u_t)\right]$$

where $\gamma \in (0,1)$ is the discount factor, a scalar constant between 0 and 1.

For sustainability, constraints are added. Define cost functions $c_j(s,u)$ (e.g., emissions, resource consumption) and thresholds d_j. The CMDP becomes:

$$\begin{aligned} &\max_{\pi} \mathbb{E}\left[\sum_{t=0}^{\infty} \gamma^t r(s_t, u_t)\right] \\ &\text{subject to} \, \mathbb{E}\left[\sum_{t=0}^{\infty} \gamma^t c_j(s_t, u_t)\right] \le d_j j = 1..k \end{aligned} \tag{7}$$

Resolution uses linear programming on occupation measures, which describe the proportion of time a policy spends on each state–action pair. Dual variables (Lagrange multipliers) associated with constraints represent shadow prices, i.e., the marginal cost of relaxing a constraint.

Example: Two states (s_1,s_2) and two actions (u_A,u_B). Action u_A yields a higher reward but more emissions, while u_B yields less reward but lower emissions. Without constraints, u_A dominates. With a strict emission constraint, however, the optimal strategy is a mixed policy that probabilistically balances between u_A and u_B. This illustrates how sustainability constraints impose explicit, computable trade-offs between performance and environmental impact.

3.1.1 Monte Carlo Simulation

Monte Carlo methods are used to estimate the performance of a strategy or system when uncertainty plays a significant role. They work by generating a large number of random scenarios and computing how the outcome changes across them.

If Z is a random variable representing a sustainability criterion (for example, *annual* CO_2*emission reduction, economic return, or resilience index*), its expected value (mean performance) can be estimated by:

$$\hat{\mu}^N = \frac{1}{N}\sum_{i=1}^{N} Z^{(i)} \tag{8}$$

where $Z^{(i)}$ are independent random samples of Z.

This estimator is unbiased, meaning that its expected value equals the true mean of Z, and it converges almost surely to the true mean as the number of samples N grows. In practice, this means that with enough random draws, the simulated average will approximate the true expected value very closely.

Beyond the Mean: Measuring Risk

Averages alone can hide risk. For sustainability and resilience studies, we often care not just about the *typical* performance but also about *worst-case* outcomes.

The Conditional Value-at-Risk (CVaR) at level α captures this. It is defined as the average of the worst $1 - \alpha$ fraction of outcomes. For instance, the mean of the worst 10% of cases when $\alpha = 0.9$. In practice:

1. Generate N samples of Z.
2. Sort them from best to worst.
3. Compute the average of the lowest $(1 - \alpha)$N values.

Example

Suppose Z uniform in [0, 10]. This means all values between 0 and 10 are equally likely. Then:

- The true mean is $E[Z] = 5$
- The CVaR at level 0.9, i.e., the average of the worst 10% of outcomes, is 9.5. This represents the mean of all values greater than the 90th percentile, which is 9.

  Now, imagine we only draw four random samples: (2, 4, 8, 10).

- The **empirical mean** (sample average) is $(2 + 4 + 8 + 10)/4 = 6$
- The **estimated $\mathbf{CVaR_{0.9}}$** (average of the worst 10% of losses) here equals the largest value, 10, since only four samples exist.

  This simple example shows how the Monte Carlo estimator approaches the true value as we increase N. The more samples there are, the closer we get to the theoretical mean and CVaR.

Variance Reduction Techniques

Naïve random sampling can require a very large N to get stable estimates. Variance reduction techniques improve efficiency by producing more informative samples.

- **Antithetic variables:** use negatively correlated samples to cancel out noise. If one simulation uses a high rainfall year, pair it with a low rainfall year to balance the estimate.

- **Stratified sampling**: divide the probability space into segments (strata) and sample from each to ensure full coverage. If modelling temperature, ensure samples represent cold, moderate, and hot years rather than random clustering.
- **Importance sampling**: over-sample rare but critical scenarios to better estimate tail risks. In resilience planning, place more emphasis on rare flood events to capture the impact of extreme climate shocks.

3.1.2 Network Resilience

Logistical and communication infrastructures are vital. Model the network as a graph G = (V,E).

Key indicators are:

- **Algebraic connectivity (λ_2):** higher values correspond to stronger robustness.
- **Effective resistance:** measures global ease of flow.
- **Max flow:** transfer capacity between critical nodes.
- **Percolation threshold:** fraction of nodes to remove before fragmentation.

This could model a power grid, transport network, or water distribution system. These indicators define optimisation problems:

- Interdiction: identify nodes/edges whose removal most degrades the network (adversarial view).
- Hardening: select reinforcements under budget to maximise robustness.

In a 4-city transport chain (A–B–C–D), removing the B–C road cuts access between A and D. Adding a diagonal link (A–C) increases redundancy and raises λ_2 .

3.1.3 Intervention Portfolios

Corrective planning can be expressed as a portfolio optimisation problem. Each intervention k has a binary decision variable y_k, cost c_k, and total budget B:

$$\sum_{k=1}^{K} c_k y_k \leq B \tag{9}$$

Sustainability indicators are functions $\phi_i(y,\xi)$ of interventions and randomness. Objective:

$$\max_{y\in\{0,1\}^K} \mathbb{E}\left[\sum_{i=1}^{M} \omega_i \phi_i(y,\xi)\right] \tag{10}$$

This is combinatorial and stochastic.

- For moderate cases: MILP.
- For many scenarios: Benders decomposition.
- For multi-objective trade-offs: evolutionary algorithms (e.g., NSGA-II).

Three interventions, each cost = 1, budget = 2.

- Intervention 1: strong emission reduction.
- Intervention 2: resilience improvement.
- Intervention 3: moderate effect on both.

With high weight on the environment, pick 1. With balanced weights, pick 3. For instance, ϕ_1 could measure CO_2 reduction, ϕ_2 could measure resilience score under uncertain disruptions ξ (e.g., floods or cyberattacks).

4 Guarantees, Algorithms, and Implementation

A rigorous methodology must go beyond models. It must guarantee theoretical validity, specify robustness conditions, provide explicit algorithms, and demonstrate computational feasibility. These guarantees ensure that sustainability assessments are not only computed but also *reliable under uncertainty*. Table 1 summarises the methods explained in the remainder of this section.

4.1 Theoretical Guarantees

A rigorous methodology must go beyond models. It must guarantee theoretical validity, specify robustness conditions, provide explicit algorithms, and demonstrate computational feasibility.

- **Well-posedness**: Convex models (e.g., additive MCDA aggregation, convex readiness dynamics) guarantee the existence and uniqueness of optimal solutions. Combinatorial problems (e.g., intervention portfolios) guarantee existence by finiteness, but not uniqueness.
- **Sensitivity**: Global sustainability scores depend on normalisation functions, indicators, and weights. If each v_i is Lipschitz with constant L_i, then a perturbation δ in the indicator yields at most $L_i\delta$ change in score. This bounds measurement error effects.
  - If L_i is **small**, the model is **stable and robust**.
  - If L_i is **large**, results are **sensitive** and require tighter calibration or data accuracy.

$$|\Delta S| \leq L_i\delta \tag{11}$$

Table 1 Summary of methods employed to achieve sustainability

Method	Domain/category	Description
Additive MCDA aggregation	Multi-Criteria Decision Analysis (MCDA)	Aggregates multiple criteria into a single score using weighted sums or additive utility functions
PROMETHEE method	Multi-Criteria Decision Analysis (MCDA)	Outranking method that compares alternatives based on preference functions and pairwise comparisons
Constrained Markov Decision Processes (CMDPs)	Stochastic Optimization	Extends MDPs to include constraints, often used for sequential decision-making under uncertainty
CVaR estimation (Monte Carlo)	Risk Management/Stochastic Optimization	Estimates conditional value-at-risk using Monte Carlo simulations to assess tail risk
Intervention portfolio optimization	Optimization/Operations Research	Optimises a portfolio of interventions (e.g., projects, policies) to maximise impact or efficiency
Multi-objective optimization (Heuristics)	Multi-Objective Optimization	Uses heuristic algorithms (e.g., genetic algorithms, simulated annealing) to solve problems with multiple conflicting objectives

- **Weight robustness**: For two options A, B we have

$$S(A) - S(B) = \sum_i \omega_i(\mathrm{v}_i(\mathrm{A}) - \mathrm{v}_i(\mathrm{A})) \tag{12}$$

The ranking of alternatives remains stable as long as the weights stay within specific regions of the weight space. These regions, defined as polyhedra, allow us to analyse the robustness of preferences and ensure conclusions hold even with slight shifts in priorities.

- **Probabilistic guarantees**: Using Hoeffding's inequality, we can bound the probability that the empirical mean $\hat{\mu}^N$ deviates from the true expected value $\mathbb{E}[Z]$:

$$P\left(\left|\hat{\mu}^N - \mathbb{E}[Z]\right| \geq \varepsilon\right) \leq 2e^{-2N\varepsilon^2} \tag{13}$$

or equivalently, to achieve a precision ε with confidence 1−δ:

$$N \geq \frac{\ln(2/\delta)}{2\varepsilon^2} \tag{14}$$

This provides a **statistical guarantee of accuracy**. By selecting an appropriate number of samples N, we can control the estimation error within a desired tolerance ε and confidence level 1−δ.

4.2 Algorithms and Complexity

The models rely on algorithms whose computational efficiency must be clearly defined. Understanding complexity ensures that the methodology is scalable and feasible for large sustainability assessments.

4.2.1 Additive MCDA Aggregation

Purpose: This aims to compute a composite sustainability score for each option.

Input:

- Normalised value matrix v_{ij}
- Weight vectorw_j

Steps:

1. For each option *i*, compute the weighted sum of criteria:

$$S_i = \sum_j \omega_i \mathrm{v}_{ij} \tag{15}$$

2. Rank options by S_i
3. Output the ranking and scores.

Output: Ordered list of options with aggregate scores.

Complexity:

- Score computation: *O(mn)*
- Sorting: *O(m.logm)*

Notes: This method is efficient for computing the sustainability with large numbers of criteria and options.

4.2.2 PROMETHEE Method

Purpose: This method performs pairwise comparisons between alternatives to obtain preference rankings [22].

Input:

- m = number of alternatives/options
- n = number of criteria
- v_{ij} = value of alternative i on criterion j
- w_j = weight of criterion j
- $P_j(a,b)$ = preference function for criterion j comparing alternatives a and b, usually normalised between 0 and 1.

$$P_j(a, b) = F_j\left(v_{aj} - v_{bj}\right) \tag{16}$$

where F_j is a chosen preference function, e.g.:

- **Usual criterion:** $F_j(d) = 1$ if $d > 0$, else 0
- **Linear criterion:** $F_j(d) = d/q_j$ if $0 < d < q_j$, else 1 if $d \geq q_j$

Steps:

Compute preference indices for all pairs of options.
Aggregate indices to obtain net flows.
Rank options based on net flows.

Output: Ranking of options considering pairwise preference dominance.

Complexity: $O(m^2n)$.

Notes: Quadratic scaling limits use for very large option sets.

4.2.3 Constrained Markov Decision Processes (CMDPs)

Purpose:

Optimise sequential decisions under uncertainty, subject to resource or policy constraints. CMDPs generalise standard MDPs by introducing cost or sustainability limits.

Input:

- State set S: possible system conditions.
- Action set $U(s)$: available actions in each state s.
- Transition probabilities $P(s'|s,u)$: probability of moving from s to s' after action u.
- Reward function $r(s,u)$: performance or utility of choosing u in state s.
- Cost functions $c_j(s,u)$: resource use or penalties for constraint j.

- Constraint bounds d_j: maximum allowed cumulative cost for each constraint.
- Initial distribution $\rho(s)$: starting probabilities of states.
- Discount factor $\gamma \in [0,1)$ optional weighting of future rewards.

Decision Variables:

$x(s,u)$ is the expected discounted frequency (occupation measure) of being in state s and taking action u.

$$x(s, u) \geq 0, \forall (s, u)$$

Problem Formulation:

$$\begin{aligned} &\text{maximize} \sum_{s,u} r(s, u)x(s, u) \\ &\text{subject to} \begin{cases} \forall j, \sum_{s,u} c_j(s, u)x(s, u) \leq d_j \\ \forall s, \sum_u x(s, u) - \gamma \sum_{s',u'} P(s|s', u')x(s', u') = \rho(s) \\ x(s, u) \geq 0 \end{cases} \end{aligned} \tag{17}$$

Steps:

1. Define states, actions, rewards, costs, and transition probabilities.
2. Express constraints and flow balance equations.
3. Solve the linear program to obtain optimal occupation measures $x^*(s,u)$.
4. Derive the optimal policy:

$$\pi^*(us) = \frac{x^*(s, u)}{\sum_{u'} x^*(s, u')}$$

5. Interpret Policy in Terms of Sustainable Decision Rules.

Solution Method: Linear Programming (LP).

Output: Optimal policy $\pi^*(u|s)$ or occupation measure $x^*(s,u)$ that maximises expected reward under all constraints.

Complexity: Polynomial time with modern LP solvers.

Notes: CMDPs are well-suited for adaptive resource management, resilient planning, and sequential optimization under sustainability or risk constraints.

4.2.4 CVaR Estimation (Monte Carlo)

Purpose: Estimate risk beyond a given quantile (tail risk).

Input:

- Samples $Z^{(1)},\ldots,Z^{(N)}$
- Confidence level α

Steps:

1. Sort the samples in ascending order.
2. Compute quantile index: $q = \lceil \alpha N \rceil$.
3. Average the $N - q$ largest samples:

$$\widehat{\mathrm{CVaR}^{\alpha}} = \frac{1}{N - q} \sum_{i=q+1}^{N} Z^{(i)} \tag{18}$$

Output: Estimated $\widehat{\mathrm{CVaR}^{\alpha}}$.

Complexity: $O(N.\log N)$ (dominated by sorting).

Notes: Statistical estimation allows control over accuracy via sample size.

4.2.5 Intervention Portfolio Optimization

Purpose: Select an optimal portfolio of interventions (projects, actions, or policies) that maximises expected sustainability impact under budgetary, operational, or risk constraints, often across multiple uncertainty scenarios.

Problem Type:

NP-hard combinatorial optimization.

Small-scale formulation: Mixed-Integer Linear Programming (MILP)

Large-scale formulation: Benders Decomposition

Sets and Indices

- $I = \{1,\ldots,n\}$: set of possible interventions
- $S = \{1,\ldots,N\}$: set of uncertainty scenarios

Parameters

- c_i: cost of intervention$i$
- b: total available budget
- r_{is}: reward (benefit) of intervention i under scenario s
- p_s: probability of scenario s

Decision Variables

- $x_i \in \{0,1\}$: 1 if intervention i is selected, 0 otherwise

Scenario Reward

$$R(x, s) = \sum_{i} r_{is} x_i \tag{19}$$

Objective Function—Expected Value Form

$$\max_{x} \mathbb{E}_s[R(x, s)] = \sum_{s \in S} \mathrm{p}_s \sum_{i \in I} r_{is} x_i \tag{20}$$

Constraints

1. Budget constraint $\sum_{i} c_i x_i \leq b$
2. Binary constraints $\forall \mathrm{i}, \mathrm{x_i} \in \{0,1\}$

Small-Scale Case—MILP Formulation

$$\begin{aligned} &\text{maximize} \sum_{s \in S} p_s \sum_{i \in I} r_{is} x_i \\ &\text{subject to} \begin{cases} \sum_{i} c_i x_i \leq b \\ \forall i, xi \in \{0, 1\} \end{cases} \end{aligned} \tag{21}$$

Solution method: Direct MILP optimization using branch-and-bound or branch-and-cut solvers (e.g., Gurobi, CPLEX).

Output: Optimal set of interventions x^*

Large-Scale Case: Benders Decomposition.

Used when the number of scenarios S is large, where direct MILP is not applicable.

Algorithm Steps

1. Master Problem: Decide a Provisional Portfolio X

$$\max_{x} \theta \text{ subject to} \sum_{i} c_i x_i \leq b \text{with} \, x_i \in \{0, 1\} \tag{22}$$

   where θ approximates expected performance.
2. Subproblem: For fixed x, Compute expected scenario performance:

$$Q(x) = \sum_{s} \mathrm{p}_s R(x, s) \tag{23}$$

   Evaluate feasibility and optimality; extract dual information.

3. Benders Cuts: Add linear constraints to refine the master problem:

$$\theta \leq \alpha + \beta^T \tag{24}$$

   where (α,β) are derived from dual variables of the subproblem.
4. Iteration: Repeat until convergence criterion is met:

$$|\theta(k+1) - \theta(k)| \leq \in \tag{25}$$

Output: Optimal or near-optimal intervention portfolio x^* satisfying all constraints and scenario conditions.

Per iteration complexity: Linear in the number of scenarios.

Scaling: Excellent scalability for large scenario sets.

Advantage: Decomposition improves the tractability of NP-hard problems by separating investment (master) and evaluation (subproblem) decisions.

4.2.6 Multi-objective Optimization (Heuristics)

Purpose: Find a set of Pareto-optimal solutions when multiple sustainability criteria (e.g., cost, impact, resilience) conflict. No single solution dominates all others.

Mathematical Formulation: Given multiple objective functions, we have:

$$\begin{aligned} f(x) &= (f_1(x), f_1(x), \ldots, f_k(x)) \\ &\text{minimize/maximize} f(x) subjecttox \in X \end{aligned} \tag{26}$$

A solution x^* is Pareto-optimal if there is no x' such that:

$$f_j(x') \leq f_j(x^*) \forall j, \text{ and} f_i(x') < f_j(x^*) \text{ for atleast one } i \tag{27}$$

The collection of all Pareto-optimal solutions forms the Pareto front.

Common Method: NSGA-II (Non-dominated Sorting Genetic Algorithm II). It is a heuristic evolutionary algorithm for approximating the Pareto front.

Algorithm Steps:

1. Initialisation: Generate an initial random population $P_0 = \{x_1,\ldots,x_N\}$
2. Evaluation: Compute objective values $f(x_i)$ for each candidate.
3. Non-dominated Sorting: Partition population into Pareto fronts F_1, F_2, … by dominance relations.
4. Crowding Distance: Within each front, calculate distances in objective space to preserve diversity.

Table 2 Comparison of methods' complexity

Algorithm/method	Purpose	Complexity
Additive MCDA	Aggregate criteria scores	O(mn) + O(mlog(m))
PROMETHEE	Pairwise preference ranking	$O(m^2n)$
CMDP	Optimise sequential decisions	Polynomial (LP)
CVaR estimation	Tail risk estimation	O(NlogN)
MILP	Small-scale intervention portfolios	NP-hard
Benders Decomposition	Large-scale portfolios	Linear per iteration
NSGA-II	Multi-objective optimization	$O(m^2)$ per generation

5. Selection: Select parents based on rank (front) and crowding distance.
6. Crossover and Mutation: Apply genetic operators to produce offspring Q_t
7. Combination and Replacement: Combine parent and offspring populations and select the next generation P_{t+1} from best-ranked individuals.
8. Termination: Repeat until the maximum number of generations or the convergence criterion is reached.

Output: Approximate **Pareto front**—a diverse set of non-dominated solutions showing trade-offs among objectives.

Complexity: $O(M^2)$ per generation due to pairwise dominance comparisons, where M = population size (Table 2).

4.2.7 Python Prototype

A Python prototype illustrates feasibility. Not operational, but transparent and reproducible.

```
Structure:
sustain_decision/
├── mcda.py        # multi-criteria methods
├── mdp.py         # CMDPs via LP
├── thc.py         # technological–human hybrid models
├── mc.py          # Monte Carlo
├── net.py         # network analysis
├── opt.py         # portfolios and optimisation
├── examples/      # minimal examples
├── tests/         # unit tests
```

Example function:

```
def additive_score(values, weights):
    """
    Compute additive score of a multi-criteria option.
    values : list of normalised values
    weights : list of weights (sum = 1)
    return : global score
    """
    return sum(w * v for w, v in zip(weights, values))
```

```
Test:
assert abs(additive_score([0.8, 0.6], [0.5, 0.5]) - 0.7) < 1e-9
```

This modular organisation ensures reproducibility and extensibility. It verifies that the methods are implementable in a clear and transparent way.

4.2.8 Quantum Perspectives for Decision-Making

Classical MCDA approaches rely on Kolmogorov probability theory. Empirical evidence shows limits: preference inconsistencies, context effects, and inefficient exploration.

Quantum formalism, used *mathematically*, extends this.

- A decision state is a normalised vector $|\psi\rangle$ in a complex Hilbert space H
- Strategies form an orthonormal basis. For two strategies A, B:

$$|\psi\rangle = \alpha|\mathrm{A}\rangle + \beta|\mathrm{B}\rangle, \text{ with} |\alpha|^2 + |\beta|^2 = 1$$

Think of α and β as expressing the organisation's *simultaneous inclination* toward both strategies before deciding.

- Each sustainability criterion is a Hermitian operator A_i on H.
- Expected performance: $\langle\psi|A_i|\psi\rangle$

For example, a city choosing between "renewable expansion" (A) and "industrial subsidy" (B) may be in a superposed state. A context shift, such as new EU funding, acts like a "phase rotation," changing preference probabilities.

Interference models context effects and preference paradoxes.

- Dynamics follow a unitary transformation U:

$$|\psi_{t+1}\rangle = U|\psi_t\rangle \quad (28)$$

Quantum walks allow faster exploration. They can model how organisations explore multiple strategies in parallel rather than sequentially.

Unlike Markov chains, quantum walks allow faster exploration.

- NP-hard sustainability problems (e.g., portfolios, CMDPs) can be encoded as Hamiltonians HH. The ground state corresponds to the optimal solution.
- Variational quantum algorithms (QAOA, VQE) approximate this ground state using parameterised quantum circuits and classical optimisation of expected energy:

$$\langle\psi(\theta)||H\psi(\theta)\rangle \cdot \langle\psi(\theta)||H\psi(\theta)\rangle \quad (29)$$

Example: Initial balanced superposition:

$|\psi\rangle = 12(|A\rangle + |B\rangle).|\psi\rangle = 21(|A\rangle + |B\rangle)$.

Probabilities: P(A) = P(B) = 0.5P(A) = P(B) = 0.5. With a context-induced phase shift:

$|\psi'\rangle = 12(|A\rangle - |B\rangle).|\psi'\rangle = 21(|A\rangle - |B\rangle)$.

Probabilities in $\{|A\rangle,|B\rangle\}\{|A\rangle,|B\rangle\}$ remain 0.5, but measurements in another basis change due to interference.

Quantum formalism enables the representation of superposed strategies, evaluation by observables, modelling of context effects via interference, and faster exploration via quantum walks. Long-term, quantum optimisation may address combinatorial problems beyond classical reach. Current processors are limited, but quantum-inspired classical heuristics already exploit this structure.

5 Conclusion

Sustainability in security and defense can no longer be treated as a peripheral concern. It has become a structural requirement that conditions the legitimacy, effectiveness, and resilience of contemporary defense systems. Mission success must now be assessed together with environmental impact, resource consumption, infrastructure resilience, and human readiness. This complexity requires analytical frameworks able to integrate heterogeneous criteria and handle uncertainty.

Five contributions structured the exposition:

Contribution 1.Reformulation of Net Assessment

The strategic method developed in the 1970s, originally based on qualitative expertise, was reframed as a quantitative multi-criteria decision problem. Additive, multiplicative, CES, and Choquet integral models enable transparent comparison of strategies, while outranking methods such as ELECTRE or PROMETHEE introduce dominance relations. The integration of risk measures like CVaR and distributionally robust optimisation strengthens the reliability of evaluations.

Contribution 2.Technological–human convergence

Operational performance was formalised as the interaction between readiness and technological capacity, rather than a simple sum. The proposed performance function, sensitive to the gap between readiness and requirements, showed that advanced technology is not always optimal. The model introduced the notion of a viability kernel, defining the trajectories that keep systems above a minimum performance threshold over time.

Contribution 3.Optimisation under uncertainty

Constrained Markov decision processes provided a rigorous framework for sequential decision-making with explicit constraints. Monte Carlo simulations enabled evaluation of strategies and the integration of risk. Network analysis offered measurable indicators of resilience, such as algebraic connectivity and effective resistance. Intervention portfolios translated corrective planning into a combinatorial optimisation problem. Together, these methods made it possible to design robust strategies in unpredictable environments.

Contribution 4.Guarantees, algorithms, and prototype implementation

Theoretical soundness was verified through convexity analysis, sensitivity bounds, and probabilistic guarantees such as Hoeffding's inequality. Algorithmic complexity was detailed: additive MCDA in $O(mn)O(mn)$, PROMETHEE in $O(m^2n)O(m^2n)$, CMDPs solvable in polynomial time by linear programming, CVaR estimation in $O(N\log N)O(N\log N)$, and multi-objective optimisation heuristics in $O(m^2)O(m^2)$. A modular Python prototype demonstrated computational feasibility, with separate modules for MCDA, CMDPs, technological–human models, Monte Carlo, networks, and portfolios. This structure ensured transparency, reproducibility, and adaptability.

Contribution 5.Quantum perspectives for decision-making

Beyond classical probability, the mathematical formalism of quantum mechanics was used as an extension. Decision states were represented as superpositions in Hilbert spaces, criteria as Hermitian operators, and outcomes as expectation values. Interference effects modelled context-dependent preferences and paradoxes. Quantum walks offered faster exploration than classical Markov chains. Hard optimisation problems were encoded into Hamiltonians, with their ground states corresponding to optimal solutions. Variational quantum algorithms (QAOA, VQE) and quantum-inspired classical heuristics illustrated possible future directions for addressing combinatorial complexity.

This work has proposed and tested a set of formal tools for addressing sustainability in defense and security. The contributions include: a reformulation of Net Assessment within multi-criteria decision analysis; a mathematical model of technological–human convergence; optimisation methods under uncertainty combining CMDPs, Monte Carlo, network analysis, and portfolio formulations; guarantees on well-posedness, sensitivity, and complexity together with a Python prototype; and finally, exploratory perspectives drawn from quantum decision models.

Taken together, these results provide a transparent, reproducible, and extensible methodological basis. They are intended to enrich strategic analysis by clarifying trade-offs, quantifying risks, and offering formal representations that complement qualitative expertise.

Acknowledgements The author would like to sincerely thank Pr. Markus Mueck and Dr. Christophe Gaie for their support throughout the development of this book chapter.

Disclaimer Note that the authors have assembled the information above to the best of their knowledge as available at the time when the present book was written. Any interpretation is the personal opinion of the authors. Also, the regulation framework can change and evolve any time. The only legally binding text is the one provided in the documents formally published in the European Union Official Journal.

References

1. COMMUNICATION FROM THE COMMISSION: The European Green Deal, COM (2019) 640 final, 11th December 2019, available at https://eur-lex.europa.eu/legal-content/EN/TXT/?uri=celex:52019DC0640
2. Ciot M-G (2021) On European green deal and sustainable development policy (the Case of Romania). Sustainability 13(21):12233. https://doi.org/10.3390/su132112233
3. Smaliukienė R (2018) Sustainability issues in the military: genesis and prospects. J Secur Sustain Issues 8(1):19–32. https://doi.org/10.9770/jssi.2018.8.1(2)
4. Kumari A (2024) Sustainability for military logistics. SSRN. https://doi.org/10.2139/ssrn.5051843
5. Kunreuther H, Meyer R, Zeckhauser R (2002) High stakes decision making: normative, descriptive and prescriptive considerations. Mark Lett 13:259–268. https://doi.org/10.1023/A:1020287225409
6. Depledge D (2023) Low-carbon warfare: climate change, net zero and military operations (Version 1). Loughborough University. https://hdl.handle.net/2134/21517611.v1
7. Lin F, Fang X, Gao Z (2022) Distributionally robust optimization: a review on theory and applications. Numeric Algebra, Contr Optimiz 12(1):159–212. https://doi.org/10.3934/naco.2021057
8. Kuhn D, Shafiee S, Wiesemann W (2025) Distributionally robust optimization. Acta Numer 34:579–804. https://doi.org/10.1017/S0962492924000084
9. Favre M, Wittwer A, Heinimann HR, Yukalov VI, Sornette D (2016) Quantum decision theory in simple risky choices. PLoS ONE 11(12):e0168045. https://doi.org/10.1371/journal.pone.0168045

10. Yukalov VI (2021) Quantum uncertainty in decision theory. In: Studies in systems, decision and control, vol 427. Springer, pp 201–218. https://doi.org/10.2139/ssrn.3843168
11. Cinelli M, Spada M, Kim W et al (2021) MCDA Index Tool: an interactive software to develop indices and rankings. Environ Syst Decis 41:82–109. https://doi.org/10.1007/s10669-020-09784-x
12. Michel G (1996) The application of fuzzy integrals in multicriteria decision making. Euro J Operat Res 89(3):445–456, ISSN 0377–2217. https://doi.org/10.1016/0377-2217(95)00176-X
13. Roy B (1996) Multicriteria methodology goes decision aiding. Kluwer Academic Publishers, Berlin. https://doi.org/10.1007/978-1-4757-2500-1
14. Mary S, Suganya G (2016) Multi-criteria decision making Using ELECTRE. Circ Syst 7:1008–1020. https://doi.org/10.4236/cs.2016.76085
15. Wim DK, Peter P (1996) A note on the use of PROMETHEE multicriteria methods. Euro J Operat Res 89(3):457–461, ISSN 0377–2217. https://doi.org/10.1016/0377-2217(94)00307-6
16. Saaty RW (1987) The analytic hierarchy process—what it is and how it is used. Math Modell 9(3–5), 161–176. ISSN 0270–0255, https://doi.org/10.1016/0270-0255(87)90473-8.
17. Yazdani M, Torkayesh AE, Santibanez-Gonzalez ED, Otaghsara SK (2020) Evaluation of renewable energy resources using integrated Shannon Entropy—EDAS model. Sustain Operat Comput 1:35–42. https://doi.org/10.1016/j.susoc.2020.12.002
18. Figueira J, Roy B (2002) Determining the weights of criteria in the ELECTRE type methods with a revised Simos' procedure. Eur J Oper Res 139(2):317–326. https://doi.org/10.1016/s0377-2217(01)00370-8
19. Pöyhönen M, Hämäläinen RP (2001) On the convergence of multiattribute weighting methods. Eur J Oper Res 129(3):569–585. https://doi.org/10.1016/s0377-2217(99)00467-1
20. Belton V, Stewart T (2012) Multiple criteria decision analysis: an integrated approach. Springer Science & Business Media. https://doi.org/10.1007/978-1-4615-1495-4
21. Rockafellar RT, Uryasev S (2000) Optimization of conditional value-at-risk. J Risk 2:21–42. https://sites.math.washington.edu/~rtr/papers/rtr179-CVaR1.pdf
22. Esfahani PM, Kuhn D (2017) Data-driven distributionally robust optimization using the Wasserstein metric: performance guarantees and tractable reformulations. Math Program 171(1–2):115–166. https://doi.org/10.1007/s10107-017-1172-1

Chapter 10
A Case Study: Sustainability in the Context of 6th Generation Mobile Communication Systems

Mattia Merluzzi, Luigi Briguglio, Serge Bories, Gilles Callebaut, Emilio Calvanese Strinati, Lucas Pereira Carwile, Emma Complido, Chaima Ghribi, Bahare Masood Khorsandi, Marja Matinmikko-Blue, Davide Montagno Bozzone, Hosein Nikopour, Carmela Occhipinti, Chathura Sarathchandra, Lars Christoph Schmelz, and Emanuela Tangari

Abstract This Chapter addresses the topic of sustainability in the context of wireless communication networks, more specifically, the case of next-generation mobile

---

M. Merluzzi (✉) · S. Bories · E. C. Strinati · E. Complido
University Grenoble Alpes and CEA-Leti, Grenoble, France
e-mail: mattia.merluzzi@cea.fr

S. Bories
e-mail: serge.bories@cea.fr

E. C. Strinati
e-mail: emilio.calvanese-strinati@cea.fr

E. Complido
e-mail: emma.complido@cea.fr

L. Briguglio · L. P. Carwile · C. Occhipinti · E. Tangari
CyberSocial Lab. Srl Impresa Sociale, Rome, Italy
e-mail: l.briguglio@cybersoclab.com

L. P. Carwile
e-mail: l.carwile@cybersoclab.com

C. Occhipinti
e-mail: c.occhipinti@cybersoclab.com

E. Tangari
e-mail: e.tangari@cybersoclab.com

G. Callebaut
Department of Electrical Engineering, KU Leuven, Ghent, Belgium
e-mail: gilles.callebaut@kuleuven.be

C. Ghribi
Nokia, Massy, France

C. Gaie and M. Mueck (eds.), *Digital Sustainability: Legislations, Standards, Solutions and Related Case Studies*, Advances in Sustainability Science and Technology,
https://doi.org/10.1007/978-981-92-0458-8_10

systems (6G). Advocating a *sustainable by design* approach, we present the challenges of sustainability from the well-established perspective of three pillars: environmental, social, and economic sustainability. All three pillars have a fundamental role, spanning ecological effects, operational costs, business opportunities, and the controversial topic of social acceptance of new technologies. Also, 6G comes with two facets: 6G as part of the problem, thus calling for actions to design a *Sustainable 6G*, and 6G as part of the solution (or, sustainability enabler), which typically referred to as *6G for Sustainability* or *6G for Sustainable Applications*. We introduce all these topics from a *value-oriented* perspective beyond performance, to then present the main technological advancement that will drive this evolution across the different network segments. With a final glimpse at standardization aspects, we aim to give insights into whether and how sustainability considerations can be brought into regulation.

Keywords Sustainability · Sustainable 6G · 6G for sustainability · Sustainable applications · Social acceptance · Telemedicine · Precision agriculture

1 Introduction

Connecting people has been one of the greatest revolutions of the last century, starting with wired access, mainly to communicate and access content, then evolving towards wireless systems that support mobility and flexibility among others, with the goal of interconnecting humans, applications, and things through the internet. This revolution greatly improved the quality of many people's lives from several perspectives, but also generated threats to our social behavior, for instance, facilitating fast (and difficult to control) circulation of malicious, harmful, and often unverified information. One key reason is that, up to now, communication networks have been designed and evolved solely to enhance performance from a purely technical perspective, namely

B. M. Khorsandi · L. C. Schmelz
Nokia, Munich, Germany

M. Matinmikko-Blue
Centre for Wireless Communications (CWC), University of Oulu, Oulu, Finland
e-mail: marja.matinmikko-blue@oulu.fi

D. M. Bozzone
Hewlett Packard Italiana SRL, Milan, Italy
e-mail: davide.montagno-bozzone@hpe.com

H. Nikopour
Intel Labs, Santa Clara, CA, USA
e-mail: hosein.nikopour@intel.com

C. Sarathchandra
Interdigital, London, UK
e-mail: chathura.sarathchandra@interdigital.com

based on Key Performance Indicators (KPIs) such as throughput, latency, reliability and energy efficiency, while standardization organizations have often considered sustainability as an implementation afterthought and left it to be managed by the device vendors and OPEX-related KPIs for network operators.. These evolved greatly during the last decades, so that today wireless systems exhibit performance levels that are close to reaching fundamental limits such as the Shannon capacity [1].

However, as the aforementioned effects on individuals and societies, but also the economy (at least to a certain extent) have not been sufficiently reflected by the design of communication networks in the past, future technical evolutions shall be accompanied by broader considerations to positively contribute to fundamental contemporary challenges, including Sustainability. 6G shall therefore focus as well on the *values* it brings to society and the economy as a whole, and make sustainability part of the early steps of the requirement setting and design processes.

Nonetheless, there is still a lot to be done technically, since next-generation systems (e.g., 6G) are being conceived to connect not only humans and things, but also different forms of intelligence, including Artificial Intelligence (AI) agents that are proliferating thanks to the recent advances in Large Language Models (LLMs) and generative AI [2]. This creates a heterogeneous ecosystem that potentially complicates its management and generates new threats in terms of trustworthiness among others. These aspects will be introduced in detail in Chap. 11.

Additionally, this is no longer sufficient to contribute positively to fundamental contemporary challenges, including *Sustainability*.[1] Technical evolutions shall also be accompanied by broader considerations. 6G shall also focus on making sustainability part of the early steps of the requirement setting and design processes. Following the United Nations (UN) Sustainability Development Goals (SDGs),[2] sustainability targets must pose their foundations on ***environmental, social and economic pillars,*** whose associated aspects (in this case for technologies) span greenhouse gases (GHG) emissions towards Net Zero networks, raw material use and extraction, security, privacy, inclusiveness, business opportunities and economic viability, bio- and geodiversity effect, bridging the digital divide (or access to digital services), and digital transformation.

These concepts (or, values) are strongly related the overarching aspects of the well-known International Telecommunication Union (ITU) framework [3] that include *sustainability, connecting the unconnected, ubiquitous intelligence, security and resilience* in a *value-oriented* "wheel" that surrounds six enabled services [4]: immersive communication, AI and communication, hyper reliable and low-latency communication, ubiquitous connectivity, massive communication, integrated sensing and communication. As highlighted in [4], the overarching aspects are in some cases mutually conflicting in nature, thus complicating the problem.

On one hand, from an environmental perspective, although today wireless communication networks represent a relatively small portion of GHG emissions [5], the

---

[1] https://www.un.org/en/academic-impact/sustainability: "meeting the needs of the present without compromising the ability of future generations to meet their own needs".

[2] https://sdgs.un.org/goals.

traffic explosion expected towards 2030 pulls the alarm to anticipate an unwanted growth in this share. Also, facilitating data sharing and introducing AI poses non-trivial dilemmas of privacy, trustworthiness, and resilience, with the latter representing a more general requirement also from an environmental perspective (e.g., in the presence of natural disasters), and at the societal level (e.g., in case of wars). Finally, economic aspects related to investments and operational costs warn operators. On the other hand, the potential positive effect[3] of 6G on other sectors plays a key role. As an example, enhancing coverage in rural areas to enable precision agriculture, or providing healthcare assistance from remote to unserved areas, might generate clear advantages from the perspective of all pillars. Therefore, fundamental questions arise:

> How will 6G contribute positively to these sustainability pillars?
>
> And how can new technologies increase the sustainability of 6G itself and help minimize its negative effects?

These complex questions require a careful analysis of what is (not) available today, what are the lessons learnt from previous generations (especially the case of 5G), and the new opportunities to achieve the objective of a *Sustainable 6G by design*, contributing to the enablement of *Sustainable applications*.[4] To do so, new concepts beyond performance must be introduced. One interesting example in the context of environmental sustainability is proposed in [6], with Decarbonization Service Agreements (DSAs) beyond Service Level Agreements (SLAs).

In this Chapter, we present an overview of the main opportunities and challenges of sustainability in the context of 6G. We first present, in Sect. 2, the key values, the sustainability pillars, and their relations with the two aspects of the problem: *Sustainable 6G* and *6G for Sustainable Applications*. These aspects are strongly related to the first-order and second-order effects of 6G, defined in [7].[5]

Further, from a social perspective, we go through the delicate problem of social acceptance, which turned out to be a very controversial topic during the 5G rollout phase. Then, we dive more deeply into the problem of Sustainable 6G (Sect. 3), considering the Radio Access Networks (RANs) and core networks. Therefore, Sect. 3 is more focused on technical aspects, which remain important for developing new technology enablers. Then, we shift the focus to 6G for Sustainability (Sect. 4), covering questions on what sustainable applications are, why 5G capabilities are insufficient, and how 6G will overcome these shortcomings. Further, insights on the currently ongoing discussions in standardization groups about 6G and sustainability are presented in Sect. 5. Finally, Sect. 6 draws the conclusions and takeaways.

---

[3] "Effect" refers to the result or consequence on the environment, economy or society that follows from ICT existence or usage [7].

[4] https://sustain-6g.eu/.

[5] A first order effect is the direct (positive or negative) effect associated with the existence of an ICT based solution. A second order effect (positive or negative) is induced by the use and application of ICT based solutions [7].

2 The Sustainability Pillars and Their Relation with 6G

This section examines sustainability throughout the entire lifecycle of 6G technology, from conception and development to deployment, adoption, and evolution. Drawing lessons from 5G and prior mobile network generations, this section proposes a broader understanding of sustainability for 6G. This approach extends beyond environmental preservation and energy efficiency to encompass critical societal and ethical challenges.

Consequently, the discussion focuses on integrating social values, ensuring long-term acceptability, and fostering responsible innovation, thereby aiming for a 6G technology that is not only efficient but also just, inclusive, and aligned with democratic principles.

2.1 *The Role of Values in the Sustainability Path*

The development of next-generation mobile networks occurs at the intersection of converging frameworks that are reshaping technology innovation. The sustainability discourse has reoriented both business design and research and development practices, shifting focus from primarily technical feasibility and economic optimization toward environmental stewardship, social responsibility, and intergenerational equity as core values for corporate responsibility. As sustainability has become established as both a business value and design principle, it now demands systematic consideration of environmental, economic, and social effects throughout the technology lifecycle.

Parallel to this sustainability integration, value-driven approaches have evolved from business practices through Responsible Research and Innovation (RRI) policy frameworks to encompass systematic consideration of stakeholder values in technology development. These dual movements intersect with a third stream: social acceptance of technology approaches to innovation, which have emerged through the reciprocal influence of user-centric design methodologies, technology acceptance models examining adoption barriers, and social acceptance frameworks pioneered in renewable energy deployment.

The rollout of 5G revealed critical gaps in technology development and governance approaches, demonstrating that technical excellence does not guarantee social acceptance without appropriate stakeholders' engagement and communication management. From antenna vandalism to municipal bans, 5G faced unprecedented, organized opposition across multiple dimensions.[6] The absence of a clear "killer application" that delivered immediate, tangible benefits to consumers partially tempered public enthusiasm [8], as well as various requirements from vertical stakeholders and supported by the Telco industry with different modes for network slices were not reflected in the final 5G specifications, leading to a certain disillusion

[6] https://www.politico.eu/article/coronavirus-5g-arson-attacks-online-theories/.

of verticals [9]. Simultaneously, geopolitical considerations regarding international technology supply chains raised questions about national security, which added complexity to deployment strategies in the EU and the US. This situation highlighted the trade-off between values of economic efficiency and technological sovereignty [9].

These challenges can be understood as manifestations of unresolved value conflicts across multiple dimensions. Legitimate concerns about electromagnetic field (EMF) exposure, raised by scientists and medical professionals since the 2017 EU 5G Appeal, reflected tensions between the values of technological progress and precautionary principles. Privacy and surveillance anxieties represented conflicts between values of connectivity and individual autonomy, as 5G's enhanced data collection capabilities clashed with growing demands for digital rights. Environmental concerns about increased energy consumption exposed between values of technological advancement and ecological responsibility, while local resistance to infrastructure deployment revealed conflicts between national competitiveness values and community self-determination. Significant intertwined economic flaws have hindered 5G's widespread acceptance. The substantial capital expenditure required for network infrastructure upgrades has often outpaced the uncertain return on investment. This uncertainty has mainly been compounded by the limited availability of compelling, monetizable end-user applications beyond enhanced mobile broadband. These factors have made it difficult for telcos to justify the high costs, highlighting a conflict between the values of technological potential and financial sustainability.

Acceptance concerns represent a signal for potential societal values at risk or compromised within a specific context, therefore, they are an insightful means to better understand how technological innovations are negotiated, contested, and legitimised through complex social processes that extend far beyond individual adoption decisions. These issues arise when technological innovations fail to align with societal and social values, when governance processes exclude affected communities, when risk communication lacks transparency, or when benefits are inequitably distributed. Unlike simple adoption or rejection, acceptance issues involve ongoing negotiations over legitimacy, governance, and competing visions of technological futures across multiple stakeholder groups and institutional levels, evolving through different phases of technological development—from conditions for implementation to consequences of implementation and institutional change. The 5G experience provides crucial lessons for developing sustainable 6G systems, where social acceptance becomes fundamental to achieving long-term environmental and social sustainability goals. The varied social perspectives and effects revealed by 5G deployment accentuate the interconnected nature of acceptance and sustainability challenges. The Social Acceptance of Technology (SAT) framework [10–12] addresses these value conflicts by conceptualizing acceptance not as a binary outcome but as a dynamic, multi-level process where diverse stakeholder values are embedded, contested, and negotiated across systemic, societal, and individual levels. This framework reveals how sustainable technology development requirements are understood and integrate diverse value systems from the earliest stages, rather than treating values as external constraints on technical solutions.

This framework was developed particularly within two SNS JU projects: 5G-SOLUTIONS [10] and 6G4Society [12], drawing from experiences with 5G deployment challenges and anticipatory analysis of 6G development needs. The 5G experience demonstrates that addressing value conflicts is fundamental to achieving sustainable 6G development across environmental, social, and economic dimensions. However, moving from recognizing value conflicts to systematically integrating values into sustainable 6G innovation requires identifying which values matter most to different stakeholders and developing mechanisms to measure progress toward value-aligned sustainability outcomes. This necessitates a structured approach to defining Key Values and establishing Key Value Indicators to evaluate the technology development, to guide responsible innovation throughout the 6G technology lifecycle, ensuring that sustainability goals are embedded from conception rather than retrofitted as external requirements.

Key values and their assessment via indicators

The value conflicts revealed by 5G's social acceptance challenges illustrate a broader methodological challenge in technology development. When approaching 6G development as a complex innovation project, one of the fundamental challenges becomes identifying and classifying social values, and consequently social effects. This challenge is particularly acute because values are intrinsically linked to the diverse interests of stakeholders. Indeed, stakeholders are "individuals and organizations who are actively involved in the project, or whose interests may be positively or negatively affected as a result of project execution or successful project completion" [13]. Therefore, differing stakeholder priorities can lead to conflicting values, making a universal or easily agreed-upon classification difficult to achieve. This inherent tension is a fertile ground for **controversies**, which erupt when these competing values and interpretations clash publicly. Such controversies not only starkly illuminate the difficulty of achieving consensus but also actively complicate efforts to embed a consistent set of social values within technological development or policymaking.

To navigate this complexity and proactively address potential conflicts, a robust methodological approach is proposed by [10] and [12]. This approach necessitates a multidisciplinary collaboration, actively involving experts from social sciences and humanities (SSH) alongside technology specialists (STEM), right from the earliest stages of defining use cases and developing proofs-of-concept (PoCs). Such early and integrated collaboration aims to collaboratively identify core social values (**Key Values—KVs**). Furthermore, throughout the development process, it enables the systematic evaluation of how effectively these values are embedded into both the technological solution itself and its anticipated usage patterns, as well as assessing the corresponding societal effects through Key Value Indicators (KVIs), which are suitable indicators designed for assessment purposes.

Key values, which refer to principles or qualities that individuals or groups deem important, desirable, or intrinsically good that may be addressed or impacted by ICT, are typically difficult to measure. This has led to the notion of so-called key value indicators (KVIs), which are metrics, either on a qualitative scale or, when possible, on a quantitative scale, and are defined within the scope of a specific use case, or in general, a specific context of a technology development [14].

According to [15], KVIs are used for monitoring and validating the effect of 6G on key social values. Differentiating between the nature of performance and values, and subsequently their assessment in a technology's context and use, enables a KVI definition that doesn't solely hinge on effects evaluation, but rather on whether and to what extent values are embedded into innovation. Therefore, another definition for KVIs could be tools designed to ensure social responsibility by identifying, monitoring, and validating the technological development, where societal responsibility includes three sustainability pillars: ***environmental, social, and economic sustainability.*** The Hexa-X-II project[7] defined KVIs as either a qualitative or quantitative indicator for assessing a key value, which in turn represents the effect of the use case and technology on the high-level human and planetary goals [16]. Two types of KVIs were identified: "Use Case KVI" which is a quantitative or qualitative indicator used to assess the effect of a use case, and "Enabler KVI" which is a quantitative or qualitative indicator to assess the effect of the technical enabler, i.e., the technology applied to a use case [16].

According to [15], environmental sustainability-related KVIs can include percentage of energy efficiency improvements, litres of water conservation, level of quality environment, percentage of waste reduction, number of initiatives contributing to climate resilience, and percentage of reduction of carbon footprint. Social sustainability-related KVIs include the social inclusivity index, percentage of improved health and well-being, workplace safety and diversity rate, improved accessibility, and percentage of population with access to digital technologies or high-speed internet. Economic sustainability-related KVIs include percentage of cost savings, percentage of increased profits, percentage of increased productivity, and increased operational efficiency ratio, among others. Hexa-X-II identified key values and KVIs for six representative use cases for the three sustainability pillars of environmental, social, and economic sustainability for both first and second-order effects [4, 16]. The identified KVs and KVIs are very specific to the use case, while some of the KVIs are more generic across several use cases. However, the work on KVs and related KVIs is still in an early stage in 6G, and there is no common methodology beyond the one presented in [14].

2.2 The Sustainability Pillars

Beyond the technical measurement of Key Value Indicators, it is crucial to understand the overarching dimensions that define sustainable technology development. While sustainability discussions in mobile communications have historically concentrated on environmental aspects like energy efficiency, the advent of 6G marks a pivotal

[7] https://hexa-x-ii.eu/.

shift. Recognizing the paramount importance of societal values and social acceptance, 6G fundamentally expands the sustainability discourse beyond mere environmental effect and performance metrics. As previously mentioned, this next generation of mobile networks fully embraces the well-known three pillars of sustainability: ***environmental, social, and economic***. These interconnected dimensions provide a comprehensive framework for assessing the 6G effect and are discussed in detail in the following sections.

Environmental sustainability

Environmental sustainability in 6G extends far beyond energy efficiency improvements to encompass comprehensive lifecycle effects, circular economy principles, and enabling effects across other sectors. The value conflicts revealed by 5G deployment, particularly around environmental claims that lacked credibility with local communities—demonstrate that environmental sustainability cannot be achieved through technical solutions alone but requires addressing the values that different stakeholders bring to environmental protection and resource use.

The challenge lies in measuring and validating environmental benefits across complex socio-technical systems while maintaining credibility with diverse stakeholder groups. Current approaches focus overwhelmingly on direct energy consumption and carbon emission while neglecting broader environmental implications, including rare earth mining for devices, electronic waste from accelerated device replacement cycles, potential ecological effects from electromagnetic fields, and biodiversity effects from expanded infrastructure deployment. The 5G experience showed that environmental promises failed to gain acceptance when they did not align with community values about transparency, participation, and precautionary approaches to environmental protection.

Understanding environmental sustainability through a Social Acceptance Framework reveals how environmental considerations operate across four objects of acceptance. At the conditions for innovation level, environmental values must be embedded in funding priorities and standard-setting processes. For conditions for implementation, environmental requirements such as circular design principles and energy-efficient infrastructure become essential for deployment acceptance. The consequences of implementation must account for both direct environmental effects and enabling effects where 6G helps other sectors reduce their environmental first-order effects. Finally, institutional change must ensure that environmental considerations are embedded in new regulatory frameworks and market structures rather than treated as external constraints.

Key priorities for environmental sustainability in 6G should include: circular design principles that address the full lifecycle from material extraction to end-of-life recycling; implementing comprehensive assessment frameworks that capture both direct technological effects and indirect societal effects; establishing credible metrics for enablement effects where 6G helps other sectors reduce environmental effects; and ensuring environmental considerations are integrated with performance metrics rather than treated as trade-offs. The opportunity exists to embed environmental sustainability values from conception, learning from 5G's failure to adequately address environmental concerns through purely technical approaches.

Social sustainability

In the literature, social sustainability is often defined from different angles and touching diverse aspects. The United Nations Global Compact states that "social sustainability is about identifying and managing business impacts, both positive and negative, on people. [...]".[8] Several definitions are provided, both as a standalone objective and in line with ecology. They also define macro and micro levels, depending on whether it is related to the basic level of physical well-being, quality of life, and equity, respectively. Additionally, in [7, 12, 15] the term "societal sustainability" is also used in the context of 6G, with the goal to focus on accessibility, security, privacy, and inclusiveness, to ensure that the technology serves diverse populations and helps prevent the reinforcement of existing inequalities, or the raise of new ones.

The 5G experience highlighted how value conflicts around privacy, surveillance, democratic participation, and community autonomy became central barriers to achieving broader societal sustainability goals. These conflicts emerged not from technical deficiencies but from misalignments between the values embedded in 5G development and the values prioritized by different social groups.

The absence of transformative applications that clearly improved citizens' lives revealed a fundamental disconnection between industry values focused on technical capabilities and public values seeking tangible societal benefits. Privacy and surveillance concerns represented deeper conflicts between values of connectivity and individual autonomy, while local resistance to infrastructure deployment reflected tensions between national competitiveness values and community self-determination. These values conflict directly undermined societal sustainability by creating social divisions, eroding institutional trust, and preventing equitable benefit distribution.

From a social acceptance perspective, achieving societal sustainability in 6G would require understanding how different stakeholder groups at different scales would evaluate and respond to the technology. At the policy and institutional level, societal sustainability depends on whether 6G development processes genuinely incorporate values like social equity and democratic participation into funding priorities and governance structures. At the community level, it requires addressing how 6G infrastructure deployment affects local quality of life, community autonomy, and social cohesion—lessons from 5G show that top-down infrastructure decisions without community input create lasting resistance. At the individual and household level, societal sustainability depends on whether people find 6G applications meaningful and aligned with their values around privacy, social connection, and personal autonomy, rather than simply technically accessible.

[8] https://unglobalcompact.org/what-is-gc/our-work/social.

Economic sustainability

According to [16], Economic sustainability in the context of 6G refers to the ability to develop ecosystems that are resilient, adaptable, and inclusive, while also being able to sustain economically, not at the expense of the environment or social equity. Economic sustainability targets long-term economic growth while respecting environmental and social sustainability objectives. Thus, economic sustainability can be seen to bring together the three sustainability pillars. Economic sustainability in the context of 6G is concerned with new business opportunities and sustainable business models that enable sustainability improvements in different sectors while at the same time ensuring 6G's economic viability (e.g., for operators).

A particular deployment model introduced in the 5G era is the notion of local 5G networks that could be deployed by different stakeholders to serve local and often vertical-specific needs [17]. This changed the business landscape in 5G, opening the market to new entries and business models, and the same development is expected to continue in 6G. These local cellular network deployments build an ecosystem around them, in which a set of stakeholders is needed to create the offerings.

Economic sustainability has been considered in the 6G context by using use cases as the unit of analysis and focusing on ecosystems and business models in [16]. The resulting business evaluation for the 6G use cases developed an ecosystem-level business model canvas, including ecosystem value proposition, ecosystem co-value creation process, ecosystem outcomes/value co-capturing, detailed stakeholder analysis of key stakeholders and their roles and value-related aspects, followed by a graphical illustration of the ecosystem pie.

2.3 *Sustainable 6G Versus 6G for Sustainability*

Beyond the sustainability pillars, one of the crucial challenges in the design of the next-generation mobile network is the integration of sustainability into the development and deployment of technologies. This is not simply about making 6G networks more energy-efficient; it is about creating a holistic approach that considers environmental, societal, and economic effects across the entire lifecycle of technology and its applications. The ambition for 6G is to fundamentally shift the paradigm, moving beyond incremental improvements to achieve a truly sustainable 6G ecosystem. The way forward to achieve this goal is two-fold.

First, aiming to develop and/or design *Sustainable 6G*, focusing on minimizing the negative effects of the technology itself. This involves a deep dive into the energy consumption of 6G networks, particularly the energy-intensive Radio Access Network (RAN). The research explores detailed energy models for various RAN components, considering factors like idle power consumption, load-dependent consumption, and the effect of different hardware and software solutions. These aspects will also be discussed in Sect. 3. The goal is to develop innovative techniques for dynamic power management, optimizing resource allocation to reduce overall energy use without compromising performance. Beyond energy, it is also important to consider addressing the material footprint of 6G, promoting the use of sustainable materials and manufacturing processes, and advocating for circular economy principles through initiatives like device reuse, repair, and refurbishment. This includes developing novel wireless services to facilitate the tracking and management of end-of-life devices, reducing e-waste.

The second, equally important aspect, is 6G for sustainability, or 6G for Sustainable Applications. This focuses on how 6G can be leveraged to drive positive environmental, societal, and economic change in various vertical sectors. This can only be possible by selecting a sector of society (a.k.a. use case) and analyzing how 6G can contribute to the sustainability goals of that particular sector. In this regard, SUSTAIN-6G,[9] a European-funded project on 6G and sustainability, has chosen three umbrella use cases (based on an analysis of the United Nations SDGs) to be further enhanced through the development of 6G technologies and improved with respect to environmental, societal, and economic aspects. These are: **agriculture, telemedicine, and smart grids** [7]. In agriculture, the focus is on providing reliable, on-demand connectivity in rural areas to support precision farming techniques, optimizing resource use, and reducing environmental impact. In telemedicine, 6G enables remote diagnostics, consultations, and rehabilitation, improving access to healthcare and reducing the carbon footprint associated with travel. For smart grids, 6G facilitates real-time monitoring and control of energy distribution, enabling better integration of renewable energy sources and improving grid resilience. These aspects will be more deeply discussed in Sect. 4.

These efforts are underpinned by a commitment to the three pillars of sustainability. Environmental sustainability is addressed directly through the reduction of 6G's energy consumption and material footprint, and indirectly through the enabling of sustainable practices in other sectors. Societal sustainability is ensured by promoting equitable access to 6G technology, bridging the digital divide while respecting privacy and other fundamental rights, and fostering trustworthy and secure communication. This includes the ethical considerations and the need for inclusive design. Economic sustainability is achieved by promoting the long-term economic viability of 6G networks and fostering sustainable business models that benefit both large corporations and SMEs. These three pillars are interconnected, and achieving true sustainability requires a balanced approach that considers the trade-offs between different objectives.

Figure 1 graphically shows a holistic perspective of 6G across the three sustainability pillars and the two aspects of 6G and sustainability. Also, the main technologies introduced in Sect. 3 and the enabled vertical use cases discussed in Sect. 4 are shown in the inner circle.

---

[9] https://sustain-6g.eu/.

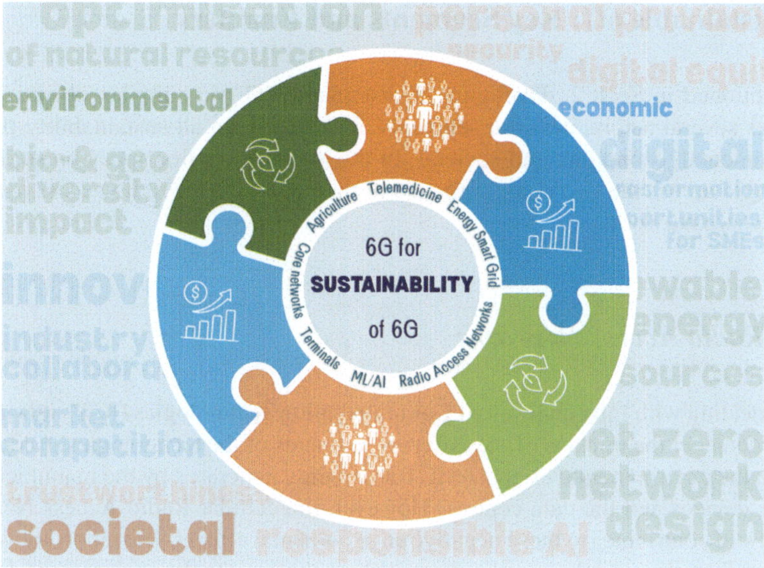

Fig. 1 The sustainability pillars

Section key takeaways: 6G technology and its applications need to be considered since their inception, the societal values that will be impacted both by the technology development (e.g., life cycle of the raw materials and products) and the usage of the technology (e.g., consumption of resources, social implications, and economic changes). For this reason, the development process has to engage a broader community of stakeholders and involve both technology and engineering (STEM) and social sciences and humanities (SSH) experts to consider the different perspectives. This value-based development approach is a prerequisite to ensure the sustainability of 6G and 6G for sustainability

Teaser on the remainder: In Sect. 3, we first present the key aspects of designing a sustainable 6G. This requires an analysis of all network segments, starting from the RAN, to then describe challenges and opportunities for the core network. Then, in Sect. 4, we will go through the main challenges and opportunities of 6G for sustainable applications, the second aspect of 6G and sustainability. This will close the loop between 6G as part of the problem and 6G as part of the solution

3 Sustainable 6G: Key Segments and Enablers

As mentioned in Sect. 2, developing a sustainable 6G is a complex objective that involves several actors, network segments, and a view on all sustainability pillars: environmental, economic, and societal. In this section, we first focus on some key challenges across the two main segments of the networks: the radio access network and the core network.

3.1 Radio Access Networks

When dealing with Information and Communication Technologies (ICT), it is important to consider the impacts of the different segments of the technologies. According to [18], the environmental effects of ICT are mainly generated by the manufacturing of user devices and the data centres. However, the cellular Radio Access Network (RAN) still contributes to these negative environmental effects. The report shows that the manufacturing and the use phases contribute the most to the environmental effect of the RAN, mainly because of the electrical energy consumption and the use of rare raw materials such as gold, copper, and other critical raw materials. Regarding the energy consumption, it is important to have in mind that the study was conducted in France, where the electricity mix is low-carbon due to nuclear plants. Consequently, the results of the study do not apply to the rest of Europe or the world.

The use of rare raw materials and critical raw materials must be addressed as well. Indeed, for instance, the use of conflict minerals such as gold, tungsten, tin, and tantalum is regulated in the EU as their extraction is associated with armed conflicts and human rights abuses,[10] in addition to the environmental effects. In the RAN, gold is usually used for connectors and plating of PCBs, tantalum for capacitors and RF filters, tin for solder, resistors, inductors, semi-conductors, and capacitors; whereas, tungsten for oscillators and integrated circuits [19].

Therefore, it is important to reduce as much as possible the diverse impacts of RAN. Regarding the use phase of the RAN, a first step is to identify the most power-consuming part of the network and try to reduce its consumption. Power models, such as the one developed by Golard et al. [20] on sub-6GHz base stations (BS), study the influence of different parameters, such as sleep modes, physical (traffic) load, bandwidth, frequency band, etc., on the power consumption of a BS sub-components.

It shows that the RF power amplifier is responsible for approximately three-quarters of the total power consumption of the macro base station at full physical load. Additionally, with high power consumption of power amplifier, comes increased cooling (between 10% [20] and 30% [21] of the total consumption) of the macro base station. Therefore, optimizing this part of the system, from the design phase

[10] https://policy.trade.ec.europa.eu/development-and-sustainability/conflict-minerals-regulation/regulation-explained_en#:~:text=The%20EU%20regulation%20covers%20tin,sense%20to%20focus%20on%20them.

to sleep mode implementation in the use phase, is an important step to reduce the global energy consumption of the base station without reducing its performance and consequent QoS and QoE.

The rest of the base station should be considered as well. The number of antennas and their topologies must be taken into account. Indeed, the analogue front-end and the digital baseband will not have the same power consumption, depending on the antenna topology, either fully analogue, fully digital, or hybrid. Therefore, the architecture of 6G's base stations should be designed along with the power model to ensure the base station power consumption is optimized globally and that optimizing one function does not lead to the overconsumption of another.

From a middle-term perspective, such a power consumption model may not only predict RAN consumption evolution but also be used during the daily operational setting of RAN components by reducing the number of monitoring points on its different components. With FR3 frequency band specificities, the overall 6G RAN objective might consist of ensuring a hardware energy-aware agility that ensures that the amount of equipment activation (its consumption) remains linear regarding the cell traffic load.

Nevertheless, the reduction of energy consumption should be correlated with the reduction of other effects as well as sustainability is not only about energy consumption. First RAN Life Cycle Analysis (LCA) allows the desired holistic approach [22]. Therefore, the base station sub-components should be a point of interest as well. For instance, reducing and reusing the amount of conflict minerals, but also components used in the base station.[11] This should reduce both environmental and social effects. However, it is crucial to have a systemic overview of the system to avoid trade-offs either between the three sustainability pillars or between a sustainability pillar's effect categories.

3.2 Core Networks: Current Status, Challenges, and Opportunities

The move to the 5G Core (5GC) is a significant advancement in network architecture away from the hardware-bound networks of the previous generations. The 5GC is designed using a Service Architecture (SBA) that drops the significant monolithic network elements in favor of small, autonomous software functions that perform specific functions and communicate with each other through open and standardized APIs. This model offers unprecedented flexibility and scalability, allowing operators to deliver and modify services with unprecedented speed in any previous network. When applied to an industry that uses Network Function Virtualization (NFV) with Software Defined Networking (SDN), SBA opens a diverse range of infrastructure and hardware resources with the potential to respond to customer needs

[11] https://www.grenoble-inp.fr/en/research/repex-new-uses-for-discarded-digital-devices.

cost-effectively based on performance requirements, including speed, capacity, and service configurations.

The 5GC also allows for the separation of the Control Plane (CP), which performs the tasks of mobility and signaling, from the User Plane (UP) that transports the data traffic. Separation enables functions, for example, the Access and Mobility Management Function (AMF) and User Plane Function (UPF), to run independently and get synchronized if needed. Simultaneously, the virtualization of the core and the continuous massive deployment of edge systems all contribute to an enormous increase in electricity consumption in both the CP and UP. To reduce the adverse effect, energy-efficient hardware, software optimization, and intelligent systems should be used to actively monitor and control energy consumption across the Radio Access Network (RAN) and Core layers, which will minimize any negative effects on the environment and energy consumption.

Such principles will be critical as the 6G Core supports speed, scale, and societal transformation. The 6G Core must support highly advanced services with ultra-low latency, high throughput, and the capacity to connect and manage billions of smart devices. As performance advances, so does the risk of uncontrolled energy consumption. The future network must be engineered considering that performance and sustainability will be equally essential when decision-making, especially regarding energy use, from hardware production to network operation to final decommissioning. To meet these demands, the architecture of 6G Core must go beyond the modular principles of 5GC and be inherently intelligent, adaptive, and self-optimizing. These features will enable real-time decisions that support technical needs and broader environmental objectives.

Environmental responsiveness will be as important as network performance in the next-generation architecture. Intelligent energy-aware network service would enable the orchestration of those customer demand-driven network functions so that they either stand by or switch off when not in use, reducing waste through its dynamic allocation approach. Practically, it would need an ecosystem of ubiquitous sensing. New analytics methods need to be located across the network to inform sensors about the state of the network, with constant monitoring of performance and energy usage. These mechanisms set the stage for an even more central role for analytics and automation in orchestrating sustainable, energy-aware networks. Nonetheless, it must be considered that the implementation of such monitoring solutions may introduce their own environmental, economic, and potentially societal costs, given the extensive hardware and sensor systems required to produce and collect the necessary data. Therefore, a thorough end-to-end assessment is required to factor in the positive contributions of these technologies while considering the potential negative consequences they impose, guaranteeing that sustainability in the operation of the network is being sought, as well as in the technologies that allow it to function.

Analytics plays a key enabling role in real-time control, prediction, and optimization. Network Data Analytics Function (NWDAF) is a 5GC function that collects information and conducts analytics on the vast range of measurements made in the network, which is a foundational aspect of enabling NWDAF to recognize user traffic patterns, user mobility over time, and performance of each function. The analytics that NWDAF can provide, in conjunction with the total amount of information used, will lead to better decision-making for operators, and allow better resource management in the network, and will be able to deliver better services and enable automated decisions that minimize a given node's energy usage. Such automation, when paired with prediction and adaptation, can push the network to be intelligent and self-aware in managing resources.

A network capable of real-time, predictive management is within reach with the evolution of NWDAF. By continuously ingesting and analyzing data, NWDAF would present data trends that would allow for predicting congestion before it occurred, releasing preventive/mitigation actions, and initiating the appropriate responses that would not only address the problem or opportunity but also do so in a sustainable way with energy efficiency considered as well. Beyond the current state, these data capabilities could help NWDAF inform future network planning, investments, and sustainability metrics reporting. Therefore, NWDAF could become a key element in green networking strategies, especially when combined with AI and machine intelligence technologies, discussed in the next section, which further enhance network management and sustainability efforts.

To anticipate part of Sect. 3.3 content, here with a focus on core networks, AI and ML are already reshaping operations to be smarter and more energy-conscious. Green AI is a critical aspect of ongoing evolution, which centers on framing computing as sustainable, energy-aware resources. Using AI/ML models, networks can predict traffic and flows, optimize resource allocation, and reduce unnecessary energy waste. Edge computing can further supplement AI/ML by performing a complete computation cycle closer to the data, thus reducing latency and energy while improving time to response. These advancements also create opportunities for newer models like LLMs to enhance future networks' analytical capabilities further.

LLMs represent an evolution in contextual analytics through structured and unstructured data integration. LLMs are utilized for their ability to use complex contextual relationships across unstructured data (text) and structured data. LLMs can use information from large log files to provide temporal analysis to identify time-based data trends while processing multiple inputs (alarms) in real-time to detect the evolving anachronous data set required to define an anomaly or performance degradation. LLMs can create a context that describes a variety of states and mine for interesting predictive or counterfactual relationships that allow for a reliable description of predicted behavior throughout network contexts, such as predicting anomalies, network demand, and planning for predicted and planned network/lifecycle management, thus enhancing the overall energy efficiency. These capabilities reflect a growing European consensus to invest in sustainable, forward-looking infrastructure. Further AI-related topics are discussed in Chap. 11.

Europe continues to lead by example in the global pursuit of network sustainability. The European Telecommunications Standards Institute (ETSI)[12] is developing technical specifications on interoperability and energy efficiency, where

[12] https://www.etsi.org/.

sustainability is a key principle.[13] The 6G Smart Networks and Services Industry Association (6G IA) is also a major player. It combines network operators, vendors, researchers, and academia.

They advocate for open, modular, and sustainable architecture and have produced position papers communicating sustainability as an obligation. They promote energy management techniques for advancing capabilities and open technologies to enable innovation using environmentally sustainable methodology. These aligned efforts underscore a broader vision of 6G as a platform for sustainable digital transformation.

In conclusion, building a sustainable 6G Core is a very complex challenge, but it is no longer an option towards contributing to the achievement of the UN SDGs.

Building a sustainable 6G Core represents a direction regarding technology, regulation, the environment, and effect. To deliver a sustainable, high-performing 6G ecosystem, we must evolve with AI-enabled management, Intelligent Orchestration and design, and new and changing standards. Next-Generation core networks will facilitate the big transformative applications ahead, such as augmented reality, autonomous mobility, Industry 4.0, and the global climate challenge. It represents a pathway to create a digital future that is technologically advanced, inclusive, and sustainable.

4 6G for Sustainable Applications

6G technologies are poised to drive not only direct efficiency improvements, but also profound **second-order sustainability effects** across key vertical domains, i.e., applications.

These effects, enabled by ultra-low latency, ubiquitous connectivity, distributed intelligence, and energy-aware design, extend beyond the infrastructure itself and catalyze **systemic environmental and societal benefits**.

What are Sustainable Applications

Sustainable applications are those that contribute meaningfully to the three sustainability pillars, as introduced in Sects. 1 and 2:

- **Environmental sustainability**: Lowering emissions, reducing waste, conserving biodiversity, and improving resource efficiency.
- **Social sustainability**: Promoting equity, inclusion, mental well-being, social cohesion, and access to essential services.
- **Economic sustainability**: Boosting operational efficiency, productivity, innovation, and business model resilience.

6G networks are uniquely positioned to act as a **catalyst** for these goals by enabling use cases that reimagine sectors such as agriculture, energy, healthcare, and, increasingly, human interaction itself.

[13] https://www.etsi.org/newsroom/news/10-events/2169-etsi-summit-on-sustainability.

Lessons learnt from 5G and opportunities for 6G

While 5G introduced capabilities such as enhanced mobile broadband, low latency, and massive machine-type communications, it falls short in supporting fully integrated, intelligent, and energy-adaptive systems at a global scale. 5G lacks:

- **Global ubiquity and vertical-native coverage**: 5G networks are limited by terrestrial deployments and exhibit inconsistent coverage, particularly in rural and remote regions. 6G is expected to integrate non-terrestrial networks (NTNs) [23, 24], including low Earth orbit (LEO) satellites and high-altitude platform systems (HAPS), enabling seamless global connectivity vital for agriculture, disaster response, and logistics. As emphasized by the Next-Generation Mobile Networks Alliance (NGMN), ensuring global coverage is not only a technological ambition but also aligns directly with the United Nations Sustainable Development Goals (SDGs), which advocate for inclusive and universal access to communication technologies for all individuals, regardless of location [23, 25].
- **Ambient IoT**: 5G falls short in enabling seamless, energy-autonomous, and massively scalable deployments of sensing infrastructure. 6G will incorporate ambient IoT systems, self-powered, cost-effective, and disposable devices capable of harvesting energy from the environment (e.g., RF, solar, thermal) [26, 27]. Standardization efforts led by ETSI ISG Europe and 3GPP Rel. 19/20 envision ambient IoT as a foundational pillar of future ultra-low power sensing infrastructures for 6G verticals [28]. These devices can be densely deployed across agricultural fields, industrial assets, or public health environments to enable pervasive monitoring without reliance on batteries or manual maintenance. Their passive nature allows for fine-grained, long-duration observation of environmental conditions, asset movements, and human activity patterns. By enabling data-rich, infrastructure-light applications, ambient IoT helps 6G bridge the scale, cost, and energy-efficiency gaps left by previous generations.
- **Integrated sensing and communication**: 5G can only support gNB-only sensing scenarios, while mostly using proprietary techniques. 6G embeds RF sensing for environmental awareness, user localization, and context-aware services. This convergence supports digital twins, autonomous navigation, and precise industrial automation [29].
- **Native AI support and distributed computing**: While 5G permits network slicing and basic edge computing, it lacks architectural support for dynamic AI workloads. 6G will enable deployment of AI/ML techniques in various parts of the system, from PHY layer optimization to orchestration and service delivery, enabling features like intent-based networking [30], federated learning, and in-network intelligence [31]. Edge-cloud continuum frameworks will enable real-time AI inference and data processing close to the source, reducing latency and data exposure across wide-area networks. Of course, unlike shared and highly optimized computing infrastructures (central clouds), distributed resources need to be carefully managed to not introduce energy consumption and material overhead, e.g., being dedicated to specific functions, and not shared and available to multiple actors, and to serve multiple purposes as in shared infrastructures.

However, thanks to recent advances in edge AI (as discussed in Sect. 3.3.3), AI processing is expected to be enabled in resource-constrained devices, without the need for high energy demand. As discussed in [9], 5G's current cloud-based C-RAN architecture suffers from performance bottlenecks and centralization issues, including high latency and energy inefficiency due to front-haul overloads and single points of failure. To address these, 6G will need to adopt decentralized architectures that support edge-native processing and intelligent handover, especially for mobile IoT and ultra-reliable applications. Furthermore, dynamic AI-native E2E slicing would allow 6G to dynamically respond to application requirements and current traffic loads. Crucially, the development and deployment of this AI-native network must consider the enacted AI Act (EU 2024/1689). This regulation establishes a framework for trustworthy AI by imposing obligations and specific requirements on high-risk AI systems, including those used in critical infrastructure (e.g., telecommunications). These requirements have to be natively integrated into 6G's architecture and AI models to ensure development of trustworthy, robust, ethical, secure, and legally compliant ecosystem. Further discussion on in-network AI is provided in Chap. 11.

- **Sustainability-by-design**: As discussed in Sects. 1 and 2, 5G KPIs were defined around throughput, latency, and reliability. 6G introduces key value indicators (KVIs) encompassing energy efficiency, carbon impact, societal benefit, and inclusivity, i.e., across all sustainability pillars. It will feature hardware and software innovations for ultra-low power IoT, energy harvesting, and dynamic energy adaptation at the RAN and core levels [32]. As highlighted [9], in current 5G networks contribute significantly to energy consumption and e-waste, due to over-deployment of base stations and a lack of coordinated energy strategies across heterogeneous networks. 6G aims to overcome these issues through ambient-powered devices, intelligent BS sleep strategies, and holistic energy management across the network stack.
- **Trustworthiness and privacy**: 5G lacks built-in mechanisms for verifiable trust, zero-trust architectures, or quantum-safe cryptography. 6G aims to integrate secure-by-design principles, including distributed ledger technologies, AI-driven threat detection, and trusted execution environments for data-sensitive verticals like e-Health and finance [33–35]. Moussaoui et al. [9] underscore the limitations of 5G SBA, which is vulnerable to protocol-level threats, side-channel attacks in NFV/SDN environments, and privacy breaches due to cloudified control of personal data. 6G must natively embed secure multi-party computation, privacy-preserving AI, and real-time threat detection.

These capabilities will not only fill existing gaps in mobile networks but also redefine how digital infrastructure contributes to long-term sustainability and societal resilience. To illustrate how 6G can unlock second-order sustainability benefits, we now focus on three sectors, i.e., agriculture, energy, and e-Health, where the convergence of advanced connectivity, sensing, and AI-enabled automation can drive measurable gains across all three sustainability pillars. These sectors were selected based on their relevance to European policy priorities (e.g., the Green Deal and Digital Decade), the availability of mature use cases from SNS-funded projects, and their alignment with use case families (e.g., massive IoT, critical IoT, immersive services) identified in 3GPP, ITU-T, and NGMN frameworks.

4.1 Agriculture

6G is set to transform agriculture into a **data-driven and precision-oriented** domain by offering massive-scale connectivity and edge AI for real-time monitoring and control. Sustainable benefits arise through:

- **Precision irrigation and fertilization**, using real-time sensor networks to reduce water and chemical use;
- **Remote crop monitoring via drones and satellites**, enabled by low-latency links and federated learning;
- **Collaborative robotics and automation** for labor-intensive tasks, improving yield and reducing waste;
- **Digital twins of farms and climate models** for simulating sustainable scenarios.

EU-funded projects and 6GIA vision papers stress that these tools will **reduce the carbon and resource intensity** of agricultural systems, especially in controlled-environment settings [36]. These capabilities also foster **resilience to climate change**, enabling real-time adaptation to weather variability and global supply disruptions. As outlined in [7], these use cases are essential to meet European Green Deal objectives and SDG-aligned sustainability goals.

Representative use cases (rUCs) defined by 3GPP RAN [27] demonstrate how A-IoT contributes to sustainable agriculture, primarily rUCs 2,6,8:

- Outdoor sensor applications, including wide-area monitoring of soil moisture, nutrient levels, and microclimate via batteryless backscatter-based devices;
- Inventory and livestock tracking, using passive tags for geolocation and condition monitoring;
- Command applications, allowing energy-autonomous actuation (e.g., triggering irrigation valves based on local sensor input);
- Integrated positioning and context-awareness, enabling optimized farm logistics and autonomous vehicle routing;

Digital twin integration, with live sensor data feeding AI-driven simulations for yield optimization and climate adaptation [26].

4.2 Energy and Smart Grids

6G is crucial for realizing **cyber-physical smart energy systems**, where sustainability is achieved through:

- **Real-time monitoring and control** of decentralized assets (photovoltaics, wind, electric vehicles, batteries);
- **Edge-based energy optimization** using distributed intelligence;
- **Digital twins for grid resilience and planning**;
- **Dynamic demand-response schemes**, balancing loads with the availability of renewable sources.

Special attention to smart grids is considered in 3GPP Rel. 19 [27] to smart grids applications (rUC #6) as it is envisioned to be one of the key applications enabled by ambient IoT.

These enable **lower transmission losses**, **decarbonized electricity flows**, and **resilient local energy islands**. Within the EU SNS framework, such deployments are envisioned as enablers of the **European Green Deal goals**, including grid decentralization and electrification [37]. The 6G-PATH and B5G-PINE initiatives, discussed in [38], underscore the importance of AI-enabled orchestration and KVIs in smart energy applications.

4.3 E-health

In the health domain, 6G enables **ubiquitous, privacy-preserving healthcare** via:

- **Federated learning** across distributed medical data sources, ensuring data sovereignty and compliance with GDPR[14];
- **Telemedicine and remote diagnostics** in underserved areas with high-resolution holographic video and tactile feedback;
- **Wearable and implantable sensors** for continuous health monitoring and preventive care;
- **AI-powered decision support** is integrated into the network infrastructure, reducing diagnostic errors and unnecessary travel.

In [39], 6G supports sustainable healthcare through:

- **Concurrent remote surgical planning** (Sect. 6.3.7 of [7]), where high-fidelity holographic data enables joint planning sessions between clinicians and engineers across borders, avoiding travel and enabling time-sensitive planning;
- **Remote rehabilitation** (Sect. 6.3.8 of [7]), using extended reality (XR) environments and haptic interaction for patients recovering at home, reducing hospital load and enabling long-term behavioral monitoring [7].

These systems support **decentralized, proactive healthcare**, which reduces emissions from patient and staff mobility, cuts over-treatment, and improves the health resilience of aging populations. Importantly, the WHO Commission on Social Connection [39] notes that digital technologies must also be deployed with caution:

[14] https://gdpr-info.eu/.

while they offer support and accessibility, they also risk displacing face-to-face interaction, exacerbating loneliness, and impacting mental health.

4.4 *Cross-Sectoral Effects*

Other verticals, such as **manufacturing, logistics, education, and smart cities**, also benefit from 6G's systemic intelligence and sustainability-by-design capabilities. Use cases include:

- **Green factories** using AI/ML for predictive maintenance and energy optimization [40];
- **Sustainable mobility ecosystems** through vehicle-to-everything (V2X) communications [41];
- **Remote learning and work platforms**, reducing emissions from commuting;
- **Urban twins for circularity** and climate-neutral planning [42] (Sec. 9.4.2);
- **Trusted immersive environments** for safe learning, healthcare, and public safety (e.g., Hexa-X-II and TrialsNet initiatives) [32];
- **Smart building and infrastructure maintenance**, leveraging in-sensor analytics to optimize energy usage and lifespan of critical assets [7].

The WHO further emphasizes that how digital technology is used, actively or passively, and by whom, determines whether it strengthens or undermines social connection [39]. 6G must therefore be designed with equity, inclusion, and psychological well-being in mind.

5 Sustainability in 6G Standardization

Sustainability in wireless networks has become an important topic as the ICT industry and its various sub-domains face rising energy demands and climate objectives. As previously mentioned, roughly 2–3% of the global energy consumption accounts for telecommunication networks, and mobile networks alone contribute about 0.2–0.4% of worldwide CO_2 emissions. In recognition of this effect, industry and governments worldwide are driving initiatives to reduce the carbon footprint of wireless infrastructure. For example, the European Union's "Fit for 55" plan calls for a 55% reduction in emissions by 2030, and the ITU has set a 45% GHG emission reduction target for the ICT sector from 2020 to 2030. In response, standardization bodies and industry alliances are integrating sustainability principles into the core of wireless network technologies. In this section, we attempt to provide an overview of state-of-the-art standardization efforts to make wireless networks more sustainable.

International Telecommunication Union (ITU): ITU's ITU-R (Radio) and ITU-T (Telecom) sets global objectives and technical guidelines. ITU-R's frameworks for next-generation networks (IMT-2020 for 5G and IMT-2030 for 6G) explicitly emphasize improved energy efficiency and support for climate goals. Likewise, ITU-T's Study Group 5 ("Environment, Climate Change and Circular Economy") develops recommendations for green ICT (from energy-efficient equipment metrics to methodologies for lifecycle assessments and e-waste reduction). For example, a set of ITU-T L-series standards covers energy efficiency, GHG emissions accounting, and circular design of telecom equipment [43, 44].

3GPP (Third Generation Partnership Project) is the primary standard development organization for cellular networks (4G, 5G, 6G and beyond). Over the years, 3GPP has progressively built energy-saving features into its specifications. Since as early as Release 10 (circa 2011), 3GPP has studied and recommended ways to reduce network power consumption, including operation recommendations, such as how to operate networks in low-power modes, and new protocol features, such as signaling to shut off unused base station components during off-peak hours. Since then, it has continued to improve energy efficiency of telecommunication systems. Fast forwarding to more recent times, in 5G 3GPP has introduced KPIs for energy efficiency (such as bits per Joule delivered, which was adopted from ETSI Technical Committee Environmental Engineering, in the ETSI ES 203 228) to quantify improvements, while the 3GPP's System Architecture and RAN groups continue to refine these metrics and features. Moreover, Release 16 also introduced optimizations like a lower-power UE mode for small IoT devices (14 dBm power class for wearables) and a "wake-up signal" mechanism to let UEs sleep deeper and only listen for specific low-power wake-up cues. 3GPP SA5 released TR 32.972 and TS 28.554, detailing how to collect energy consumption data from RAN and core equipment via standardized counters, and how to compute network-wide energy efficiency. In release 17, energy efficiency remained an important consideration, despite its focus was primarily on performance enhancements. However, release 18 (5G-Advanced) places an elevated emphasis on reducing network energy consumption. It may, for example, manifest in the form of cross-node AI/ML techniques to intelligently manage resources and traffic for energy savings (e.g., AI-driven sleep mode orchestration).

Another related advancement in Release 18 is the introduction of the RedCap (Reduced Capability) device category. RedCap devices are simplified 5G devices, used primarily in 5G IoT use cases, designed with the goals of significantly cutting device complexity and power usage. Ongoing work is investigating energy-harvesting and wake-up radio features for these devices. In one example, a tiny secondary receiver that can wake the main radio only when needed, allowing the primary radio to remain off most of the time. Such wake-up receivers (being studied in 3GPP and IEEE) could let future devices recharge from ambient sources or run on coin-cell batteries, potentially for months, or even for years.

Massive IoT standards are inherently focused on low power. 3GPP's IoT technologies—NB-IoT and LTE-M (Cat-M1)—were designed to support devices that may need to run for 5–10 years on a single battery. To achieve this, 3GPP introduced features like Power Saving Mode (PSM) and extended Discontinuous Reception (eDRX). PSM allows a device to essentially "sleep" for long durations (hours or days) without deregistering from the network, drawing virtually no power during sleep, until it needs to periodically check in. eDRX extends the interval a device can avoid listening for downlink messages to save energy.

6G is currently being conceived with sustainability at the core of the principles that drive the development from the outset. The ITU-R's IMT-2030 Vision Framework (approved in 2023) explicitly lists "Environmental sustainability" as a key objective, alongside technical performance and societal goals. One of the four design pillars in the IMT-2030 framework is sustainability, ensuring that 6G requirements will include drastic improvements in energy efficiency, carbon footprint reduction, and even features to aid other sectors in reducing emissions.

6 Conclusions

In this Chapter, we discussed the main challenges and opportunities to make 6G a sustainable technology, with a potential of also being sustainability-enabling for other sectors. To ensure a holistic view of sustainability, the latter must be treated jointly from three perspectives related to environmental, social, and economic pillars, as promoted by the UN SDGs. This means not only focusing on energy consumption aspects, but also on economic viability and the impact of new technologies on people, including the controversial topic of economic sustainability.

After discussing these complex aspects, we described the main technical challenges related to the main segments of the network. Further, we discussed the impact of 6G on other sectors, with use cases related to agriculture, telemedicine, and energy with smart grids. Similarly, these use cases require a holistic approach across the three sustainability pillars.

Finally, we provided a description of the current status of 6G standardization, with a perspective on sustainability. Overall, we believe these and other aspects, e.g., related to AI and discussed in Chap. 11, will contribute to a sustainable and value-oriented 6G by design, enabling sustainable services and applications.

Acknowledgements This work is co-funded by the European Union under Grant Agreement 101191936. Views and opinions expressed are, however, those of the authors only and do not necessarily reflect those of all SUSTAIN-6G consortium parties nor those of the European Union or the SNS JU (granting authority). Neither the European Union nor the granting authority can be held responsible for them. The authors would like to thank the whole SUSTAIN-6G consortium.

The authors thank the editors, Dr Christophe Gaie and Dr Markus Mueck, for their thorough review and detailed corrections.

Disclaimer Note that the authors have assembled the information above to the best of their knowledge as available at the time when the present book was written. Any interpretation is the personal opinion of the authors. Also, the regulation framework can change and evolve at any time. The only legally binding text is the one provided in the documents formally published in the European Union Official Journal.

References

1. Shannon CE (1948) A mathematical theory of communication. Bell Syst Tech J 27:379–423
2. Raiaan MAK et al (2024) A review on large language models: architectures, applications, taxonomies, open issues and challenges. IEEE Access 12:26839–26874
3. International Telecommunication Union (2023) Framework and overall objectives of the future development of IMT for 2030 and beyond. Recommendation ITU-R M.2160–0
4. Merluzzi M et al (2025) Towards sustainable 6G: a holistic view of trade-offs and enablers. In: IEEE Communications Magazine (submitted)
5. Malmodin J, Lövehagen N, Bergmark P, Lundén D (2024) ICT sector electricity consumption and greenhouse gas emissions—2020 outcome Telecommun Policy 48
6. Bolla R, Bruschi R, Lombardo C, Siccardi B (2023) 6G enablers for zero-carbon network slices and vertical edge services. IEEE Netw Lett 5:173–176
7. SUSTAIN-6G (2025) Deliverable D2.1: Sustainability baseline, UCs, and baseline requirements. [Online]. Available: https://doi.org/10.5281/zenodo.16684673
8. Espinoza J (2023) EU considers mandatory ban on using Huawei to build 5G. [Online]. Available: https://www.ft.com/content/a6900b0f-08d5-433d-bfb0-f57b6041e381.
9. Moussaoui M, Bertin E, Crespi N (2022) 5G shortcomings and beyond-5G/6G requirements. 6GNet
10. Briguglio L, Nesse P, Di Giglio A, Occhipinti C, Durkin P, Markopoulos I (2021) Business value and social acceptance for the validation of 5G technology. In: 2021 IEEE international Mediterranean conference on communications and networking (MeditCom), Athens, Greece
11. Occhipinti C, Carnevale A, Briguglio L, Iannone A, Bisconti P (2022) SAT: a methodology to assess the social acceptance of innovative AI-based technologies. J Inf Commun Ethics Soc 21:94–111
12. 6G4Society (2024) Deliverable D1.1: societal aspects in 6G technology: concerns, acceptance models and sustainability indicators. [Online]. Available: https://doi.org/10.5281/zenodo.147 47637
13. Project Management Institute (PMI®) (1996) A guide to the project management body of knowledge (PMBOK® Guide). Project Management Institute, Newtown Square, PA
14. Wikström G et al (2024) Key value indicators: a framework for values-driven next-generation ICT solutions. Telecommun Policy 48
15. Uusitalo M et al (2024) European vision for the 6G network ecosystem. White paper
16. Hexa-X II (2025) Deliverable D1.4. 6G Value, requirements and ecosystem. Available: https:// hexa-x-ii.eu/wp-content/uploads/2025/05/D1.4-final.pdf.
17. Ahokangas P, Matinmikko-Blue M, Yrjölä S, Hämmäinen H (2021) Platform configurations for local and private 5G networks in complex industrial multi-stakeholder ecosystems. Telecommun Policy 45
18. ADEME-ARCEP (2023) Evaluation de l'impact environnemental du numerique en France et analyse prospective. [Online]. Available: https://www.arcep.fr/uploads/tx_gspublication/note-synthese-au-gouvernement-prospective-2030-2050_mars2023.pdf.
19. Ericsson (2019) What do you know about conflict minerals?. [Online]. Available: https://www.ericsson.com/en/blog/2019/6/what-do-you-know-about-conflict-minerals
20. Golard L, Agram Y, Rottenberg F, Quitin F, Bol D, Louveaux J (2024) A parametric power model of multi-band sub-6 GHz cellular base stations using on-site measurements. In: IEEE 35th international symposium on personal, indoor and mobile radio communications (PIMRC), Valencia, Spain
21. SK Telecom and NTT DOCOMO (2023) Green mobile network: energy saving efforts by SK telecom and NTT DOCOMO. [Online]. Available: https://www.docomo.ne.jp/english/binary/pdf/corporate/technology/rd/docomo6g/GreenMobileNetworksWhitePaper_22February2023.pdf
22. Golard L, Dethienne R, Louveaux J, Bol D (2024) A parametric life-cycle model for assessing environmental impacts of 4G and 5G cellular base stations. Int J Life Cycle Assess

23. Next Generation Mobile Networks Alliance (2022) White Paper: 6G use cases and analysis. [Online]. Available: https://www.ngmn.org/work-programme/6g-use-cases-and-analysis.html
24. Harounabadi M, Heyn T (2023) Toward integration of 6G-NTN to terrestrial mobile networks: research and standardization aspects. IEEE Wirel Commun 30:20–26
25. Next Generation Mobile Networks Alliance White Paper: 6G Drivers and Vision [Online]. Available: https://www.ngmn.org/wp-content/uploads/NGMN-6G-Drivers-and-Vision-V1.0_final_New.pdf.
26. 3rd Generation Partnership Project (3GPP) (2023) TR 38.848 V18.0.0: Study on ambient IoT (Internet of Things) in RAN (Release 18)
27. 3rd Generation Partnership Project (3GPP) (2024) TR 38.769 V0.0.1: Study on solutions for ambient IoT (Internet of Things) RAN (Release 19)
28. 3rd Generation Partnership Project (3GPP) (2025) Ambient IoT: Redefining Wireless Communication for Industry 4.0. [Online]. Available: https://www.3gpp.org/technologies/anbient-iot-tsdsi
29. Kaushik A (2024) Toward integrated sensing and communications for 6g: key enabling technologies, standardization, and challenges. IEEE Commun Stand Magaz 8:52–59
30. Gharbaoui M, Martini B, Castoldi P (2023) Intent-based networking: current advances, open challenges, and future directions. In: 23rd international conference on transparent optical networks (ICTON), Bucharest, Romania
31. Letaief K, Shi Y, Lu J, Lu J (2022) Edge artificial intelligence for 6G: vision, enabling technologies, and applications. IEEE J Sel Areas Commun 40:5–36
32. 6G Smart Networks and Services (2025) White Paper: 6G KVIS – SNS projects initial survey results 2025. [Online]. Available: https://smart-networks.europa.eu/wp-content/uploads/2025/05/sns-ju-white-paper-6g-kvis-survey-2025_final-1.pdf
33. Wang S (2024) Explainable AI for 6G use cases: technical aspects and research challenges. IEEE Open J Commun Soc 5:2490–2540
34. Blika A (2025) Federated learning for enhanced cybersecurity and trustworthiness in 5G and 6G networks: a comprehensive survey. IEEE Open J Commun Soc 6:3094–3130
35. Siddiky M (2025) A comprehensive exploration of 6G wireless communication technologies. Computers
36. 6G-IA (2024) White Paper: The role of 6G in agriculture. [Online]. Available: https://aioti.eu/wp-content/uploads/6GIA-AIOTI-White-Paper-Agriculture-Final.pdf
37. 6G-IA (2024) Sustainability of 6G: Ways to reduce energy consumption. [Online]. Available: https://6g-ia.eu/wp-content/uploads/2025/01/sustainability_of_6g-path_forward_v1.2.2.pdf.
38. Maglogiannis I, Iliadis L, Macintyre J, Avlonitis M, Papaleonidas A (2024) Artificial intelligence applications and innovations. In: 20th IFIP WG 12.5 International conference, AIAI 2024, Corfu, Greece
39. World Health Organization (2025) From loneliness to social connection: charting a path to healthier societies. [Online]. Available: https://www.who.int/publications/i/item/978240112360
40. Fisher WP (2024) How deterministic chaos defines natural laws, Springer LNME
41. Roger SA (2024) Sustainable mobility in B5G/6G: V2X technology trends and use cases. IEEE Open J Veh Technol 5:459–472
42. Raes L, Ruston McAleer S, Croket I, Kogut P, Brynskov M, Lefever P (2025) Decide better: open and interoperable local digital twins. Springer Nature
43. International Telecommunication Union (2012) L.1420: Methodology for energy consumption and greenhouse gas emissions impact assessment of information and communication technologies in organizations. [Online]. Available: https://www.itu.int/rec/T-REC-L.1420.
44. International Telecommunication Union (2019) L.1451: Methodology for assessing the aggregated positive sector-level impacts of ICT in other sectors. [Online]. Available: https://www.itu.int/rec/T-REC-L.1451

Chapter 11
Artificial Intelligence: Sustainability Challenges, Opportunities, and the Case of 6G

Mattia Merluzzi, Luigi Briguglio, Serge Bories, David Brayford, Gilles Callebaut, Emilio Calvanese Strinati, Lucas Pereira Carwile, Emma Complido, Tim Dykes, Olivier Forceville, Chaima Ghribi, Utz-Uwe Haus, Bahare M. Khorsandi, Wei Mao, Marja Matinmikko-Blue, Davide Montagno Bozzone, Hosein Nikopour, Carmela Occhipinti, Chathura Sarathchandra, Lars Christoph Schmelz, Omid Semiari, Emanuela Tangari, Lili Wei, and Shu-Ping Yeh

Abstract From our daily life to complex industrial environments in almost all sectors, Artificial Intelligence (AI) undoubtedly joined the list of the greatest revolutions of human history. However, like many of them, it comes with opportunities and

---

M. Merluzzi (✉) · S. Bories · E. C. Strinati · E. Complido
University Grenoble Alpes and CEA-Leti, Grenoble, France
e-mail: mattia.merluzzi@cea.fr

S. Bories
e-mail: serge.bories@cea.fr

E. C. Strinati
e-mail: emilio.calvanese-strinati@cea.fr

E. Complido
e-mail: emma.complido@cea.fr

L. Briguglio · L. P. Carwile · C. Occhipinti · E. Tangari
CyberSocial Lab. Srl Impresa Sociale, Rome, Italy
e-mail: l.briguglio@cybersoclab.com

L. P. Carwile
e-mail: l.carwile@cybersoclab.com

C. Occhipinti
e-mail: c.occhipinti@cybersoclab.com

E. Tangari
e-mail: e.tangari@cybersoclab.com

D. Brayford
Hewlett Packard Enterprise EMEA Research Lab Germany, Ratingen, Germany

C. Gaie and M. Mueck (eds.), *Digital Sustainability: Legislations, Standards, Solutions and Related Case Studies*, Advances in Sustainability Science and Technology,
https://doi.org/10.1007/978-981-92-0458-8_11

threats. From the environmental impact of data centers to economic and social risks in labor markets, the sustainability of AI systems must progress hand in hand with their technical development. In this Chapter, we provide an overview of the main challenges of sustainability in the context of AI. Considering all sustainability pillars related to environmental, social, and economic challenges, we introduce the main aspects of *Responsible AI.* Then, we focus on the fundamental question of where to host AI-related processing, with a thorough discussion on data centers and edge computing. Finally, in continuity with Chap. 10, we consider the use of AI in 6G

G. Callebaut
Department of Electrical Engineering, KU Leuven, Ghent, Belgium
e-mail: gilles.callebaut@kuleuven.be

T. Dykes
Hewlett Packard Enterprise EMEA Research Lab, Wokingham, UK

O. Forceville
Hewlett Packard France SAS, Puteaux, France

C. Ghribi
Nokia, Massy, France

U.-U. Haus
Hewlett Packard Enterprise EMEA Research Lab, Meyrin, Switzerland

B. M. Khorsandi · L. C. Schmelz
Nokia Germany, Munich, Germany

W. Mao
Santa Clara, USA

M. Matinmikko-Blue
Centre for Wireless Communications (CWC), University of Oulu, Oulu, Finland
e-mail: marja.matinmikko-blue@oulu.fi

D. M. Bozzone
Hewlett Packard Italiana SRL, Milan, Italy
e-mail: davide.montagno-bozzone@hpe.com

H. Nikopour · O. Semiari · S.-P. Yeh
Intel Labs, Santa Clara, USA
e-mail: hosein.nikopour@intel.com

O. Semiari
e-mail: omid.semiari@intel.com

S.-P. Yeh
e-mail: shu-ping.yeh@intel.com

C. Sarathchandra
Interdigital, London, UK
e-mail: chathura.sarathchandra@interdigital.com

L. Wei
Quebec, Canada

networks, with its double role of enabling and enabled technology, with a specific focus on sustainability aspects.

Keywords Sustainability · Responsible artificial intelligence · Sustainable artificial intelligence · Artificial intelligence for sustainability · Artificial intelligence for networks · Networks for artificial intelligence · AI act

1 Introduction

This Chapter is intimately related to Chap. 10, whose focus is on sustainability in the context of 6G. In the present Chapter, we instead focus on Artificial Intelligence (AI), but with a final special focus on the use of the latter in the context of 6G. There are multiple reasons why these topics are inevitably and strongly interrelated.

First, in Chap. 10, we extensively discussed the dichotomic role of 6G as part of the problem (i.e., how to enable a *Sustainable 6G*), and as part of the solution (i.e., how 6G can help other sectors to be sustainable, or *6G Sustainable Applications*). AI shares the same role of sustainability-enabled and sustainability-enabling technology. We can refer to it as *Sustainability of AI* and *AI for Sustainability* (also referred to as AI4Good[1]) for the former and the latter, respectively [1]. The sustainability of AI aims at ensuring that the development, usage, and lifecycle of AI adhere to sustainable principles across all sustainability pillars: ***environmental, social, and economic***. AI for sustainability refers to the opportunities enabled by AI to promote sustainable development in other sectors, for example, thanks to process automation and human safety in industrial environments. Both aspects, as for 6G and as advocated in Chap. 10, cannot be treated separately or as optional, but rather in tandem to extrapolate first and second order effects jointly.[2] In [1], it is also argued how sustainable AI principles must be considered across the entire lifecycle, including training, tuning, deployment, maintenance, and decommissioning. This is illustrated in Fig. 1.

Second, AI plays a dichotomic role also in 6G: (i) ***AI for networks***, by which AI is used as a powerful tool to design, optimize, and self-configure networks, and (ii) ***networks for AI,*** by which wireless networks are used as an efficient platform to enable heterogeneous agents to share data over wireless and perform complex cooperative tasks. This comes with fundamental challenges on data representation techniques, to lighten the communication burden generated by different devices such as sensors and machines. These challenges will be discussed in Sect. 4.2, introducing the paradigm of semantic and goal-oriented communications [2].

Besides wireless networks' evolution, it is clear today that AI is becoming pervasive in almost all sectors, thus raising fundamental questions on its positive and negative impacts on the environment, society, and economy. With the advancement

[1] https://ai4good.org/

[2] See Chapter 10 and [30] for the definitions.

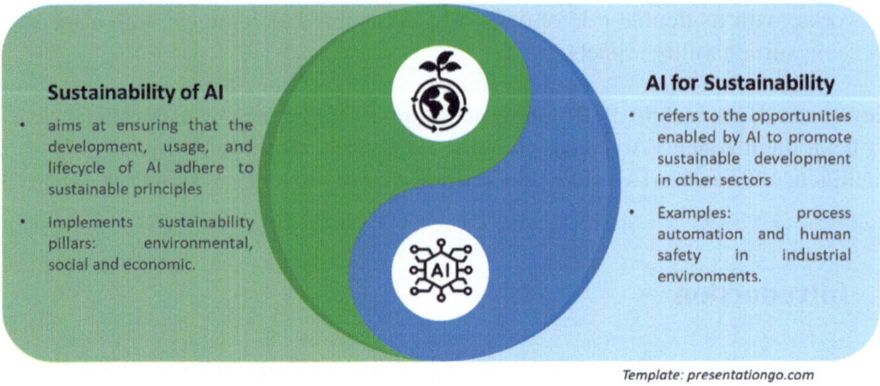

Fig. 1 Interconnections between artificial intelligence and sustainability

of Large Language Models (LLMs) [3], AI is becoming more and more a popular and widely adopted product that is not only used by highly specialized individuals but by entire businesses and citizens whose use differs considerably from one to the other. Despite the differences in using these tools, the common aspects of potentially sharing private and sensitive information represent a social sustainability issue to be carefully integrated into the lifecycle assessment.

Therefore, it is not coincidental that the European Commission recognized since 2019 the disruptiveness of AI technology with respect to ethics and social aspects and promoted primarily guidelines for ensuring trustworthiness [4], and later enacted the AI Act in 2024 [5], providing obligations to mitigate or address potential risks within the whole lifecycle of such technology. These obligations are further supported by existing and forthcoming harmonized standards [6] to ensure their wider implementation.

In this context, and toward the goals of considering AI and sustainability across the three pillars and with its double facet, in this chapter, we will first introduce the concept of *Responsible AI* (Sect. 2), also in reference to 6G systems. Then, we will focus on the sustainability of AI, from the perspective of its deployment, discussing benefits and drawbacks of data centers and edge computing deployments, from environmental and economic perspectives (Sect. 3). The introduction of these topics naturally leads to the use of AI in Information and Communication Technology (ICT), and more in particular in wireless networks, with the abovementioned dichotomic role. Although the 6G case is present in the whole Chapter (in relation to Chap. 10), we also dig into it as a potential use case and enabling technology for sustainable AI, discussing detailed implications of these technologies in Sect. 4. This covers AI used for Radio Access Networks (RANs) optimization and in (and for) semantic and goal-oriented communications, therefore in strong relation with Chap. 10. Further, in continuity with Chap. 10, we will discuss the main challenges and opportunities of AI in standardization (Sect. 5). Finally, we will draw the conclusions in Sect. 6.

2 Responsible Artificial Intelligence

Responsible AI refers to the design, development, and deployment of AI systems that prioritize sustainability across environmental, social, and economic pillars, while adhering to ethical standards and regulatory requirements. It ensures that AI technologies are environmentally friendly, ethical, transparent, accountable, and aligned with societal goals.

In the wake of AI's recent exponential growth, embedding responsibility into AI products and services has become increasingly challenging yet critical. As we transition into the 6G era, Responsible AI integration is a key challenge moving forward, especially as AI becomes a foundational component of the 6G architecture, both as enabler and enabled technology, as mentioned above.

For these reasons, a need is emerging to define common principles and guidelines to ensure the responsible development of this technology. It is not coincidental that the European Commission recognized since 2019 the disruptiveness of AI technology with respect to ethics and social aspects and promoted primarily guidelines for ensuring trustworthiness [4], and later enacted the EU Regulation 2024/1689 "AI Act" [7] in 2024, providing obligations to mitigate or address potential risks within the development, deployment, operation and maintenance of such technology. These two regulatory principles represent milestones that are triggering the process for a corpus of global AI regulations, which are yet fragmented and still evolving.

At the worldwide level, many similar initiatives are following the seed launched by Europe. Namely, we may remind the harmonized standards with AI Act [8] from ISO/IEC JTC1/SC42 and CEN/CLC JTC21, and the Governing AI for Humanity of the United Nations [9]. All of them offer valuable direction; however, there are initiatives that vary significantly across jurisdictions, leading to regulatory disparities. For instance, the U.S. approach is fragmented, relying on executive orders [10] and state laws [11], which may widen as 6G deployment accelerates worldwide, especially towards 2030. These guidelines set out three core pillars of trustworthy AI (i.e., lawful, ethical and robust) and seven key requirements that AI systems should meet to be considered trustworthy: **(1) human agency and oversight, (2) technical robustness and safety, (3) privacy data and data governance, (4) transparency, (5) diversity non-discrimination and fairness, (6) societal and environmental well-being, (7) accountability.** These requirements are illustrated in Fig. 2.

Several companies are adapting and aligning their vision to this common effort to establish principles for responsible and trustworthy AI. As an example, Nokia has articulated this vision through its Six Pillars of Responsible AI framework: **fairness, reliability and safety, privacy, transparency, environmental sustainability, and accountability** [12]. Further, Amazon and Ericsson are contributing to the definition of a standard "AI Trustworthiness Framework" in the CEN/CLC JTC21/WG4 [13], aligning with the seven key requirements defined in [4, 5]. This shows a clear movement of the 6G (or, more generally, the ICT) ecosystem with its industries, academies, and small and medium-sized enterprises towards integrating these values into technology development.

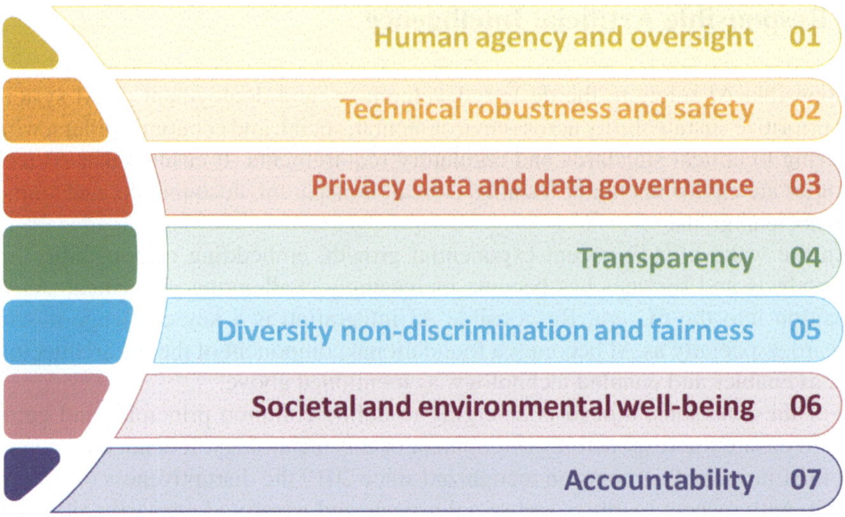

Fig. 2 Key requirements of AI for trustworthiness

These principles provide a structured approach to ensuring that AI technologies remain ethical, trustworthy, and aligned with environmental and societal goals.

Responsible AI can be explored through different angles -environmental, social, and economic—highlighting how the pillars guide ethical and sustainable AI in practice.

Environmental sustainability in AI focuses on minimizing the ecological second-order effects of AI systems throughout their lifecycle, from training and deployment to operation. With AI models growing exponentially in size and complexity, their energy and resource consumption are rising concerns. Organizations such as ISO/IEC's JTC 1/SC committee and the ITU-T are leading the development of standards targeting "green AI," aiming to measure and reduce AI's energy use, water consumption, and carbon emissions. For example, ISO/IEC TR 20226 [14] establishes frameworks for assessing AI environmental effects, while ITU-T [15] promotes energy-efficient algorithms and green computing standards for data centers and networks (more details will be provided in Sect. 3).

Within telecommunications, the 3rd Generation Partnership Project (3GPP) has several working groups addressing the use of AI and ML within the context of 5G and beyond. R20 is advancing AI-native architectures and sustainable communication as a part of early 6G preparations.[3] 3GPP SA1[4] Release 20 study on 6G use cases incorporates responsible AI considerations. Meanwhile, SA2 is advancing AI-native 6G architecture, ensuring sustainability is embedded as a core design principle.

Industry contributions, such as Nokia's ENVIAA framework [16], provide transparent environmental effect assessments by distinguishing AI training emissions

[3] https://www.3gpp.org/news-events/3gpp-news/ai-ml-2025.

[4] https://www.3gpp.org/3gpp-groups/service-system-aspects-sa/sa-wg1.

from hardware manufacturing, allowing targeted reduction strategies. Similarly, the Next G Alliance [17] emphasizes reducing embodied emissions, promoting renewable energy, and optimizing AI models using techniques like compression, pruning, quantization, and federated learning.

Social sustainability in AI entails fostering fairness, inclusivity, and positive societal effects. It encompasses protecting fundamental rights of users from biases in data and AI results, ensuring transparency and explainability of the AI behavior, and maintaining accountability throughout AI operations [18]. Frameworks such as ISO/IEC 42001 and UNESCO's Recommendation on the Ethics of AI provide guiding principles but currently lack standardized metrics or mature regulations to quantify social effects reliably [19]. This is the reason why CEN/CENELEC and ISO/IEC are involved in a joint effort with many working groups and projects to define harmonized standards before all the obligations of the AI Act enter into force (i.e., August 2026).

Core principles underpinning social sustainability include security, safety, fairness, transparency, explainability, and data quality, each vital to trustworthy and ethical AI use. Indeed, the AI Act is adopting a risk-based approach to preserve the centrality of human beings and their fundamental rights, building on top of seven key requirements, as outlined in [4], where the most paramount is arguably **Human Agency and Oversight**. This requirement acts as the foundation for the others by ensuring that AI systems are developed and used as a tool to **serve people**, upholding **human dignity, autonomy, and fundamental rights**.

Without effective human control, especially the capacity for human override and meaningful understanding of the AI's functions, all other safeguards (e.g., technical robustness, fairness, and privacy) could be undermined, leading to potential significant harm or loss of control over critical decisions. This human-centric focus is central to the entire EU approach to AI governance, especially when AI is applied to support decision-making for healthcare use cases. For healthcare applications, specific standards [20] are going to be defined following the challenges identified with respect to care ethics and bioethics principles.

Moreover, social sustainability faces critical challenges: existing guidelines often overlook the long-term societal shifts AI may provoke, such as changes in power dynamics, workforce displacement, and the erosion of human connections. All of them rely on communication, public engagement, social perception, and expectation of AI technology, leading to controversy and social acceptance.

Nevertheless, recent research highlights key social risks like bias and discrimination, privacy violations, and job displacement while suggesting mitigation strategies like ethical AI design, worker reskilling, AI literacy, and stronger regulation [21–23]. Trustworthy AI, with requirements such as technical robustness, fairness, and privacy, serves as a practical foundation to realize social sustainability in AI systems [4, 5, 24].

Within telecom ecosystems, these concerns manifest across different stakeholders: vendors, service providers, and end users. Efforts by 3GPP to incorporate trustworthy AI features into network standards continue, with more robust inclusion anticipated in forthcoming releases aligned with 6G's AI-native vision [25].

Economic sustainability in AI emphasizes its potential to enhance productivity, spur innovation, and drive equitable growth while mitigating risks such as job displacement and rising inequality. Though there is no universally accepted definition, frameworks from OECD [26] and the European Commission [27] underscore AI's role in fostering inclusive economic development.

Studies by the International Monetary Fund (IMF) [22], McKinsey [28], and the World Economic Forum [29] highlight AI's transformative effect on labor markets and economies globally, pointing out vulnerabilities among older and lower-skilled workers, while also noting opportunities for digitally advanced nations to lead AI adoption. Key metrics for economic sustainability include AI profitability and Return on Investment (ROI), adoption rates, cost-to-revenue ratios, job market effects, and investment growth. However, standardization of these indicators across industries remains challenging. To conclude, and towards Sect. 4, the integration of Responsible AI into 6G presents both unprecedented opportunities and challenges. This demand coordinated global action, anticipatory regulation, and a commitment to balancing innovation with sustainability and equity. Without this, disparities may deepen, and the transformative potential of 6G-powered AI could be limited.

For other details on Responsible AI, the interested reader can refer to [30].

Section key takeaways: Sustainability in the context of AI comes with the double facet of *Sustainable AI* and *AI for Sustainability*. These two aspects must be balanced to make the best use of this technology while mitigating its negative impacts across environmental, social, and economic pillars. With the European AI Act, many industries are adapting to fundamental principles that have the objective of ensuring an ethical use of AI from a social and societal perspective. A Responsible use of AI refers to these principles, as well as environmental and economic considerations

---

Teaser on the remainder: In Sect. 3, we first present an overview of the problem of Sustainable AI, particularly related to the processing burden it brings and the associated challenges in data centers deployment and operations, as well as edge processing typical of low-latency applications. Then, in continuity with Chap. 10, in Sect. 4, we analyze the potential use of AI for Sustainability, for the case of 6G. More specifically, Sect. 4.1 relates to the use of AI to optimize 6G RAN (AI for Sustainability and AI for Networks), while Sect. 4.2 focuses on the use of 6G as a sustainable platform to enable AI (Sustainable AI and networks for AI), with semantic and goal-oriented communications as an emerging paradigm in 6G

---

3 From Cloud to Edge Processing for Artificial Intelligence

One of the biggest challenges of AI is the computational burden that it brings. Although computing capacity has progressed greatly over the last decade, especially with new hardware solutions specifically adapted to AI workloads (e.g., Graphical, Tensor, and Neural Processing Units—GPUs, TPUs, NPUs), generative AI is exploding the need for more processing power. The latter requires new hardware (thus materials) and, during operations, a huge amount of energy.

3.1 AI and Sustainability in Data Centers

The Sustainability Impact of AI Scaling

The performance of AI models is strongly dependent on *scale*, where the scale consists of three primary factors: the number of model parameters, the size of the training dataset, and the amount of compute used for training [31]. As such, the ongoing race for AI supremacy and industrial competitive advantage has led to rapid and exponential growth in scale. It is currently estimated that the number of parameters of AI models has doubled every year since 2010, with current large language models reaching over 1 trillion parameters. Similarly, training datasets and required compute are estimated to double in size approximately every six months**, with current models reaching over 100 trillion training tokens** [32] **with the equivalent of thousands of tons of estimated carbon emissions during training** [33].

The increasing scale of AI models places new demands on the availability and capability of the data centers where model training and deployment predominantly occur. Global investment in data centers is projected to increase to USD 6.7 trillion by 2030, with the scale of both training and inference workloads contributing to infrastructure growth and motivated by factors including mass adoption of AI-based workloads, enterprise integration, industrial competition, and geopolitical priorities [34]. This investment has led to an explosion in demand for AI hardware infrastructure.

The commercial sector is at the forefront of data center scaling to support training frontier AI models, with systems such as xAI Colossus (est. 200,000 GPUs, 280 MW), Meta 100 k and OpenAI/Microsoft Goodyear (each est. 100 k GPUs, 143 MW). National infrastructures are following with supercomputing systems and AI Research Resource programs instituted in the US and the UK, AI factory programs funding data centers across Europe, and various similar programs found globally.

Such a growth in data center infrastructure has an inevitable impact on sustainability. The power requirements of leading AI supercomputers are doubling every 13 months [35], and in 2025, data centers accounted for roughly 1.5% of the world's electricity consumption and are expected to more than double by 2030. It is estimated that, on average, a 100 MW hyperscale data center in the US consumes ~ 2 m liters of water per day, equivalent to 6500 households, with global water consumption for data centers estimated at 560 billion liters per year and rising to 1200 billion liters by 2030 [36]. The two highest performing systems on the Top500 list from June 2025 consume a combined 54.2 MW of power. **This is equal to the energy consumption of around 224,497 single-family homes in Germany, each consuming 2105 kWh of electricity per year** [37]**.**

Sustainable Data Center Design, Construction, and Operation

Improving data center sustainability through design, construction, and operation is strongly motivated by multiple factors such as *regulatory pressure*, as governmental policy and international regulatory bodies impose stricter standards around energy, carbon, and water usage. *Cost reduction,* as metrics for sustainability are often tied to operational efficiency and correlates strongly with cost, while operational expense reduction for customers also acts as a competitive advantage for infrastructure providers. *Market demand*, as the previous two factors impact customers, alongside power and water availability and usage constraints, procurement criteria on efficiency and sustainability push vendors to innovate to compete.

The ISO/IEC 30134 standard, similarly the European EN50600-4, provides specifications and key performance indicators (KPIs) for the design, construction, and operation of data centers. These include various efficiency metric definitions, such as Power Usage Effectiveness (PUE), measuring the ratio of total facility energy to IT equipment energy, with similar metrics for water and carbon usage, as well as KPIs on energy reuse (ERF), renewable energy use (REF), and more. These KPIs support quantification of sustainability and are increasingly in use as mechanisms to apply regulatory and procurement pressure on data centers. The International Energy Agency report on Energy and AI highlights the evolution of global policy on data center energy efficiency, showing national strategies primarily incorporating PUE mandates in the near-term 1.2–1.5 region [36].

Location has a strong impact when designing a sustainable data center, where factors such as close proximity to clean energy and availability of brownfield site building can greatly improve KPI metrics such as REF or metrics on material reuse and embodied carbon. The LUMI data centre in Kajaani, Finland, is an excellent example of such, with a PUE of 1.04, or 1.24 with excess heat utilization due to the use of heat pumps to pass heat to the local district heating networks. This highlights that PUE does not tell the whole story, as the excess heat utilization raises PUE, however offsets CO_2 emission and supports energy reuse, providing 20% of the heating requirements for the local town in the LUMI case [38]. Hence, the importance of utilizing a wider scope of KPIs, including those such as the energy reuse factor.

Alternate approaches to data center construction can further improve sustainability. Modular data centers (MDC) are factory-integrated standardized, shipping-sized container deployments that integrate IT, power, and cooling. An MDC may be constructed wherever an appropriate supply of power and water is available in a matter of weeks. Such deployments reduce the carbon footprint of construction, while also enabling high material reuse, with the MDC typically repurposed after 5–10 years of operation. The UK national AI research infrastructure, Isambard AI, was constructed in such a fashion and estimates 90% material recyclability for the entire data center [39].

The overconsumption of power is a major concern in the design, procurement, and operation of large high-performance computing (HPC) systems [40, 41], especially in the Exascale era. **Concerns over the total cost of ownership (TCO) have moved the focus from peak computational performance, in terms of total FLOPS, to improving power and energy efficiency of the HPC systems and data centers** [42]. As such, beyond the initial design and construction, ongoing sustainable data center operation is an active topic of research. For example, to help reduce the amount of energy used by a cluster during its operational period, we can tune the hardware setup for specific objectives during the application's execution. These objectives include minimizing the energy to solution (EtS), using the least amount of power, minimizing time to solution (TtS), and any combined trade-off metric. The EtS deals with energy savings, which have a direct impact on the costs of running HPC systems. Power reduction strives to keep power demand below a given limit in order to avoid penalties imposed when power demand spikes or falls below a certain limit defined by the power company. EtS and TtS trade-off optimization takes into account the costs associated with time and thus addresses the productivity of data centers. An example can be seen in the UK national supercomputing service, Archer2, case study, where an average of 21% power reduction was achieved by introducing low-overhead, system-wide default settings for BIOS and CPU frequency [43].

In addition, the use of a dynamic power management tool such as PowerSched [44] employed at the top-tier German national supercomputing centre HLRS enables the system to balance performance and energy efficiency by dynamic power capping. That is achieved by controlling the amount of power sent to each node based on the dynamic performance characteristics of the running jobs. So that nodes running jobs that require more power to attain optimal performance receive more power, and jobs that require less power to achieve optimal performance receive less power. This has been shown to **reduce up to 20% in energy consumption** in experiments at HLRS without degrading the performance of the running jobs. Thus, reducing CO_2 emissions and operational costs, which results in a more sustainable system in terms of environmental and economic pillars. PowerSched is employed at HLRS for overprovisioning management to ensure that the desired power budget of the system is enforced, and that the system remains stable.

Beyond system configuration and management software-based approaches, operational sustainability efforts are also underway using advanced data analytics. Digital Twins are an active area of research, both for improving operational efficiency for the systems running AI workloads and for using AI-based analytics to do so [45]. Data center digital twins have been shown to be beneficial for energy efficiency studies [46], while allowing operators to build a holistic view of the liquid-cooled supercomputing data centers that are required for large-scale AI. Whether as a provider of infrastructure for AI models to support 6G network analysis, or indeed as a user of 6G technology, data center operators could also benefit from **6G-focused digital twin use cases, as surveyed in** [47]**, from an operational efficiency and sustainability perspective**.

Data Movement as a Cross-Cutting Driver of Resource Efficiency in Data Centers

One of the cross-cutting concerns is the impact of data movement on energy consumption. In the data center, this is common wisdom at multiple levels: In a classic data center not moving data, or using intelligent storage tiering, improves responsiveness and performance, reducing idle times; in supercomputers, good computational code is only limited by the unavoidable I/O, i.e., data layout is optimized for the application, and storage I/O is overlapped with other computation. But data movement is in fact a bigger factor: normalizing to total worldwide installed data center energy usage, **workload doubles every 3 years (indicating better efficiency), but interconnectivity traffic increases at double that factor**, as shown in the IEA reports

[48], and interconnectivity energy cost thus requires special attention in overall IT industry sustainability metrics. With federated infrastructures becoming increasingly common—for organizational, resilience, and political reasons—this problem will only increase. 6G networks may, in fact, play a favorable role in increasing efficiency here.

At the same time, data centers can be considered as industrial plants—and in the spirit of Industry 4.0 plants, can utilize **6G campus networks**—from coordinating automated hardware management (robotic automation of component replacement, monitoring, operational support of on-site personnel), to site security surveillance, and finally out-of-band customer and workload observation. In this context, AI models of the respective component behavior, network behavior, and data movement consequences are critical. This includes the full gamut of usage of LLMs for human-focused AI usage, as well as data center local retraining or specialization of generic AI models. The fact that 6G networks will in general be providing a second, independent, layer of networking, is useful in resilient data center management, even in extreme situations, where no spare capacity in the managed resources would be available.

Finally, 6G edge AI inference enables ultra-low latency, real-time applications such as autonomous driving, logistics routing, and emergency service routing based on dynamic traffic shaping based on city-scale sensing. This allows moving computation closer to the data source, processing it on devices or local servers instead of distant data centers, supported by techniques like distributed and split inference. This approach reduces latency, improves privacy by processing sensitive data locally, and increases efficiency by minimizing data transfers, as discussed in the next Section.

3.2 Edge Computing

Edge computing brings computation, storage, and artificial intelligence (AI) close to the user devices and terminals at which the data is created. Rather than being in a centralized data center with all the challenges and opportunities described above, edge computing takes advantage of distributed nodes closer to the data source, for example, in gNB [49], eNB [50], (for 4G–5G continuity), and local data centers. These distributed nodes run virtualized network functions (VNF) [51] and applications on ETSI MEC Platforms [52], which will be standardized and integrated into the 5G Core (5GC) [53]. The proximity to the device enables lower latencies, reduced traffic back to the Core, and more secure handling of sensitive information, and opens new real-time services for application in areas such as autonomous driving, collaborative robotics, telemedicine, and extended reality (XR) [54]. For this reason, edge computing is a key element of 5G networks and the expected 6G future (cf. Chap. 10). These distributed endpoints eventually create a cognitive network that authenticates and coordinates connectivity, computing, and AI. As networks become more complex as services become based on performance, there are other complexities around latency, data traffic, and security that will need to be managed.

A main goal for 5G and 6G networks is to support ubiquitous applications (low latency, high trust), through a variety of use cases. For example, applications with a trust layer for connected mobility (V2X) [54], industrial robotics, telemedicine, and XR experiences, generate large amounts of data with limited possibility for real-time, centralized transfer, due to bandwidth and energy costs. Moreover, some of the data includes some sensitive information that needs to be protected due to privacy and security concerns, and this can happen in centralized applications. Traditional centralized (pure) systems for processing cannot solve this on their own or, if they can, are cumbersome. To help with the problems identified in the previous paragraphs and in Chap. 10. 5G applications support edge processing alongside AI and analytics that will allow for local processing, improved and faster decisions, and predictive handling of data traffic in the network.

In 5G, the 5GC facilitates edge logic using a Service-Based Architecture [53] and cloud-native principles, enabling functions to be placed on geographically separated nodes. For example, the User Plane Function (UPF) [53] can be repositioned and function under the RAN [49] so that it can handle traffic that breaks out locally instead of always going back to the core. There are control functions such as Session Management Function (SMF) [53] and, in multi-site architectures, the Access and Mobility Management Function (AMF) [53], that can also be placed locally and require less signaling latency. MEC platforms allow applications to reside connected to the 5GC to the network, in which applications can use the Network Exposure Function (NEF) [53] to connect to the network and the Application Function (AF) [53], through which applications can consume network data. The lifecycle of edge functions is facilitated by NFV MANO [51] or O-RAN SMO orchestrators [55]. Network slicing creates a separation of traffic through functional layers to isolate and prioritize traffic-related considerations (performance/data priority/ latency). The Network Data Analytics Function (NWDAF) [56] could also be hosted on the edge, to collect, analyze and process live network metrics (including energy) to assist with localized predictive decisions to drive improved service quality and efficient network operations. In a V2X use case, for example, if there is an accident on the highway, UPF and MEC nodes can process the live data immediately for analysis and consumption, and relay alerts in reduced timing than if the data were processed centrally. Each of these elements advanced by the edge facilitates 5G edge architecture to address current issues while preparing networks for next-generation capabilities, especially including environmental, social, and economic sustainability aspects.

Looking forward, 6G will pursue this strategy with a complete cloud–edge–device continuum. The edge will only become more cognitive and self-adaptive with tasks such as local AI inference, federated learning with collaborating nodes, as well as a working layer for dynamic orchestration while considering some of the factors like latency, energy efficiency, and sustainability. Computational tasks will more than likely be allocated dynamically across core–edge-devices as tasks emerge. Future technologies may also include neuromorphic hardware or edge-aware computing. Finally, with 6G, AI will become integrated into the control and management of the network, allowing network functions and processes to dynamically and effectively move among core, edge, and devices to create ultra-reliable, autonomous, and sustainable services. Standardization, as well as industrial and collaborative projects like 6G-IA [57], 6Green [58], and SUSTAIN-6G [59, 60], etc., will define guidelines and tools that optimize energy consumption, lower environmental impact, and promote conscientious use of computing resources. This, accompanied by a responsible use of AI, as introduced in Sect. 2. Nevertheless, with these changes, energy efficiency management, energy consumed by AI workloads, and privacy will continue to represent a critical challenge in edge computing, AI, and 6G.

AI processing locally at the edge also advances privacy and sustainability. Local processing avoids having to send the data that includes all the sensitive information to a central core or cloud servers and subsequently eliminates the privacy issues while processing and using that data. In actual V2X use cases, NWDAF will be local to the edge and identify if congestion or another anomaly is occurring, with respect to a vehicle flow pattern, allowing traffic management through UPF and user transports, and alerting nearby vehicles of needed actions through local guidance MEC applications. Besides privacy and acknowledgement issues, this scenario of local protocol presents all the challenges previously discussed about latency in performance, consumption of energy, and an increase in overall reliability of the network. This represents a typical example of second-order effects (as introduced in Chap. 10), in which the technology (in this case, edge computing, AI processing, and 6G) enables sustainability aspects in other sectors (or verticals).

Using this kind of application, predictions and AI inference can use the data and applications that are enabled through the edge while preserving application performance and securing sensitive data in real-time and at the edge for location. This technique of combining edge analytics with AI localization and analytics provides solutions to many of the modern networks' performance challenges, energy challenges, and security challenges.

In summary, edge computing, AI, and data analytics technologies, including the NWDAF, will be important technologies, also in driving intelligent and efficient networks in future 6G and 5G networks. The potential advantages of edge computing significantly impact data latency, data and privacy security, and network resiliency, among others. Networks will be able to leverage predictive analytics, and the use of local AI will empower networks to act efficiently and intelligently, as well as provide significant assistance with critical real-time actions related to applications. In conjunction with sustainability-related research and initiatives (e.g., 6G-IA, 6Green, SUSTAIN-6G), these technologies will create a foundation for a trusted high-performance network, allowing AI to be deployed sustainably, in an environmentally socially responsible way, and service reliable more sustainable digital ecosystems more equitable globally that are working to enhance the complexities of real-world balance against the progressive need for complex services for future networks.

4 A Usage Scenario: In-Network Artificial Intelligence with 6G

Radio access networks (RANs) and core networks are the two fundamental components of mobile networks, with the former representing today the highest source of energy consumption, as discussed in Chap. 10. Core networks are still to be optimized to minimize their negative effects. However, these apparently disjoint segments can become strongly interrelated in the case of other technology enablers, including edge computing, among others.[5] In Sect. 2, we discussed the main consequences of the choice of AI-based workload placement, spanning central cloud data centers and edge facilities. From an environmental perspective, central clouds still represent a great solution thanks to resource sharing among others. However, edge facilities guarantee low latency and less risk of data leakage, among others.

With wireless networks and edge computing convergence, end-to-end enablers are more and more bonding the two worlds, especially due to the great advancements of AI, both as a powerful enabler of smarter decisions in complex environments, and as part of the problem with the possibly increased use of resources and data-hungry applications, as discussed in Sect. 1.

We separate the discussion between the two dichotomic aspects of AI: **AI for Networks** (i.e., AI as part of the solution in 6G) and **Networks for AI** (i.e., AI as part of the problem in 6G). For the former (Sect. 4.1), we present some recent advances in using AI for optimizing different aspects of the RAN, while for the latter (Sect. 4.2), we consider different cross-layer aspects within the domain of semantic and goal-oriented communication. This double facet of AI runs in parallel to the double role of 6G as part of the problem (designing a Sustainable 6G) and part of the solution (6G for Sustainable Applications) introduced in Chap. 10.

4.1 Artificial Intelligence for Networks

The telecommunications industry is increasingly adopting machine learning (ML) to optimize next-generation RANs due to their growing complexity. Modern RAN systems support heterogeneous network nodes, diverse services, multi-carrier communications, large antenna arrays, and dense cell deployments, leading to high energy consumption, despite extremely optimized energy efficiency, which has improved by at least an order of magnitude at every generation [61]. This necessitates advanced network energy saving (NES) solutions to reduce operating expenses (OPEX) and enable large-scale wireless deployments. Emerging use cases, such as industrial automation, automotive (V2X), and edge/cloud services (e.g., 4 K/8 K video streaming, virtual/augmented reality), impose strict quality-of-service (QoS) requirements, including high data rates, low per-packet delay, and high reliability.

[5] https://www.etsi.org/technologies/multi-access-edge-computing.

Current RAN implementations rely on simple rule-based policies for QoS management, focusing on maximizing coverage and throughput with limited emphasis on NES. ML-based solutions can address complex multi-objective RAN challenges, including NES, resource scheduling, network slicing, interference mitigation, load balancing, and mobility management, to optimize QoS under bandwidth and energy constraints. The 6G architecture can integrate optimized software stacks to leverage ML, pushing RAN performance to theoretical limits and enabling data-driven ML tailored for specific verticals such as industrial automation.

One of the primary goals of 6G research is to develop a Hierarchical Multi-agent RAN Management framework that integrates wireless domain knowledge with ML tools, such as reinforcement learning (RL), graph neural networks (GNN), recurrent neural networks (RNN), transformers, and generative models, to address complex RAN management challenges. This framework aims to: (1) break down high-level RAN tasks into simpler sub-tasks, (2) apply suitable ML tools to solve these subtasks, (3) proactively detect and mitigate policy conflicts by ensuring robust design. As an example, shown in Fig. 3—RAN optimization via HVI and HHI frameworks., the RAN optimization problem is divided into three sub-tasks: (A) Network energy saving (NES), (B) Connection management, and (C) RAN slicing, which collaboratively optimize energy consumption, QoS, and service level agreements (SLAs) for specific verticals.

The framework employs two integration levels:

- **Vertical Integration (HVI)**: An intra-task multi-agent ML approach where agents align on objectives with minimal conflict risk. For example, NES task uses multiple ML agents for cell on/off, cell discontinuous transmission/discontinuous reception (DTX/DRX), and multiple-input multiple output (MIMO) antenna reconfiguration, operating on different time scales and node impacts, trained jointly to achieve NES goals.
- **Horizontal Integration (HHI)**: An inter-task multi-agent ML framework optimizing RAN performance across potentially conflicting tasks, such as NES and connection management. For instance, cell on/off policies require load balancing

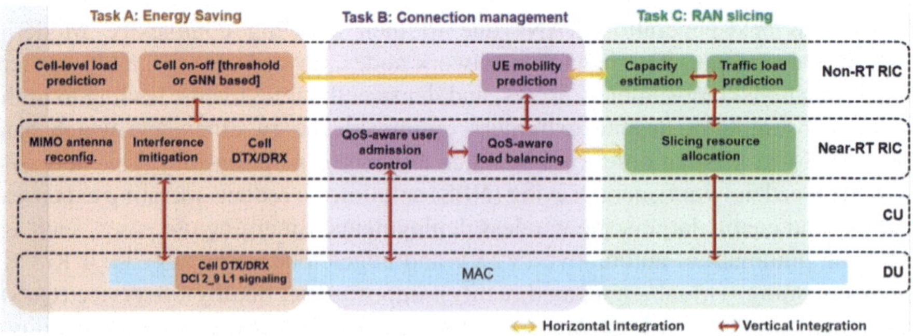

Fig. 3 RAN optimization via HVI and HHI frameworks

to reassign users, while load balancing can maximize cell shutdown opportunities. HHI minimizes redundant ML workloads by using common data embeddings across tasks.

Both HVI and HHI leverage temporal and functional hierarchies in 6G architecture, incorporating a design loop for joint ML model training and mechanisms to resolve policy conflicts and robustness issues.

Solutions for Network Energy Saving Task

Energy saving in RAN is vital for reducing operating costs, meeting environmental requirements, and ensuring service level agreements. For a major operator, a 15% RAN energy reduction could save $165 M annually [62]. In remote areas with limited energy, NES ensures service continuity where constant monitoring of remote infrastructure is critical. A hierarchical framework enables a comprehensive NES solution, focusing on three domains: frequency, time, and spatial.

ML-based Cell on/off switching: As introduced in Chap. 10, cell on/off switching is one of the most promising solutions to reduce energy consumption while maintaining acceptable QoS and/or QoE. In this sense, AI/ML can certainly help managing the optimization of transitions between sleep and active states. Existing ML methods, like threshold-based approaches that switch cells on/off based on bandwidth utilization, often overlook temporal and spatial load dependencies. Switching off a cell requires transferring its traffic to neighboring cells, risking overload in dense deployments with QoS-sensitive traffic, necessitating methods that account for these dependencies.

A novel Graph Reinforcement Learning (GRL) approach addresses this by integrating GNN and RL. GRL models RAN states as graphs, with cells as nodes and logical links as edges, using GNN for scalable, permutation-invariant processing of RAN data. The cell on/off switching problem is formulated as a Markov Decision Process (MDP), with a Deep Q-learning (DQN) algorithm using a GNN-based Q-network. Inputs include cell features like user count and bandwidth utilization. In a small-scale scenario with three cell sites, the GRL method improved energy efficiency by 30% compared to threshold-based policies, as shown in Fig. 4.

To adapt the design for a specific vertical, data traffic dynamics in the vertical's RAN must be analyzed to identify load distribution and energy-saving opportunities. Load dependencies among nearby cells and the frequency of cell on/off switching are determined based on this analysis. As an extension, the known locations of devices such as sensors and robots can be utilized with RNN load predictors to enhance GNN design. Domain-specific safeguards for coverage, throughput, and energy can be implemented by masking invalid actions, such as cell switch-offs causing coverage holes, during GRL training to ensure safe exploration.

Cell DTX/DRX: Cell DTX/DRX mechanism is introduced in 3GPP Rel-18 as an important NES feature. As shown in Fig. 5, within one cell DTX/DRX periodic cycle, there are cell transmission/reception active periods and non-active periods. The cell operates normally during the active period, but there is no data transmission/reception together with corresponding Layer 1 (L1) signals/channels during non-active period,

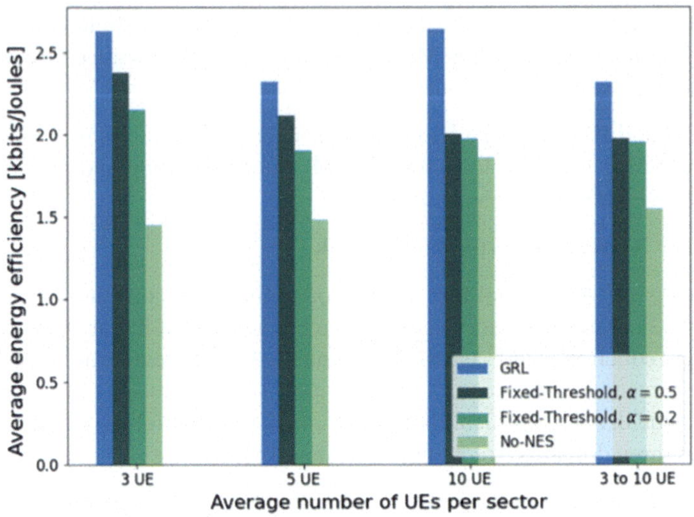

Fig. 4 Cell on/off switching performance

allowing the activation of advanced sleep modes (SM), which amounts to more power saving.

Compared with cell on/off switching, which operates at a relatively large time scale (on order of at least tens of minutes) when the load of a cell is low, cell DTX/DRX operates at a much smaller time scale, e.g., from symbol-level to up to hundreds of milliseconds (ms). Besides, no UE handover is needed for cell DTX/DRX as the cell is still maintained with corresponding procedures. Hence, cell DTX/DRX provides an opportunity for energy saving at such timescales if completely switching off a cell is not possible. It has been observed that cell DTX/DRX can achieve 47.8% ~ 71.3% energy-saving gain under the different system loads [63]. With the decrease in system load, more NES gain is achieved.

Depending on how long the cell is inactive (during cell DTX/DRX non-active period), the cell can shut off/deactivate certain parts, or even most parts of its hardware components, to enable different SMs and save energy. 3GPP defines three different levels of SMs and provides an example of the relative power consumption, as well as the total transition (entering and leaving) time needed for each sleeping mode [64].

Fig. 5 Cell DTX/DRX illustration

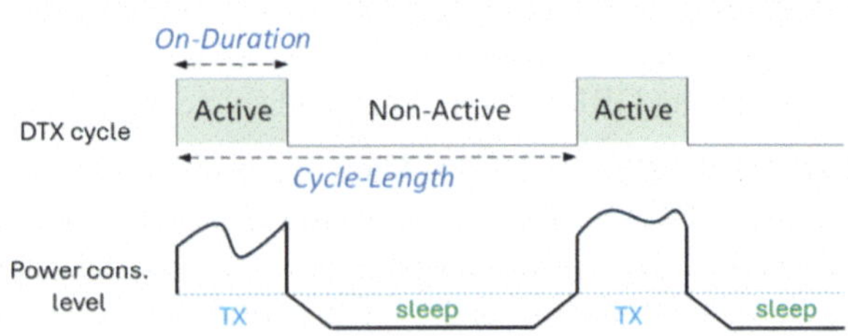

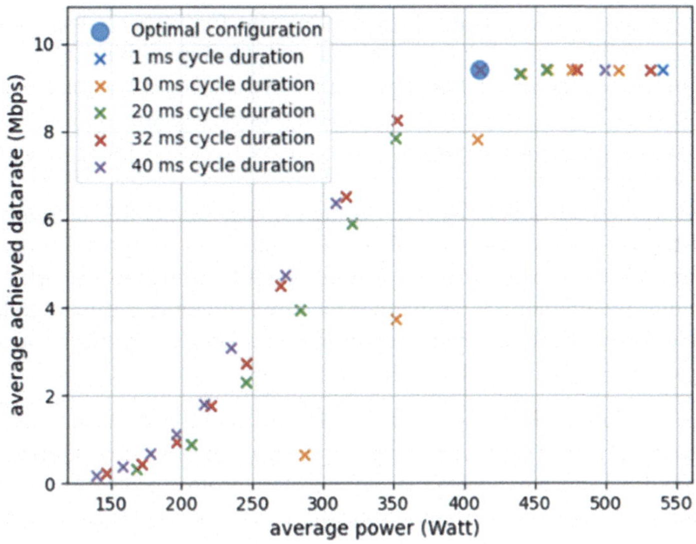

Fig. 6 Cell DTX/DRX, impact of cycle length and duration [65]

Cell DTX/DRX delays packets arriving during non-active periods, with delay increasing based on the non-active duration, creating a trade-off between QoS and energy savings for delay-sensitive traffic. Long cycle lengths or short on-durations may hinder timely packet delivery, while short cycles or long on-durations limit deep sleep, reducing energy efficiency. Figure 6 shows optimal DTX/DRX parameters maximizing energy savings while meeting QoS, i.e., delivering packets within delay constraints without data rate loss. These parameters vary across cells due to differing traffic loads and delay requirements, making optimization complex.

RL methods, including contextual bandits (CB) and Deep Q-Network (DQN), train an AI agent to select optimal cell DTX/DRX configurations. The agent monitors traffic and transmission status every ~ 1 s, applies a configuration, and uses performance and power data to compute a reward for training its Q-network. Configurations are updated periodically based on traffic load changes. Simulations on small-scale, low-interference networks show up to 33% energy savings with less than 1% data rate degradation compared to no cell DTX/DRX, outperforming fixed configuration policies.

For industrial verticals like manufacturing, high-speed, low-latency networks are critical. To enable efficient cell DTX/DRX in dense networks with high inter-cell interference, the design, originally suited for small, low-interference networks, incorporates a *start-offset* parameter in the RL scheme to add a time offset to cell DTX/DRX cycles, improving performance by reducing interference. Initial random offset results show RAN performance gains, with potential for further energy savings via AI-coordinated policies.

QoS Management Task: Private RAN deployment for verticals faces challenges in managing stringent mixed data traffic and high QoS sensitivity. QoS violations in private networks, unlike public ones, can have severe consequences, such as disrupting URLLC-dependent remote control or robot navigation, or impacting guaranteed bit rate (GBR) for surveillance camera feeds. Thus, QoS management is critical for vertical applications.

The HHI framework addresses QoS through two tasks: (1) RAN slicing, orchestrating end-to-end resources to meet SLAs, forming slices for typical vertical traffic demands, and (2) QoS-aware load balancing, handling dynamic scenarios like cell congestion during emergencies or maintenance by offloading traffic to underloaded cells. A multi-agent ML approach leverages RAN architecture, with slicing focusing on long-term patterns and load balancing addressing short-term dynamics.

QoS-aware load balancing: Supporting QoS traffic requires provisioning resources and managing cell traffic to prevent congestion, ensuring cells meet QoS constraints for assigned UEs. Load balancing (LB) offloads UEs from congested to underloaded cells, freeing resources without compromising performance. However, QoS-aware LB in vertical deployments is complex due to mixed QoS requirements, fading channels, interference, UE mobility, and traffic variations. Existing LB solutions, including model-based and AI/ML-based approaches, often focus on cell-level data, neglecting per-UE QoS constraints. Additionally, deep neural networks in prior work lack scalability, as fixed-sized inputs fail to generalize across diverse RAN topologies.

A multi-agent ML solution (Fig. 7) for QoS-aware load balancing includes two agents [66]: (1) A CB agent, called the action masking (AM) ML agent, predicts QoS before UE handovers, excluding actions that risk cell overloads, related to QoS-aware admission control [67], (2) A load balancing module, modeled as a GRL agent, uses GNN and RL to represent the RAN as a graph, framing load balancing as a sequential decision-making process that adjusts UE-to-cell edges.

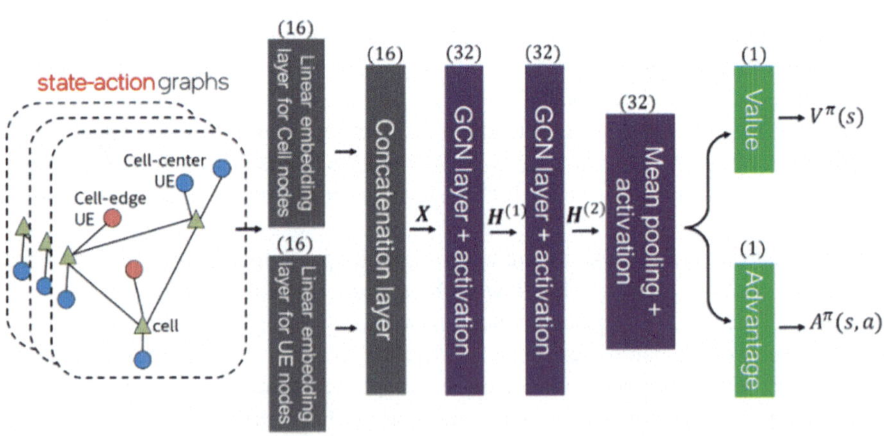

Fig. 7 Multi-agent QoS-aware load balancer

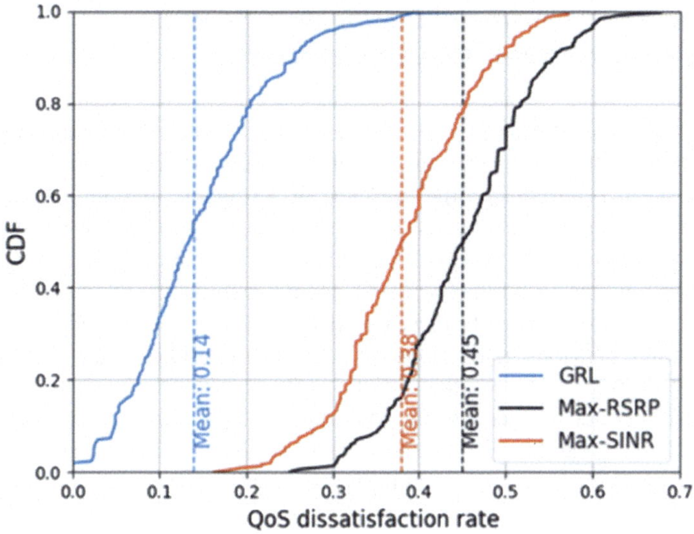

Fig. 8 QoS-aware load balancing performance

An LB decision to offload a cell-edge UE involves choosing between state-action graphs differing only in UE cell association. This dual-agent HVI design, where the AM agent shapes the GRL agent's action space, enables faster training convergence and reduced inference overhead.

The study applies a framework to optimize QoS-aware LB for mixed traffic, including GBR and best-effort (BE) downlink transmissions. Figure 8 illustrates the GNN-based Q-network, which takes graphs representing LB actions as input and outputs state-action values. The GRL method's performance is compared to max-SINR and max-RSRP baselines, with Fig. 8 showing the QoS dissatisfaction rate for GBR traffic (ratio of UEs meeting QoS to total QoS UEs).

The design can be extended to various QoS traffic types (non-GBR, GBR, URLLC) using an HVI framework integrating a CB action masking agent and a GRL-based LB agent, considering RAN slicing constraints. A joint training algorithm for AM and LB agents incorporates reward signals for QoS satisfaction, throughput, and coverage. HHI integration combines LB with RAN slicing via hierarchical RL, where the slicing agent sets goals for the LB agent, adjusting resources based on performance feedback.

Overall, several AI-based methods can be leveraged to reduce the environmental and economic cost of radio access networks, while not sacrificing QoS and QoE (thus potentially related to societal aspects) below acceptable levels.

4.2 Networks for Artificial Intelligence, Semantic, and Goal-Oriented Communications

As highlighted from different angles in Sect. 4.1, AI is a fundamental technical enabler for more flexible, zero-touch, and lean network management. However, in-network AI also pertains to the presence of AI agents making use of network resources to exchange data towards cooperation to perform complex tasks, e.g., in industrial environment (i.e., Networks for AI). This, coupled with the surge of traffic generated by humans to interact with AI agents such as Large Language Models (LLMs) [3], might dramatically increase AI-related traffic, especially in the uplink direction, unlike previous generations, or for side-link and device-to-device (D2D) communication (e.g., collaborative robots—cobots [68]). Projections suggest that AI-related (direct and indirect) traffic may reach 1000 exabytes/month globally by 2033 [69]. This extends the role of in-network AI, which becomes part of the problem, besides being part of the solution, similarly to 6G in the context of sustainability (cf. Chap. 10).

To avoid network overwhelming, especially of RANs (cf. Section 4.1), classical countermeasures such as network densification, higher frequency, and efficient antenna technologies can be employed. These would certainly lead to capital expenditure for operators and potentially more hardware deployment. The main issue is that networks and applications typically follow a "***separation of concerns***" paradigm, as defined in [70]. This principle is based on a clear distinction between network and application roles. From one perspective, this paradigm is very convenient as it allows operators to optimize their networks purely based on network-centric metrics, which are more easily measurable and controllable. Also, they do not need to expose sensitive operational data. On the one hand, application providers do not expose sensitive data, and they do not need to care about network-centric metrics. Their only concern is for their application to work. Despite these positive aspects, this principle also represents a bottleneck for network and application optimization. If both actors could expose relevant information, more informed decisions could be taken. From the perspective of the application, **semantic and goal-oriented communication** is recently gaining great attention as a promising paradigm to conceive communication not only based on bit-level reliability, but rather by considering the meaning (semantics) and the effect (goal/task-oriented) that the conveyed information brings [2, 71]. This requires an *understanding-before-transmit* paradigm, also based on the a priori knowledge between the transmitter and the receiver.

If a transmitter is aware of the AI capabilities and the background knowledge of the receiver(s), the amount of information and the needed quality can potentially be greatly reduced, with (possibly degraded) communication achieving the same desired effect [72]. If next-generation wireless systems open the way for bridging the gap between networks and applications, a huge amount of unnecessary data transmissions can potentially be avoided, thus reducing the traffic explosion, especially in the case of AI agents communicating and collaborating across the network. Obviously, there exists a plethora of dark sides and drawbacks to this paradigm. The most trivial one to be considered is the **trade-off between communication and computation**. As a fact, the semantic and goal-oriented paradigm is based on (local) relevant information extraction, usually based on deep neural networks that require hardware and energy resources to run. It is fundamental to analyze and consequently optimize the balance between communication and computation resource consumption, to not fall into the trap of greatly optimizing a system (e.g., 6G), while dramatically increasing the effect of other segments, e.g., (far) edge computing. In this direction, recent advances in edge AI [73] can help keep this powerful technology as frugal as possible, thanks to the processing of AI-related workloads either in edge computing facilities (cf. Section 3.2) or directly by end devices such as resource-poor sensors and robots. Further, recent advances in implementing feature extractors in circuits exhibit huge energy savings [74]. Some of the main challenges and trade-offs in this context are discussed in what follows.

Implementation challenges for semantic and goal-oriented communication

Moving to an AI-native based communication introducing the semantic and goals-oriented approaches [2] is tightly linked to the evolution of the embedding AI processing technologies, particularly those dedicated to far-edge equipment (cf. Section 3.2). As most of the semantic source and channel encoders use Deep Learning (DL) models like Convolutional Neural Networks (CNN) or Vision Transformers (ViT) [75, 76], one major technical enabler for their adoption is definitely the capability to optimize their resource usage and so energy consumption during inference phase. Ongoing improvement of techniques like DL model pruning in a static way [76–79], token selection in a more dynamic manner [80] and mixing both practices already proves this resource usage optimization capability when deploying models on various categories of devices. Some existing small and tiny semantic compression models are already trained to fit with the energy constraint of inference execution at the far edge when performance like throughput, latency, energy storage, or refueling capacity are the most constrained. Moreover, the rapid improvement of AI accelerators and chipsets offering adequate energy efficiency as much as latency and throughput performance will permit the adoption of more advanced AI-native encoders trained for larger media content and high-speed transmission while guaranteeing their adaptation to the device's low energy constraints. Currently, for most devices, inference execution and optimization frameworks, benchmarks using semantic encoders like ones based on CNN or ViT model architectures without proper optimization can highlight not negligible energy consumption when compressing large media content compared to the energy cost induced by the compressed data transmission over the air on a regular transmission technology (LTE or 5G).

This statement is particularly observed when deploying inferences on existing high-performance Graphics Processing Unit (GPU) or Central Processing Unit (CPU) models. Recent benchmarks on lower-power AI accelerators on new Neural Processing Unit (NPUs) [81] demonstrate better energy efficiency for inference execution, but still with limitations in terms of throughput and latency.

Moving to model specialized chipsets or Field-Programmable Gate Arrays (FPGAs) is another alternative to solve this energy balance aspect but tends to restrict the use of semantic encoders to very specific tasks and communication goals. However, this also strongly depends on several factors, including radio equipment. As an example, discontinuous transmission and reception can help further reduce energy consumption thanks to data compression, with the opportunity to more quickly transmit inference data and then enter low-power sleep modes [82].

This trade-off between a wide spectrum of use and tailored energy efficiency by the device running the inference is a key point to perfectly reach the promises of semantic and goals-oriented communication as a replacement for the standard way of compression and transmission for a wide variety of use cases. Another direction to address a larger spectrum of communication goals is to multiply the number of models loaded on the device dedicated to multiple usage or to respond to a large diversity of transported data semantic representations. This requires flexible mechanisms for models' selection and operation that on-device "pre-coded model" solutions cannot easily allow. This also implies smart recommendation mechanisms to assist the model's selection based on constant user data semantic evaluation, but also on foreseeable impact on the semantic communication quality of service, radio resources usage, latency, throughput, and consequently energy efficiency. Those impacts are correlated with the evolutionary conditions of the radio link, requiring near or real-time mechanisms and so adequate protocols and architecture to support them in a guaranteed time. Different ways to address those adaptative semantic compression models' selection and settings, as well as radio access and resources allocation are currently explored [83].

However, as already mentioned, adaptive model selection requires the devices to store multiple models, thus incurring costs in terms of storage resources that could be instead dedicated to other services and applications. In this case, a multi-dimensional trade-off between flexibility, communication, computation costs, and storage resources arises.

Modeling the semantic encoder/decoder inference execution and radio transmission resource's usage to apply numerical optimization is one feasible way to prove already relevant results. Alternatively, adopting a data-driven approach following analytics-based closed-loop principles is another interesting solution as soon as all data collection, machine learning framework, decision-making capacity, and protocols for its activation are in place on both user equipment and radio access network components. This research leads to a new architecture proposal able to integrate all these features into existing frameworks, such as the one proposed by Open Radio Access Network (O-RAN) [84] and as insights for future 6G standardization. Figure 9 summarizes the main inputs, optimization techniques, and outputs for an overall energy saving applied to a semantic and goals-oriented communication system.

A third alternative to amplify the energy savings offered by the semantic and goal-oriented communication paradigm consists of mixing model-based optimization and data-driven techniques to mitigate respectively models' approximation issue of the first approach, lack or inaccuracy of collected data for the second one. To go further into an "energy-aware" approach, enabling the capability on cellular systems to measure and report the energy consumption of the diverse components and processing parts all along the semantic and goal-oriented communication path can be consider

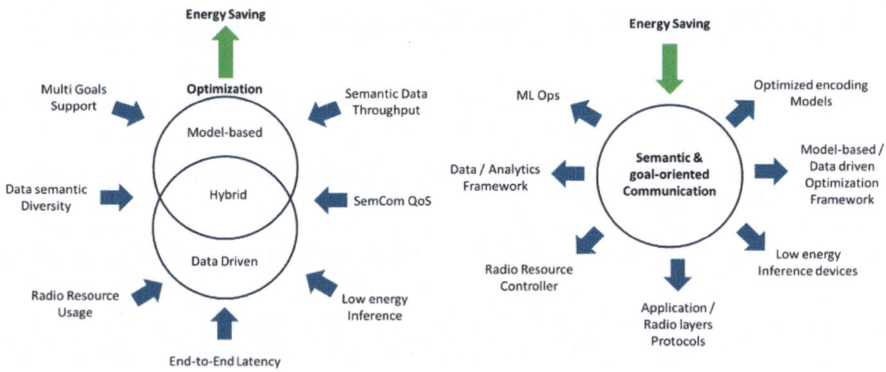

Fig. 9 Semantic communication and goal-oriented optimization for overall energy saving: optimization inputs, techniques, outputs

as a solid alternative to address the energy efficiency optimization goal as well as providing the ability to evaluate and monitor the effectiveness of energy-saving principles put in place. This energy consumption measurement capability has already been developed to monitor model inference on various DL execution frameworks, operating systems, and devices. Same approaches to measuring energy consumption on the radio transmission part are a more challenging effort requiring the proper granularity level of reported metrics related to radio resources' usage and channel conditions exposed by radio appliance APIs. This is particularly key to deducing the share of each semantic and goal-oriented communication session handled by the radio equipment in the overall energy measurement. Adopting a proper mix of energy consumption measurements on the different blocks involved in the radio transmission, energy prediction models trained with machine learning techniques, and numerical modeling leveraging known energy estimate formulas is an under-investigation topic with ongoing evaluation on proof-of-concept testbeds.

The adoption of semantic and goal-oriented communication for a wide panel of transmission goals will also introduce Machine Learning Operations (ML Ops) concepts as native parts of the radio system, not only for network operations but widely applied to the processing all along the communication path itself. Defining a generic framework covering the models' training, deployment over the air as well as monitoring their performances and selecting them for inference at the edge are important incoming challenges. Introducing proper interfaces between the application and radio lower layers and new control functions to support those new AI-native communication modes is a joint effort between academics, industry, and telco operators to infuse all essential innovations and required enablers into the next wireless communication systems' standards.

5 Sustainability, AI, and Standardization in 6G and Beyond

A thorough discussion on standardization in the context of 6G and sustainability is provided in Chap. 10. Across the telecom industry, there is a growing consensus that 6G must deliver order-of-magnitude improvements in energy and sustainability metrics compared to 5G. Multiple roadmaps and white papers have set aggressive targets. While specific numbers are not yet standardized, a general sentiment is emerging across the industry, which is indicated through various industry white papers highlighting the significant improvements 6G must achieve compared to 5G (e.g., Nokia Bell Labs and others envision "zero power" devices and AI-optimized networks that cut energy use dramatically). **Such gains hinge on innovations at all levels, from hardware to AI-driven network management.**

ITU-R has launched the process for 6G with Recommendation M.2160 (IMT-2030 framework) and will set detailed performance requirements by 2027. In parallel, 3GPP and other SDOs plan to initiate 6G specification work in the latter half of this decade. The 3GPP's Release 19 discussions (2025+) are expected to include studies for beyond-5G, ensuring alignment with IMT-2030 goals.

Standards for 6G will likely mandate advanced energy-saving features (extreme sleep modes, cognitive radio for efficient spectrum use, use of renewable energy in infrastructure, etc.), making technical progress synonymous with green progress. Moreover, 3GPP is advancing work on AI-enabled network architecture (for example, Study Items in SA2 on AI-native 6G architecture) and on AI-driven services with sustainability focus (in SA1). Moreover, earlier efforts like ITU-T's focus groups and ISO/IEC SC42 have begun creating standards for Green AI—for instance, ISO/IEC TR 20226:2025 provides a framework to assess the environmental impact of AI systems across their lifecycle, and an upcoming ITU-T standard is addressing energy-efficient AI best practices.

Considering the social dimension of sustainability, CEN/CLC JTC 21 and ISO/IEC JTC1/SC42 are defining a comprehensive set of harmonized standards with AI Act. These forthcoming standards are crucial for operationalizing social and ethical requirements of the AI Act, and for this reason, they will need to be completed before the AI Act goes into full effect (i.e., August 2026), especially for high-risk systems and critical infrastructures (e.g., telco). Key standards addressing the ethics dimension include:

Guidance on addressing societal concerns and ethical considerations (ISO/IEC 22443), Example template for documenting ethical issues of an AI system (ISO/IEC NP TS 25571), Objectives and approaches for explainability of ML models and AI systems (ISO/IEC TS 6254), AI system impact assessment (ISO/IEC 42005), Guidance for human oversight of AI systems (ISO/IEC CD 42105), technical reports on bias in AI systems (ISO/IEC TR 24027). This standardization work ensures AI systems not only meet technical robustness but are also developed and deployed in a manner that aligns with values and protects citizens' fundamental rights.

In parallel, industry alliances like the Next G Alliance in North America have dedicated groups (e.g., the "Green G" Working Group) focused on 6G sustainability.

These groups promote collaboration on metrics, methods, and roadmap recommendations so that 6G standards encompass worldwide convergence on ambitious eco-friendly objectives. The Next G Alliance's recent white paper, "Sustainable AI in Telecom: Promises and Challenges in 6G," emphasizes reducing embodied emissions in networks (e.g., manufacturing impact), using renewable energy sources, and optimizing AI workloads via techniques like model compression and federated learning, all of which inform future standards.

With reference to Sect. 4.2, Semantic Communication is a paradigm shift from conventional data transmission towards meaning-aware information exchange. By reducing redundant data transmission, networks can achieve significant energy savings per bit, particularly in scenarios involving massive machine-type communications, XR, or AI-driven IoT, where bandwidth and energy are traditionally strained. While specific details on semantic communications are evolving, and 3GPP has not yet formalized specifications for semantic communication, it has recently been discussed as part of the 6G use cases and candidate technologies in preparation for Release 20 and beyond.

AI-related standardization for 6G is firmly intertwined with sustainability objectives. Quantitatively, the aim is for 6G to be vastly more energy-efficient (on the order of 10×–100× per unit data) and to lower absolute network power despite higher traffic demands [85]. Qualitatively, the standards community is embracing concepts like semantic communication, AI-native orchestration, and environmental lifecycle accountability as integral parts of 6G. Sustainability is becoming a first-class requirement in 6G standardization, ensuring that the introduction of AI and advanced capabilities in networks also drives greener, more socially responsible outcomes rather than just technical performance.

6 Discussion and Conclusions

In this Chapter, we presented an overview of sustainability aspects and pillars in the context of AI. These span across the three pillars of environmental, social, and economic sustainability, covering aspects of energy consumption, carbon footprint, security, privacy, trustworthiness, as well as the sensitive topic of the transformation of the labor market.

Although AI is exploited in many sectors, this Chapter mostly focuses on a unifying theme within one of them: AI in 6G networks, in line with Chap. 10. After a thorough discussion on the central topic of Responsible AI, which considers all sustainability pillars, we focused on the challenges of AI processing itself, answering fundamental questions on where to place the workload and how to improve environmental sustainability aspects.

Then, we detailed our analysis on AI in 6G systems, presenting recent results on AI for RAN optimization, and AI in semantic and goal-oriented communications, with questions on communication and computation trade-offs. Finally, we discussed the ongoing standardization efforts, also with a glimpse at semantic communications and the AI Act. Overall, there is still a lot to be done to make AI a sustainable and sustainability-enabling technology, for ICT and beyond.

Acknowledgements This work is co-funded by the European Union under Grant Agreement 101191936. Views and opinions expressed are, however, those of the authors only and do not necessarily reflect those of all SUSTAIN-6G consortium parties nor those of the European Union or the SNS JU (granting authority). Neither the European Union nor the granting authority can be held responsible for them. The authors would like to thank the whole SUSTAIN-6G consortium.

The work was also partly funded by SNS JU project 6G-GOALS under the EU's Horizon program Grant Agreement (No 101139232), and by the European Commission Horizon Europe SNS JU 6G-XR project (GA 101096838).

Acknowledgment: The authors would like to sincerely thank Dr. Christophe Gaie and Dr. Markus Mueck for their invaluable support and insightful contributions throughout the development of this book chapter. Their expertise, guidance, and constructive feedback greatly enhanced the quality and clarity of this chapter.

Disclaimer Note that the authors have assembled the information above to the best of their knowledge as available at the time when the present book was written. Any interpretation is the personal opinion of the authors. Also, the regulation framework can change and evolve at any time. The only legally binding text is the one provided in the documents formally published in the European Union Official Journal.

References

1. Van Wynsberghe A (2021) Sustainable AI: AI for sustainability and the sustainability of AI. AI Ethics 1:213–218
2. Calvanese Strinati E et al (2024) Goal-oriented and semantic communication in 6G AI-native networks: The 6G-GOALS approach. In: Joint European conference on networks and communications & 6G summit (EuCNC/6G Summit), Antwerp, Belgium
3. Shao M, Basit A, Karri R, Shafique M (2024) Survey of different large language model architectures: trends, benchmarks, and challenges. IEEE Access 12:188664–188706
4. European Commission (2019) EU ethics guidelines for trustworthy AI [Online]. Available: https://digital-strategy.ec.europa.eu/en/library/ethics-guidelines-trustworthy-ai. Accessed 2025
5. European Parliament and the Council (2024) Regulation (EU) 2024/1689 AI Act [Online]. Available: https://eur-lex.europa.eu/eli/reg/2024/1689
6. CEN/CLC (2024) TR 18115:2024—aata governance and quality for AI within the European context [Online]. Available: https://standards.cencenelec.eu. Accessed 2025
7. European Parliament and the Council (2024) Regulation (EU) 2024/1689 AI Act [Online]. Available: https://eur-lex.europa.eu/eli/reg/2024/1689. Accessed 2025
8. CEN/CENELEC (2024) TR 18115:2024—aata governance and quality for AI within the European context [Online]. Available: https://standards.cencenelec.eu
9. United Nations (2025) Governing AI for humanity [Online]. Available: https://www.un.org/sites/un2.un.org/files/governing_ai_for_humanity_final_report_en.pdf. Accessed 2025
10. The President of the United States of America (2023) Executive order 14110 of October 30, 2023: safe, secure, and trustworthy development of artificial intelligence [Online]. Available: https://www.govinfo.gov/content/pkg/FR-2023-11-01/pdf/2023-24283.pdf. Accessed 2025
11. Governor of California (2025) AB 853. California AI transparency Act [Online]. Available: https://leginfo.legislature.ca.gov/faces/billNavClient.xhtml?bill_id=202520260AB853. Accessed 2025

12. Introducing Nokia's 6 Pillars of Responsible (2022) Nokia [Online]. Available: https://www.nokia.com/blog/introducing-nokias-6-pillars-of-responsible-ai/
13. CEN/CENELEC (2025) Working Groups & Projects | CEN-CENELEC JTC 21 [Online]. Available: https://jtc21.eu/working-groups/
14. ISO Standards (2025) Information technology—artificial intelligence—environmental sustainability aspects of AI systems [Online]. Available: https://www.iso.org/standard/86177.html. Accessed 09 2025
15. International Telecommunication Union (2024) AI and the environment—international standards for AI (Report) [Online]. Available: https://www.itu.int/dms_pub/itu-t/opb/env/T-ENV-ENV-2024-1-PDF-E.pdf. Accessed 09 2025
16. Nokia (2024) A transparent and standards-based way to assess the environmental impact of AI systems [Online]. Available: https://www.nokia.com/blog/a-transparent-and-methodical-framework-to-assess-the-sustainability-impact-of-ai. Accessed 09 2025
17. NextG Alliance Report (2025) Sustainable AI in telecom: promises and challenges in 6G [Online]. Available: https://nextgalliance.org/white_papers/sustainable-ai-in-telecompromises-and-challenges-in-6g/. Accessed 09 2025
18. Malmio I (2024) Artificial intelligence and the social dimension of sustainable development: through a security perspective. Discov Sustain 5
19. UNESCO (2023) Recommendation on the ethics of artificial intelligence [Online]. Available: https://www.unesco.org/en/articles/recommendation-ethics-artificial-intelligence. Accessed 09 2025
20. ISO/IEC (2025) ISO/IEC CD TR 18988—application of AI technologies in health informatics [Online]. Available: https://www.iso.org/standard/85602.html
21. Paes V, Silveira F, Akkari A (2021) Social impacts of artificial intelligence and mitigation recommendations: an exploratory study. In: Proceedings of the 7th Brazilian technology symposium (BTSym'21)
22. Cazzaniga M et al (2024) Gen-AI: artificial intelligence and the future of work (White Paper) [Online]. Available: https://www.imf.org/en/Publications/Staff-Discussion-Notes/Issues/2024/01/14/Gen-AI-Artificial-Intelligence-and-the-Future-of-Work-542379. Accessed 09 2025
23. Al-Emran M, Abu-Hijleh B, Alsewari AA (2024) Exploring the effect of generative AI on social sustainability through integrating AI attributes, TPB, and T-EESST: a deep learning-based hybrid SEM-ANN approach. IEEE Trans Eng Manage 71:14512–14524
24. ETSI (2025) ETSI AI conference—how standardization is shaping the future of AI [Online]. Available: https://www.etsi.org/events/2451-etsi-ai-conference-2025. Accessed 09 2025
25. 3GPP (2021) TR. 28.908: study on artificial intelligence/machine learning (AI/ ML) management. [Online]. Available: https://portal.3gpp.org/desktopmodules/Specifications/SpecificationDetails.aspx?specificationId=3965. Accessed 09 2025
26. OECD Digital Economy Papers (2022) Measuring the environmental impacts of artificial intelligence compute and applications: the AI footprint [Online]. Available: https://www.oecd.org/en/publications/measuring-the-environmental-impacts-of-artificial-intelligence-compute-and-applications_7babf571-en.html. Accessed 09 2025
27. European Commission (2024) Artificial intelligence: economic impact, opportunities, challenges, implications for policy [Online]. Available: https://economy-finance.ec.europa.eu/publications/artificial-intelligence-economic-impact-opportunities-challenges-implications-policy_en. Accessed 09 2025
28. McKinsey & Company (2025) AI in the workplace: a report for 2025 [Online]
29. World Economic Forum (2025) The future of jobs report 2025 [Online]. Available: https://www.weforum.org/publications/the-future-of-jobs-report-2025/. Accessed 09 2025
30. SUSTAIN-6G (2025) Deliverable D2.1: sustainability baseline, use cases, and baseline requirements [Online]. Available: https://zenodo.org/records/16684673
31. Kaplan J, McCandlish S, Henighan T, Brown TB, Chess B, Child R, Gray S, Radford A, Wu J, Amodei D (2020) Scaling laws for neural language models, p 19. arXiv:2001.08361[cs.LG]

32. Epoch AI (2025) Epoch AI compute trends: data on AI models [Online]. Available: https://epoch.ai/data/ai-models. Accessed 17 Oct 2025
33. HAI Human-Centered Artificial Intelligence (2025) The 2025 AI index report. Stanford University, Stanford
34. Noffsinger J, Goodpaster M, Patel M, Chang H, Sachdeva P, Bhan A (2025) The cost of compute: A $7 trillion race to scale data centers. Mckinsey & Company, 28 Apr 2025 [Online]. Available: https://www.mckinsey.com/industries/technology-media-and-telecommunications/our-insights/the-cost-of-compute-a-7-trillion-dollar-race-to-scale-data-centers. Accessed 18 Oct 2025
35. Pilz KF, Rahman R, Sanders J, Emberson L, Heim L (2025) Power requirements of leading AI supercomputers have doubled every 13 months. Epoch AI, 05 06 2025. [Online]. Available: https://epoch.ai/data-insights/ai-supercomputers-power-trend. Accessed 18 Oct 2025
36. IEA (2025) Energy and AI. IEA, Paris
37. Federal Statistical Office (2023) Electricity Consumption of households by household size. Destatis 27 Sept 2023. [Online]. Available: https://www.destatis.de/EN/Themes/Society-Environment/Environment/Environmental-Economic-Accounting/private-households/Tables/electricity-consumption-private-households.html. Accessed 20 Oct 2025
38. Leinonen V-A (2025) Kajaani data center: improving eco-efficiency. CSC, 25 Apr 2023 [Online]. Available: https://csc.fi/en/news/kajaani-data-center-improving-eco-efficiency/. Accessed 18 Oct 2025
39. McIntosh-Smith S, Alam S, Woods C (2025) Isambard-AI: a leadership-class supercomputer optimised specifically for artificial intelligence. In: Cray user group (CUG '24), New York (2025)
40. Feng X, Rong G, Cameron KW (2005) Power and energy profiling of scientific applications on distributed systems. In: 19th IEEE international parallel and distributed processing symposium, Denver
41. Tiwari A, Laurenzano M, Carrington L, Snavely A (2012) Modeling power and energy usage of hpc kernels. In: IEEE 26th international parallel and distributed processing symposium workshops, Shanghai
42. Borghesi A, Bartolini A, Milano M, Benini L (2019) Pricing schemes for energy-efficient HPC systems: design and exploration. Int. J. High Perform. Comput. Appl. 33(4):716–734
43. Jackson A, Simpson A, Turner A (2023) Emissions and energy efficiency on large-scale high performance computing facilities: ARCHER2 UK national supercomputing service case study. In: SC '23 Workshops of the international conference on high performance computing, network, storage, and analysis, Denver
44. Simmendinger C, Marquardt M, Mäder J, Schneider R (2024) PowerSched—managing power consumption in overprovisioned systems. In: 2024 IEEE international conference on cluster computing workshops, Kobe
45. Athavale J, Bash C, Brewer W, Maiterth M, Milojicic D, Petty H, Sarkar S (2024) Digital twins for data centers. Computer 57(10):151–158
46. Brewer W, Maiterth M, Kumar V, Wojda R, Bouknight S, Hines J, Shin W, Greenwood S, Grant D, Williams W, Wang F (2024) A digital twin framework for liquid-cooled supercomputers as demonstrated at exascale. In: Proceedings of the international conference for high performance computing, networking, storage, and analysis, Atlanta
47. Han B, Habibi MA, Richerzhagen B, Schindhelm K, Zeiger F, Lamberti F, Patticò FG, Upadhya K, Korovesis C, Belikaidis I-P, Demestichas P, Yuan S, Schotten (2023) Digital twins for industry 4.0 in the 6G Era. IEEE Open J Veh Technol 4:820–835
48. IEA (2021) Global trends in internet traffic, data centres workloads and data centre energy use, 2010–2020. IEA, 04 Nov 2021 [Online]. Available: https://www.iea.org/data-and-statistics/charts/global-trends-in-internet-traffic-data-centres-workloads-and-data-centre-energy-use-2010-2020. Accessed 19 Oct 2025
49. European Telecommunications Standards Institute (ETSI) (2018) 5G; NG-RAN; Architecture description. 3GPP TS 38.401 version 15.2.0 Release 15

50. European Telecommunications Standards Institute (ETSI) (2015) LTE; evolved universal terrestrial radio access (E-UTRA) and evolved universal terrestrial radio access network (E-UTRAN); overall description; stage 2 (3GPP TS 36.300 version 12.7.0 Release 12)
51. European Telecommunications Standards Institute (ETSI) (2020) 5G; management and orchestration; architecture framework. 3GPP TS 28.533 version 15.3.0 Release 15
52. European Telecommunications Standards Institute (ETSI) (2024) 5G; Study on alignment with ETSI MEC for edge computing management. 3GPP TR 28.903 version 18.0.1 Release 18
53. European Telecommunications Standards Institute (ETSI) (2024) System architecture for the 5G System (5GS). Technical Specification 23.501, V18.6.0
54. European Telecommunications Standards Institute (ETSI) (2024) LTE; 5G; Vehicle-to-Everything (V2X) application enabler (VAE) layer; protocol aspects. Stage 3 (3GPP TS 24.486 version 18.4.0 Release 18)
55. European Telecommunications Standards Institute (ETSI) (2024) Publicly available specification (PAS); O-RAN architecture description. O-RAN.WG1.OAD-R003-v08.00
56. European Telecommunications Standards Institute (ETSI) (2020) 5G; 5G system; network data analytics services. Stage 3 (3GPP TS 29.520 version 16.5.1 Release 16)
57. Erkens H et al (2024) 6G smart networks and services industry association (6G-IA). White paper Open RAN and 6G future networks development
58. 6GREEN (2023) 6Green, project website [Online]. Available: https://www.6green.eu/. Accessed 01 Oct 2025
59. SUSTAIN-6G (2025) SUSTAIN-6G project website [Online]. Available: https://sustain-6g.eu/. Accessed 01 Oct 2025
60. Merluzzi M, Bouchet O, Balador A, Callebaut G, Gavras A, Van der Perre L, Banchs A, Boldi MR, Strinati EC, Khorsandi BM, Matinmikko-Blue M (2026) Toward sustainable 6G: a holistic view of trade-offs and enablers. In IEEE Commun Mag. https://doi.org/10.1109/MCOM.001.2500386
61. Merluzzi M et al (2025) Towards sustainable 6G: a holistic view of trade-offs and enablers. IEEE Communications Magazine (submitted)
62. Open Networking Foundation (2023) SMaRT-5G project-sustainable mobile and RAN transformation (SMaRT) [Online]. Available: https://opennetworking.org/wp-content/uploads/2023/07/ONF-SMaRT-5G-White-Paper-v11.pdf. Accessed 09 2025
63. CATT-R1-2211210 (2022) Network energy saving techniques in time. 3GPP TSG RAN WG1#111, Toulouse, France
64. Islam T, Lee D, Lim S (2023) Enabling network power savings in 5G-advanced and beyond. IEEE J Sel Areas Commun 41:1888–1899
65. Mao W, Wei L, Semiari O, Yeh S, Nikopour H (2025) Deep reinforcement learning-based cell DTX/DRX configuration for network energy saving [Online]. Available: https://arxiv.org/pdf/2507.21385. Accessed 09 2025
66. Semiari O, Nikopour H, Talwar S (2025) Graph reinforcement learning for QoS-aware load balancing in open radio access networks [Online]. Available: https://arxiv.org/pdf/2504.19499. Accessed 09 2025
67. Semiari O, Nikopour H, Talwar, S (2024) Reliability-optimized user admission control for URLLC traffic: a neural contextual bandit approach. In: IEEE international conference on machine learning for communication and networking (ICMLCN), Stockholm, Sweden
68. Pires R et al (2024) Closed-loop automation in 6G for minimum downtime task continuity in surveillance cobots. In: Joint European conference on networks and communications & 6G summit (EuCNC/6G Summit), Antwerp, Belgium
69. Nokia (2024) Global network traffic report: understanding the growing impact of advancing technologies on future networks [Online]. Available: https://www.nokia.com/asset/213660. Accessed 09 2025
70. Lampin Q, Dufrène L, Larue G (2025) Semantic communications services within generalist operated networks. In: IEEE 25th international workshop on signal processing advances in wireless communications (SPAWC), Lucca, Italy

71. Getu TM, Kaddoum G, Bennis M (2024) A survey on goal-oriented semantic communication: techniques, challenges, and future directions. IEEE Access 12:51223–51274
72. Merluzzi M, Filippou M-C (2025) Goal-oriented spectrum sharing: trading edge inference power for data streaming performance. In: EURASIP EUSIPCO, Palermo, Italy
73. Mao Y et al (2024) Green edge AI: a contemporary survey. Proc IEEE 112:880–911
74. Miro-Panades I et al (2024) A 772μJ/frame imagenet feature extractor accelerator on HD images at 30FPS. In: 2024 IEEE Asia Pacific conference on circuits and systems (APCCAS), Taipei, Taiwan
75. Vaswani A et al (2017) Attention is all you need. Adv Neural Inf Process Syst 30
76. Dosovitskiy A et al (2021) An image is worth 16 x 16 words: transformers for image recognition at scale [Online]. Available: https://arxiv.org/abs/2010.11929
77. Fang G et al (2024) Isomorphic pruning for vision models [Online]. Available: https://arxiv.org/abs/2407.04616
78. Steiner A et al (2022) How to train your ViT? Data, augmentation, and regularization in vision transformers [Online]. Available: https://arxiv.org/abs/2106.10270.
79. Li J, Cotterell R, Sachan M (2021) Differentiable subset pruning of transformer heads. Trans Assoc Comput Linguist 9:1442–1459
80. Devoto A, Pomponi J, Merluzzi M, Di Lorenzo P, Scardapane S (2025) Adaptive semantic token communication for transformer-based edge inference. [Online]. Available: https://arxiv.org/abs/2505.17604
81. Millar J, Huang Y, Sethi S, HaddadiI H, Madhavapeddy A (2025) Benchmarking ultra-low-power μNPUs. [Online]. Available: arXiv:2503.22567v2
82. Peschiera E et al (2025) On optimizing time-, space- and power-domain energy-saving techniques for Sub-6 GHz massive MIMO Base Stations. In: 2025 IEEE 226th international workshop on signal processing and artificial intelligence for wireless communications (SPAWC), Surrey, United Kingdom
83. Mohammad-Ali M et al (2021) An energy-efficient inference method in convolutional neural networks based on dynamic adjustment of the pruning level. CM Trans Des Autom Electron 26
84. O-RAN Work Group 3 (Near-real-time RAN intelligent controller and E2 interface), "Near-RT RIC Architecture," O-RAN.WG3.TS.RICARCH-R004-v07.00
85. Ahmadi H et al (2025) Towards sustainability in 6G and beyond: challenges and opportunities of open RAN. IEEE Communications Standards Magazine
86. European Union, The EU artificial intelligence act [Online]. Available: https://artificialintelligenceact.eu/. Accessed 09 2025
87. European Telecommunications Standards Institute (ETSI) (2021) 5G; Extended Reality (XR) in 5G. 3GPP TR 26.928 version 16.1.0 Release 16
88. Euopean Commission (2019) EU ethics guidelines for trustworthy AI [Online]. Available: https://digital-strategy.ec.europa.eu/en/library/ethics-guidelines-trustworthy-ai
89. CEN/CLC (2024) TR 18115:2024—data governance and quality for AI within the European context [Online]. Available: https://standards.cencenelec.eu.
90. United Nations (2024) Governing AI for humanity [Online]. Available: https://www.un.org/sites/un2.un.org/files/governing_ai_for_humanity_final_report_en.pdf

Chapter 12
A Case Study: Sustainability in the Context of Ambient-Powered IoT

Riku Jäntti, Mateen Ashraf, Gilles Callebaut, Jona Cappelle, Lukas D'angelo, Miltiadis Filippou, Kalle Koskinen, Guus Leenders, Onel López, Ulrich Muehlmann, Daniel Poehl, Priyesh Pappinisseri Puluckul, Lyssa Ramaut, Ritesh Kumar Singh, Bikramjit Singh, Jarne Van Mulders, Jenny Wiklund, and Boxuan Xie

Abstract While previous chapters have explored sustainability objectives at the level of large-scale Information and Communications Technology (ICT) systems, this chapter examines sustainability from the perspective of device-level design and

---

R. Jäntti (✉) · K. Koskinen · J. Wiklund · B. Xie
Department of Information and Communications Engineering, Aalto University, Espoo, Finland
e-mail: riku.jantti@aalto.fi

K. Koskinen
e-mail: kalle.koskinen@aalto.fi

J. Wiklund
e-mail: jenny.wiklund@aalto.fi

B. Xie
e-mail: boxuan.xie@aalto.fi

M. Ashraf · O. López
Center of Wireless Communication, University of Oulu, Oulu, Finland
e-mail: mateen.ashraf@oulu.fi

O. López
e-mail: onel.alcarazlopez@oulu.fi

G. Callebaut · J. Cappelle · L. Ramaut · J. Van Mulders
Department of Electrical Engineering, KU Leuven, Ghent, Belgium
e-mail: gilles.callebaut@kuleuven.be

J. Cappelle
e-mail: jona.cappelle@kuleuven.be

L. Ramaut
e-mail: lyssa.ramaut@kuleuven.be

J. Van Mulders
e-mail: jarne.vanmulders@kuleuven.be

C. Gaie and M. Mueck (eds.), *Digital Sustainability: Legislations, Standards, Solutions and Related Case Studies*, Advances in Sustainability Science and Technology,
https://doi.org/10.1007/978-981-92-0458-8_12

deployment. We focus on the concept of ambient-powered Internet of Things (IoT), where Energy-Neutral Devices (ENDs) harvest energy from their environment, such as light, radio frequency, or thermal gradients, operate autonomously without wired power or frequent battery replacement, and communicate wirelessly with the infrastructure. This chapter proposes a categorization of ENDs based on their energy storage and operational profiles, ranging from ultra-passive devices with no energy buffering to continuously operating systems with rechargeable storage. We discuss associated design trade-offs and illustrate their use through selected prototypes and case studies. Furthermore, we analyze the sustainability implications of END adoption across economic, societal, and environmental dimensions, and outline a preliminary Life Cycle Assessment (LCA) framework tailored for such devices. The chapter concludes by discussing how energy harvesting principles embodied in ENDs can be extended to support more broadly sustainable ICT infrastructures.

Keywords Internet of Things (IoT) · Ambient-powered IoT · Backscatter communications · Energy-neutral devices · Wireless power transfer · Life cycle assessment

L. D'angelo
Graz University of Technology, Graz, Austria
e-mail: lukas.dangelo@tugraz.at

M. Filippou
WINGS ICT Solutions S.A., Athens, Greece
e-mail: mfilippou@wings-ict-solutions.eu

G. Leenders
Quicksand, Roeselare, Belgium
e-mail: guus.leenders@quicksand.be

U. Muehlmann · D. Poehl
NXP Semiconductors Austria GmbH & Co KG, Gratkorn, Austria
e-mail: ulrich.muehlmann@nxp.com

D. Poehl
e-mail: daniel.poehl@nxp.com

P. P. Puluckul · R. K. Singh
IDLab, University of Antwerp – imec, Antwerp, Belgium
e-mail: priyesh.pappinisseripuluckul@uantwerpen.be

R. K. Singh
e-mail: ritesh.iiit10@gmail.com

B. Singh
LMF Ericsson, Kirkkonummi, Finland
e-mail: bikramjit.b.singh@ericsson.com

J. Wiklund
Department of Bioproducts and Biosystems, Aalto University, Espoo, Finland

1 Introduction

Emerging Internet of Things (IoT) technologies and applications are poised to transform future networks into systems comprising a massive number of low-power, heterogeneous devices. A critical challenge in this evolution is ensuring dependable and energy-efficient connectivity while simultaneously addressing broader sustainability objectives. In this context, sustainable connectivity denotes a paradigm in wireless communications, wherein devices operate autonomously, are environmentally benign, and can still realize their intended societal and economic impact.

Central to this vision are energy-neutral IoT Devices, which operate by harvesting ambient energy (EH) or receiving power via Wireless Power Transfer (WPT) to replenish their energy storage. By reducing or entirely eliminating the need for battery replacement or wired power connections, such devices enable the deployment of scalable, maintenance-free, and long-lived IoT infrastructures that support overarching sustainability objectives [1, 2].

The emerging vision of ambient-powered IoT has garnered increasing momentum, not only in academic and industrial research but also within standardization bodies. The 3rd Generation Partnership Project (3GPP) has introduced the concept of the Ambient Internet of Things (A-IoT) in its study items and specification work [3, 4], aiming to incorporate ambient-powered devices into future 5G-Advanced and 6G networks. A-IoT emphasizes ultra-low-power communication protocols, ambient energy utilization, and support for massive numbers of intermittently active devices. Complementary efforts are taking place within the IEEE 802.11 working group, where the Ambient Power Topic Interest Group and the subsequent AMP Study Group explore enabling ultra-low-energy and long-life cycle IoT devices using ambient power communication techniques [5].

Importantly, these technical and standardization efforts resonate with broader sustainability objectives, particularly those set forth in the European Green Deal. As part of the European Union's strategy to achieve climate neutrality by 2050, the Green Deal advocates for sustainable digital transformation, resource efficiency, and circular economy principles. Ambient-powered IoT devices reduce the need for batteries, enable maintenance-free operation, and minimize energy consumption. These features make them a foundational technology for environmentally responsible ICT systems and a meaningful contributor to Europe's ecological transition.

This chapter provides a case study perspective on sustainability in the context of ambient-powered IoT. It is divided into four sections. The rest of Sect. 1 introduces the key technical components of energy-neutral IoT devices and briefly introduces the sustainability aspects. Different device classes are introduced and discussed in Sect. 2 along with example device designs. The ecological assessment regarding each individual device class is conducted in Sect. 3. Finally, this chapter is concluded in Sect. 4.

1.1 Energy-Neutral Devices

IoT devices can be broadly categorized as either active or passive, depending on their radio architecture and transmission capabilities. Active devices are equipped with Radio Frequency (RF) oscillators and can generate and modulate their own carrier signals, enabling them to initiate data transmissions autonomously. A typical example of an active IoT device is a Bluetooth peripheral, which actively transmits data using its onboard RF circuitry. In contrast, passive devices lack RF oscillators and rely entirely on external RF sources for their operation. These devices employ backscatter communication, where data transmission is achieved by modulating the reflection coefficient of the device's antenna, thereby encoding information onto an incident carrier wave emitted by another device or infrastructure node. These two operation principles of the devices are illustrated in Fig. 1.

Figure 2 presents a classification of low-power IoT devices, distinguishing between two principal categories: conventional battery-powered devices and Energy-Neutral Devices (ENDs), also referred to as Zero Energy Devices (ZEDs) in the literature, which either harvest energy from ambient sources or receive power via Wireless Power Transfer (WPT). ENDs can be further classified according to their energy storage and operational characteristics:

- **Ultra-Passive ENDs (UP-ENDs)** lack dedicated energy storage beyond intrinsic circuit-level capacitance and are only capable of operating when illuminated by an external power source, typically using backscatter communication.
- **Duty-Cycled Barebone ENDs (DB-ENDs)** contain minimal short-term energy storage, such as capacitors or small supercapacitors, allowing for intermittent operation by accumulating small amounts of energy.
- **Duty-Cycled Persistent ENDs (DP-ENDs)** feature larger rechargeable storage elements (e.g., supercapacitors), enabling energy accumulation over longer durations, but still operate in a duty-cycled manner due to limited energy reserves.
- **Continuous Operation ENDs (CO-ENDs)** are equipped with long-term rechargeable storage, such as batteries, allowing for continuous or near-continuous

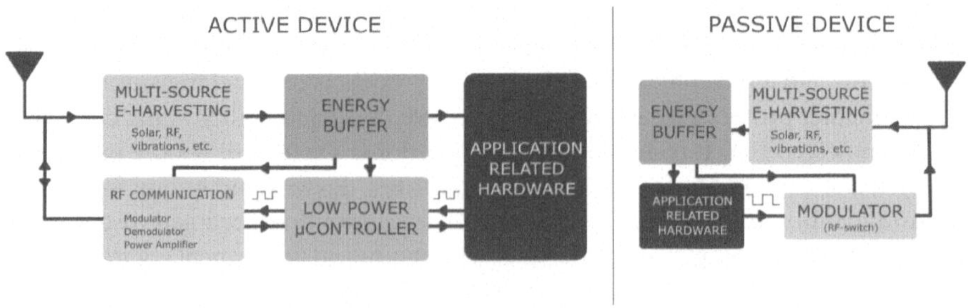

Fig. 1 General architecture of active and passive ENDs. The application-related hardware can include memory and interfaces

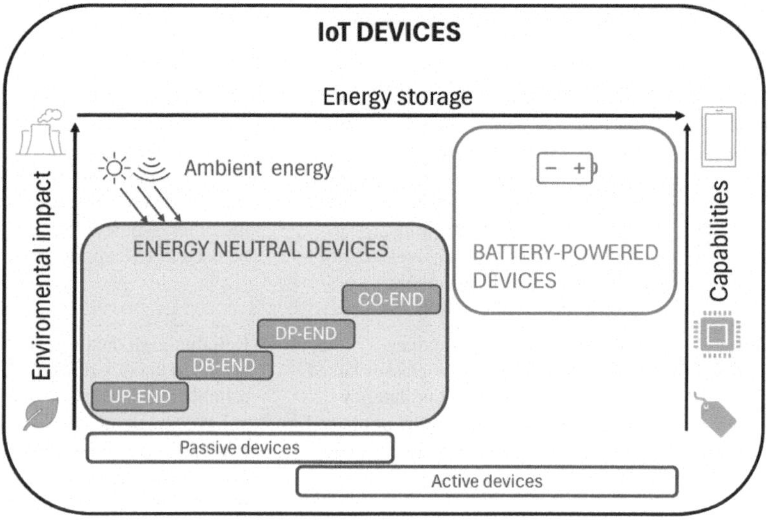

Fig. 2 Classification of low-power IoT devices

operation while remaining energy neutral through ongoing ambient energy harvesting.

This classification illustrates the diversity of ambient-powered IoT device designs and their applicability across a wide range of deployment scenarios. Alternative classifications of IoT devices presented in the literature are summarized in Table 1.

1.1.1 Key Technical Components of END

Designing ENDs, those capable of sustained operation without battery replacement or external recharging, requires careful consideration of each architectural component. These devices typically operate within stringent energy budgets, harvesting energy from ambient sources and adapting their functionality to match available power. The fundamental building blocks of such devices can be grouped into six key domains: energy storage, energy harvesting, processing, communication, interfaces, and memory.

Energy Storage: Energy storage is central to sustaining operation during intermittent power availability. Options include electrostatic (e.g., capacitors, supercapacitors), hybrid (e.g., lithium-ion capacitors), and electrochemical (e.g., conventional batteries) technologies, each differing in power and energy densities, form factor, and lifetime characteristics [11]. Electrostatic storage devices, such as Electric Double-Layer Capacitors (EDLCs), are favorable for short burst operations and high-power tasks due to their rapid charge/discharge capabilities, but they exhibit lower energy density. Hybrid and electrochemical storage solutions provide extended endurance,

Table 1 Summary of existing IoT device classifications

Classification name	Description	Classes names and applicability to IoT networks
5G device classification [6]	Service type-based classification such as eMBB, mMTC, and URLLC. This classification is based on communication QoS aspects such as data rate, reliability, and ubiquitousness. The devices are not classified according to their architectural components	(3) eMBB 5G devices, fixed wireless access 5G devices, industrial 5G devices. Apart from the devices supporting eMBB service types, the other two device classes can be used in IoT applications
6G device classification [7]	Enhancement of 5G device classifications and new classes based on energy, reliability, and latency needs. The classification is not based on the architectural components of the device	(4) Reliable high data rate with bounded latency devices, highly reliable low latency device class, enhancements of massive machine type communication device class, and END. Except for reliable high data rate with bounded latency devices, the remaining classes can be used in IoT applications
A-IoT device classification by 3GPP [3, 8, 9]	Classification of low-power/energy devices based on their energy storage capacity and their ability to generate RF signals for transmissions. Although this classification considers architectural components of the devices, it is limited to energy and communication aspects only	(3) Device A, Device B, Device C. All device classes can be used in IoT applications
LoRaWAN device classification [10]	Devices are categorized on the basis of transmit power, synchronous/asynchronous communication, and deterministic/lowest latency capabilities	(3) Class A, Class B, Class C. All device classes can be used in IoT applications

though often at the cost of slower charge dynamics, limited cycle life, and greater environmental impact. The selection of a suitable energy storage technology must therefore balance application-specific power profiles, autonomy requirements, and sustainability constraints—particularly when considering the environmental costs and disposal implications of electrochemical systems.

Energy Harvesting and Conversion: ENDs depend on energy harvesting mechanisms that exploit various physical phenomena. Scavenged energy sources include electromagnetic radiation (e.g., photovoltaic or RF), mechanical vibrations, acoustic waves, thermal gradients, and fluid flow [1]. The choice of energy harvester and conversion circuit depends on the operating environment and the anticipated power density. RF-powered solutions, including backscatter and inductive/capacitive coupling techniques, offer compact form factors suitable for wireless tags and low-duty cycle sensors [12]. Coupled wireless power transfer provides controlled delivery of energy, typically with higher power transfer efficiency over short distances.

Power Management Unit: The power management unit forms one of the fundamental building blocks of any active END design. A PMU is required due to the wide electrical characteristics of different converters, making it difficult to power the electronics directly using the harvester's output. Therefore, a PMU takes care of this and manages the output of the harvester so that the remaining component of the END can be powered. Normally, the PMU performs Maximum Power Point Tracking (MPPT) to extract maximum power from the harvester, translate the voltage to the required level using a DC-DC converter and charge an energy buffer such as a battery or a capacitor. The DC-DC converter could be an inductor-based switching regulator or an inductor-less switching capacitor converter.

Processing Subsystems: Processing capabilities in ENDS vary from logic-less implementations, e.g., chipless Radio Frequency Identification (RFID) [13], to ultra-low-power microcontrollers capable of rule-based decisions, signal filtering, and even edge AI workloads. Processing power is often quantified in terms of Million Instructions Per Second (MIPS), with higher-performance nodes enabling local analytics or inference [14]. However, increased processing capability imposes additional energy and thermal constraints, mandating tight coupling with efficient power management and low-leakage silicon processes.

Communication Interfaces: Communication is often the most energy-intensive function in an IoT device. Passive communication, such as backscatter, avoids active RF transmission and enables sub-microwatt operation. Semi-passive schemes allow limited amplification, and active schemes provide full duplex and long-range transmission at the cost of higher power draw [15]. The physical medium (RF, optical, acoustic, magnetic) and achievable data rate must be chosen to balance energy use, latency, and application needs. In ambient-powered contexts, trade-offs between range, throughput, security, and complexity are critical to sustainability.

Peripheral Interfaces: Interfaces to sensors, actuators, or localization modules define the operational context of the device. These can be digital (e.g., GPIO, SPI/I2C buses), analog (e.g., ADC/DAC for physical sensing), or RF-based (e.g., for positioning or radar sensing). Peripheral selection affects both energy consumption and processing requirements, with more complex sensing (e.g., positioning or radar) necessitating greater support from the processing and memory subsystems.

Memory and Data Storage: Memory requirements in ambient-powered devices must reconcile the constraints of both volatile and non-volatile storage technologies. Volatile memory, such as Static Random-Access Memory (SRAM), offers fast access speeds but requires continuous standby power to retain data. In contrast, non-volatile memory technologies, such as flash memory, Ferroelectric Random-Access Memory (FRAM), and Magnetoresistive Random-Access Memory (MRAM), retain data without power, though they differ significantly in endurance, write latency, and energy efficiency [16]. Among these, emerging options like FRAM and MRAM are particularly promising for intermittently powered systems due to their combination of low write energy, high endurance, and suitability for frequent state preservation under energy-constrained operation.

Security: The security of ENDs is ensured through a layered approach that leverages material-based identification, dynamic impedance modulation, and progressive cryptographic techniques. Physical Unclonable Functions (PUFs) are exploited for device identification, while impedance states are manipulated to enable basic challenge-response authentication under minimal energy budgets. In devices with larger energy reserves, simple computations are enabled and cryptographic keys are retained in non-volatile memory, facilitating more robust authentication and lightweight encryption. Devices with extensive energy reserves support secure firmware update of the device, mutual authentication with network gateways, and real-time intrusion detection through anomaly based or AI-driven monitoring.

1.1.2 Sustainability Dimensions of END for IoT

The proliferation of IoT devices has introduced multifaceted sustainability challenges and opportunities [1]. These can be broadly categorized into economic, societal, and environmental dimensions.

Economic Sustainability: Traditional IoT deployments often rely on battery-powered devices, necessitating periodic maintenance and battery replacements. This maintenance not only incurs significant labor and material costs but also disrupts service continuity. For instance, large-scale deployments may require substantial investments over time solely for battery maintenance [17]. In contrast, ENDs can operate autonomously without the need for battery replacements. This autonomy reduces operational expenditures and enhances the economic viability of long-term IoT deployments.

Furthermore, passive backscatter-based IoT devices, which modulate and reflect existing RF signals instead of generating their own, possess simpler architectures with fewer components. This simplicity translates to lower manufacturing costs, making such devices economically attractive for large-scale applications [18].

Societal Sustainability: The reduced cost and maintenance requirements of ENDs and passive IoT devices can facilitate broader adoption across various sectors, including agriculture, health care, and urban infrastructure. By enabling affordable and scalable solutions, these technologies can bridge digital divides, providing underserved communities with access to essential services and information. For example, low-cost environmental sensors can empower communities to monitor local air and water quality, fostering community engagement and informed decision-making [19].

Environmental Sustainability: The environmental impact of IoT devices is a growing concern, particularly regarding electronic waste (e-waste) and battery disposal. According to the United Nations' Global E-waste Monitor 2024 [20], global e-waste generation reached 62 million tons in 2022, with projections estimating an increase to 82 million tons by 2030. Alarmingly, only 22.3% of this waste was formally collected and recycled in 2022, indicating a widening gap between e-waste generation and recycling efforts.

Batteries, especially non-rechargeable ones, contribute significantly to this problem. Estimates suggest that by 2025, approximately 78 million batteries will be discarded daily, exacerbating environmental pollution and resource depletion [21]. ENDs, particularly those without batteries, offer a sustainable alternative by minimizing hazardous waste and reducing reliance on finite resources.

Moreover, the simplified design of passive backscatter devices not only lowers production costs but also reduces material usage, further mitigating environmental impact. Some ENDs can be manufactured using environmentally friendly materials and processes, enhancing their sustainability profile [22].

2 Device Classes

This section introduces four broad END classes and discusses their distinguishing feature, architectural and operational properties, and some existing prototypes and case studies.

2.1 Ultra-Passive END

Distinguishing Feature(S)

No meaningful energy storage beyond transient circuit-level capacitance.

2.1.1 Architectural and Operational Properties

UP-END represents the simplest form of Energy-Neutral Devices. These devices operate exclusively under the presence of an external energy or signal source, without any means of energy storage. As a result, their operation is strictly instantaneous and fully dependent on the availability of external power or excitation, ceasing immediately once the stimulus is removed. Their functional capabilities are inherently limited to basic identification, signal reflection or modulation, and minimal sensing tasks, relying on passive physical phenomena such as electromagnetic backscattering, structural resonance, or reflection.

Representative examples of UP-END includes basic RFID tags, PUF devices whose hardware elements exploit uncontrollable manufacturing variations to produce unique and repeatable responses, and metasurfaces, such as passive reflective structures used in wireless communication and sensing. Owing to their minimal component count and passive operation, UP-END devices offer near-zero maintenance, extreme longevity, and minimal fabrication complexity. A high-level block diagram of a representative UP-END architecture is shown in Fig. 3.

Energy: The energy needs of UP-ENDs, if any, are met mainly by RF-EH. As these devices do not have an energy storage capability, the harvested energy cannot be used at a later stage. To stabilize the use of available energy for their instantaneous operations, it is possible that UP-END are equipped with tiny transient capacitors.

Typical Functions: UP-END can only provide limited functionality, mainly relying on reflection and scattering. Other useful functions, such as PUF and passive sensing, can be realized by utilizing the material properties.

Communication: UP-END devices communicate by modulating a secondary physical signal, most commonly an RF carrier, through mechanisms such as Backscatter Communication (BC). Alternatively, they may leverage their inherent physical structure or material properties to encode information, such as an identifier, which can be extracted through RF, optical, or other interrogation methods.

Security: The security features for UP-END rely on their structural/material properties and hence require precise fabrication techniques. At a basic level, PUF can be used for the identification of these devices. However, challenge-response-based authentication mechanisms can also be implemented by dynamically altering the device's impedance state.

Applications: UP-END mainly provides passive support to IoT applications. For instance, UP-END in the form of passive reflectors can improve the channel quality for the link between transmitter and receiver through passive beamforming. Other common applications for these END include localization, RCS tuning, identification/authentication, and environmental and structural monitoring.

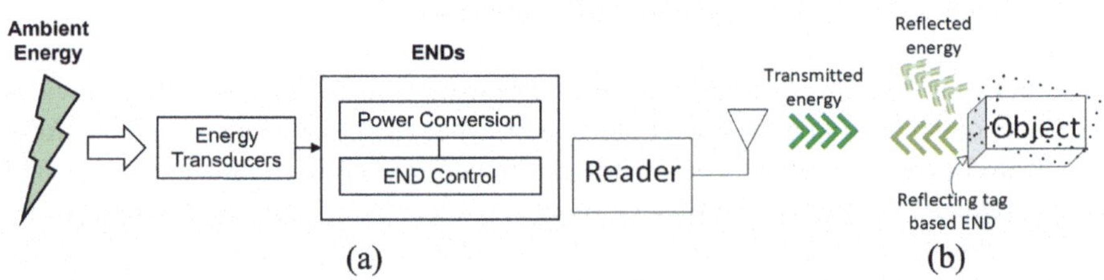

Fig. 3 High-level block diagram of a UP-END: (left) END with transient level energy storage capability; and (right) END with no energy storage at all, where the presence/absence of reflected energy can, e.g., convey orientation of the object

2.1.2 Prototypes and Case Studies

"Iridescent" Reflective Tags to Enable Radar-Based Orientation Estimation:

The concept of tags with a frequency-selective angular response property is proposed and discussed in [23]. These tags have maximal reflected energy in a sub-band of the received wavefront spectrum. Such "iridescent" reflective tags are UP-END and can be used to assist robots to detect obstacles in visually challenging environments, e.g., in situations where visibility is limited due to, e.g., smoke or fog, as shown in Fig. 4. The tags merely reflect the impinging RF waves, and the radiating entity (mobile robots in this case) can infer relevant information, such as position and orientation from the reflected waves.

The working principle of an "iridescent" reflective tag is shown in Fig. 5: a radar (e.g., equipped on the robots in Fig. 4) illuminates the tags with different frequency transmissions and based on the reflected signals decides on the orientation of the tag, and potentially that of the object it is attached to. These tags can be constructed using Leaky-Wave Antenna (LWA) technology. In such a case, some minimal RF-EH circuitry is needed for feeding.

NFC-Based Passive Flow Sensor for Smart Silos

To refill a silo with animal feed, a trailer with multiple compartments is connected to the silo using a filling tube. The feed is then blown through the tube and into the silo. Such a system, illustrated in Fig. 6, is deployed to verify both the correct feed type is dispensed from the appropriate trailer compartment into the correct silo and that the feed flow is properly tracked. The system utilizes Near-Field Communications (NFC) technology, consisting of an NFC reader mounted on the trailer and a passive NFC sensor embedded in the silo, which is an UP-END.

The NFC sensor operates without an internal power source, relying entirely on the electromagnetic field generated by the NFC reader for activation. When the reader positioned on the trailer is aligned with the sensor embedded in the silo,

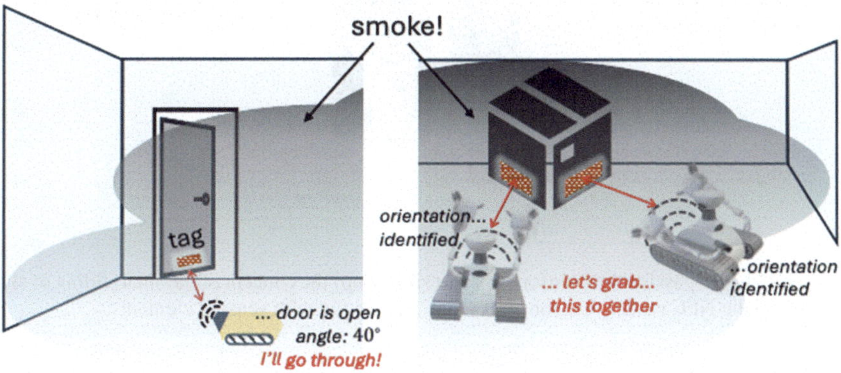

Fig. 4 Use case for "Iridescent" reflective tags [23]

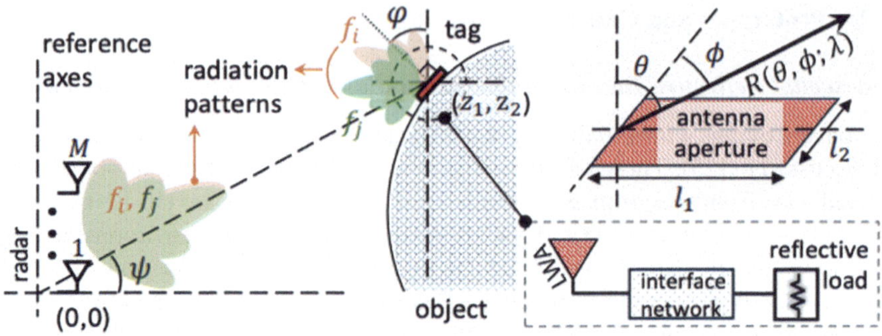

Fig. 5 Working principle of "Iridescent" reflective tags [23]. Herein, M refers to the number of radar antennas; fi and λ_i are the i-th radiation frequency and wavelength (which activate a tag's distinctive response), (z_1, z_2) are the tag location coordinates, ψ and φ are orientation angles, θ and ϕ are the radiation pattern response angles of the tag, and (l_1, l_2) are the tag dimensions

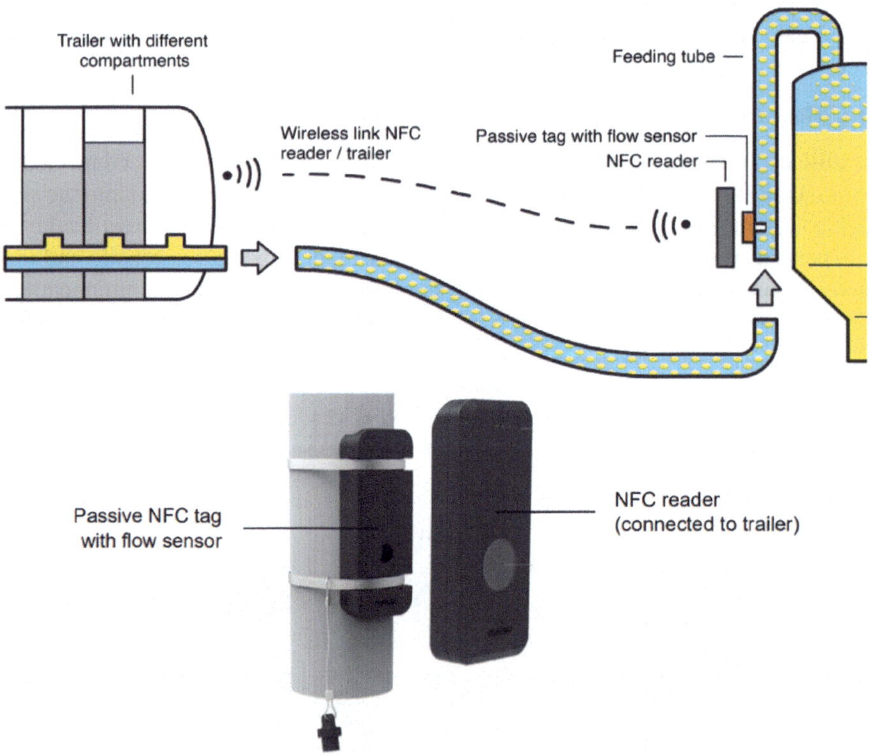

Fig. 6 NFC-based passive flow sensor for smart silos: (top) the concept of the smart filling of silos; and (bottom) the NFC reader that connects to the passive tag with sensing element

it energizes the system, allowing the sensor to function as an NFC tag and power up its onboard Microcontroller Unit (MCU). Upon activation, the MCU writes a unique identification code into the NFC memory buffer, enabling the system to verify both the silo being filled and the corresponding trailer compartment. In addition to identification, the prototype integrates an analog flow sensor, which the MCU monitors using its built-in Analog-to-Digital Converter (ADC). The ADC begins sampling as soon as feed flow is detected, and the resulting data is stored in the NFC memory buffer. This information becomes immediately accessible to the NFC reader, allowing real-time tracking of flow rate, while the system remains powered. By eliminating the need for energy storage and operating entirely via NFC-based power delivery, the embedded sensor is maintenance-free and can be seamlessly integrated into the silo structure in a weatherproof manner. The absence of batteries and exposed power contacts further enhances the system's durability and resilience to environmental conditions.

RFID Sensor System Using a Tag Antenna as Sensing Device

The prototype created by Görtschacher et al. [24] presents a RFID sensor in the Ultra High Frequency (UHF) domain that helps detect the fill level of a water canister based on the backscattered transponder signals of the sensor tag, as illustrated in Fig. 7. This is achieved by exploiting changes in the tag's impedance coming from external stimuli, like variations in water level in this example. These impedance changes affect the backscattered signal received by the RFID reader. By processing these signal variations, the system can deduce the corresponding environmental conditions. This approach maintains the passive nature of the RFID tags, ensuring low-maintenance and cost-effective sensing capabilities.

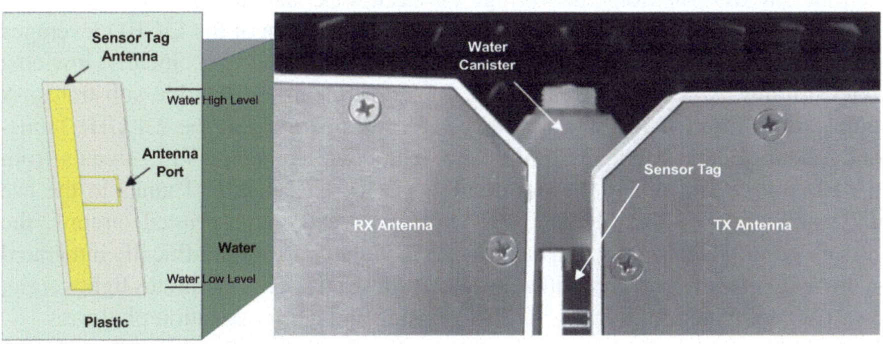

Fig. 7 Schematic overview and measurement setup of the RFID sensor tag for detecting the water level of a water canister based on the backscattered transponder signal of the sensor tag (adapted from [24])

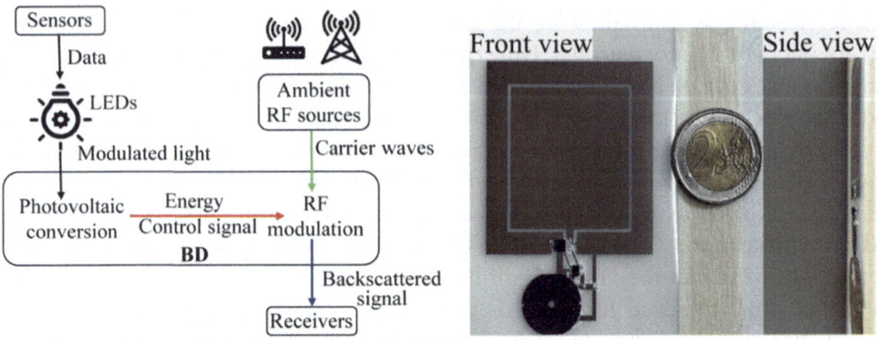

Fig. 8 Light-controlled BD for a hybrid VLC/RF network

Light-Controlled Backscatter Device

A light-controlled BC system is proposed in [25] to enable seamless integration of Visible Light Communication (VLC) and RF-based backscatter for future ambient-powered IoT applications. The key component, a Light-controlled Backscatter Device (LiCBD), utilizes light signals for both energy harvesting and modulation control. The device comprises a light harvester and a backscatter modulator.

The operation principle of the LiCBD is illustrated in Fig. 8. The light harvester captures incident light, typically from an LED, and converts it into electrical energy used both to power the device and to control the modulation state. Under illumination, the LiCBD modulates the reflection coefficient of incoming ambient RF carrier waves, thus encoding information onto the backscattered signal. These modulated signals are then detected by receivers equipped with RF front ends. In this way, the LiCBD functions as a bridge between VLC and RF networks, enabling hybrid communication and power delivery. The fabrication of the LiCBD leverages advanced packaging and additive manufacturing techniques. Specifically, low-cost inkjet printing was used to realize the RF components on flexible substrates. A compact thin-film LiCBD was developed in [26], operating at the 2.4 GHz Industrial, Scientific, and Medical (ISM) band. The device receives VLC signals from an LED source and relays the information via RF backscatter. Thanks to the use of Commercial Off-The-Shelf (COTS) materials and inkjet-printed circuits, the antenna, control electronics, and backscatter modulator are monolithically integrated on flexible Polyethylene Terephthalate (PET) substrates. This facilitates lightweight, low-cost deployment of hybrid VLC/RF sensing and communication platforms.

2.2 *Duty-Cycled Barebone END*

Distinguishing Feature(s)

Presence of short-term energy storage to power operations in a duty-cycled manner.

2.2.1 Architectural and Operational Properties

DB-END extends the basic functionality of the UP-END class by introducing a small, short-term energy buffer in the form of capacitors or tiny supercapacitors. This allows the device to accumulate minimal amounts of power and operate intermittently in short bursts, e.g., performing basic on-device sensing, rudimentary data processing, or brief communication tasks on a duty-cycled schedule. Their operation is not necessarily instantaneously triggered by RF illumination. Examples include smart labels that log environmental conditions (e.g., temperature or humidity), simple wearable health tags that periodically check vital signs, and ultra-low-power sensor nodes for industrial or agricultural monitoring. While slightly more complex than UP-END, DB-END is still low-maintenance and has long operational lifetimes and modest fabrication costs, making them ideal for cost-effective, semi-autonomous IoT solutions. A high-level block diagram for DB-END is provided in Fig. 9.

Energy: In contrast to UP-END where energy could not be stored for later use, a short-term energy storage either using small capacitors or supercapacitors is available for DB-END. Thus, energy does not have to be consumed instantaneously but it can be rather accumulated for short burst operation. Typical energy sources for this END class include RF, small solar cells, piezoelectric harvesters, and thermoelectric gradients.

Typical Functions: DB-END can perform some basic functions, such as sensing and data processing. Examples of sensing can include, among others, temperature and humidity measurements. On the other hand, data processing examples can include basic filtering or threshold detection. These functions can be activated either by some event (event-driven) or periodic signal (periodic-activation). Indeed, energy preservation mechanisms are used to keep the device at a low or no-power state most of the time while waking it up briefly to perform a task.

Communication: Although energy storage is possible for devices in this class, the communication is still predominantly passive, i.e., backscattering-based, or semi-passive, i.e., supporting low-power active communication, depending on energy buffer and energy dimensioning. Additionally, the supported transmission has low-throughput and uses communication in bursts, i.e., transmit quickly and go dormant.

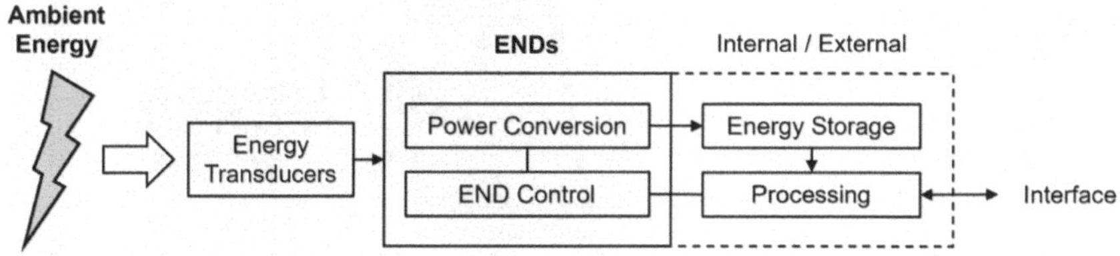

Fig. 9 High-level block diagram of a DB-END

Security: Since DB-END can implement some basic data processing functions, the challenge-response mechanisms in this class do not need to rely solely on the device's impedance state, as for the UP-END. Indeed, they can adopt simple challenge-response mechanisms if the hardware allows brief computation. Furthermore, DB-END may also store small cryptographic keys in NVM for authentication.

Applications: Similar to UP-END, DB-END mainly provides passive support to IoT applications, but wherein the former failure due to their rather stringent constraints. This includes DB-END as smart labels for inventory and logistics, wearable health tags, and environmental sensors in industrial or agricultural monitoring.

2.2.2 Prototypes and Case Studies

RF-Powered Electronic Shelf Label

A battery-less Electronic Shelf Label (ESL) powered by RF-WPT using COTS components is investigated in [12]. The proposed ESL architecture is validated through a dedicated testbed and is accompanied by a detailed analysis of design trade-offs, including energy consumption, power conversion efficiency, and storage strategy. A leaded aluminum electrolytic capacitor is employed as the primary energy storage element, selected for its favorable balance between cost, size, and energy density. Experimental results show that the ESL is capable of performing display updates within 4 to 120 min, depending on the incident RF power and operating frequency, with energy harvesting efficiencies reaching up to 30. An illustration of the prototype is shown in Fig. 10.

Hybrid RF Indoor Positioning System

A low-power hybrid positioning device that combines RF backscatter and acoustic sensing is presented in [27]. The system employs RF signals for synchronization and communication via frequency-shifted, load-switching backscatter modulation,

Fig. 10 Prototype of the proposed ESL architecture. The black box in the upper right corner includes two connectors, allowing real-time voltage monitoring of the capacitor

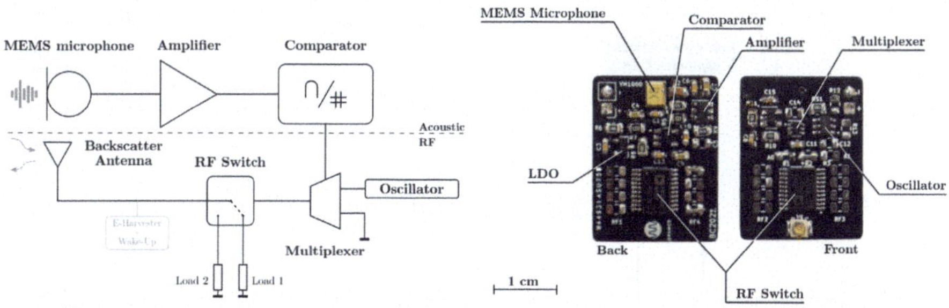

Fig. 11 Block diagram and PCB layout of the backscatter module

while distance estimation is achieved by pulse compressing reflected acoustic chirp signals. The device includes an acoustic front end based on a microelectromechanical systems (MEMS) microphone, which captures ultrasonic signals that are digitized using a zero-crossing circuit. These digitized signals are then modulated onto an ambient RF carrier by dynamically switching the impedance of the antenna load.

The current prototype is powered by an external energy buffer and stabilized using a Low-Dropout (LDO) voltage regulator. This setup provides a local power supply for both the acoustic and RF subsystems and enables intermittent operation under constrained energy conditions. A Printed Circuit Board (PCB) implementation integrates the acoustic acquisition chain, local oscillator, and RF switching logic into a compact module. A block diagram and layout of the backscatter hardware are shown in Fig. 11.

To achieve full energy-neutral operation, the system is designed for future integration with an RF energy harvester (e.g., AEM40940) and a wake-up detector to enable synchronized activation. These components are not yet integrated but represent a clear development path toward a self-sustaining indoor positioning platform. Such a hybrid approach holds promise for enabling maintenance-free location-aware IoT devices in environments, where conventional wireless localization methods are unreliable or energy-prohibitive.

RFID-Based Passive Distributed Control for a Reconfigurable Intelligent Surface

A Reconfigurable Intelligent Surface (RIS) is a programmable electromagnetic surface capable of dynamically altering the propagation characteristics of incident radio waves. RISs are increasingly investigated for applications in wireless communication and WPT systems. A prototype RIS device, classified as a DB-END, is presented in [28] and shown in Fig. 12.

The prototype features an energy harvesting power supply based on ambient light, which charges a supercapacitor with a storage capacity of approximately 2.2 J. A RFID interface is used to wirelessly receive control signals that configure the surface. The control logic is implemented on a microcontroller unit (MCU, Atmega4809), which decodes the RFID commands and controls an array of RF switches to dynamically reconfigure the RIS antenna elements.

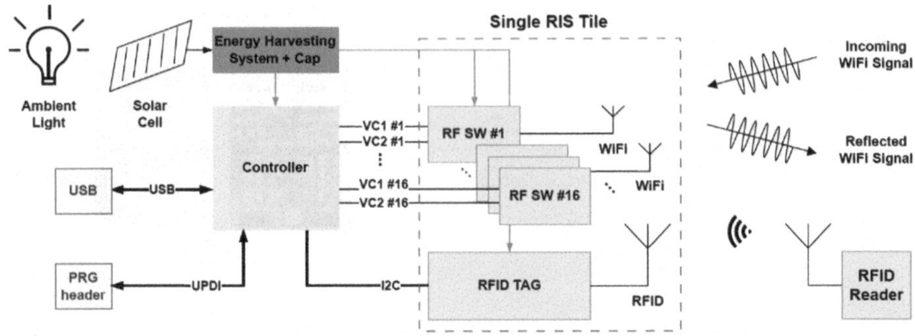

Fig. 12 Schematic overview of the RIS END, highlighting the ambient light energy harvesting supply and RFID-based control logic

This architecture demonstrates the feasibility of low-power, battery-less RIS configurations using commercial off-the-shelf components. The use of an RFID interface for distributed control enables remote, infrastructure-assisted reconfiguration without continuous powering or data links, making it well-suited for future ambient-powered intelligent wireless environments.

GHz RF Energy Harvesting Test Device

The device shown in Fig. 13 serves as a platform for evaluating an RF-EH integrated circuit (IC), originally designed for the UHF RFID band around 900 MHz. This evaluation focuses on its performance at 2.45 GHz, a frequency situated within the ISM band commonly used by wireless communication protocols such as Wi-Fi, Bluetooth, and Zigbee [29]. The platform includes a custom-designed dipole antenna matched for 2.45 GHz and is used in conjunction with the microPowerProfiler (μPP) tool for harvested power analysis.

Although not an END prototype in itself, this platform demonstrates the charge pump functionality of the RF-EH IC, which is critical for converting received RF energy into usable Direct Current (DC) power. A Power Conversion Efficiency (PCE) of 11.6% was measured using the default antenna setup, while efficiencies up to 25.9% were achieved with improved impedance matching [29]. These results confirm the IC's suitability for integration into future RF-powered ENDs.

The same device was subsequently used in real-world RF-WPT experiments in an indoor laboratory environment. As illustrated in Fig. 14 and Fig. 15, the test setup involved beamformed downlink power transmission from a Multiple-Input Multiple-Output (MIMO) Base Transceiver Station (BTS) using a Time-Division Duplexing (TDD) Long-Term Evolution (LTE) signal at 2.44 GHz with 20 MHz bandwidth. Uplink pilot signals from an LTE user equipment (UE) were used to steer the BTS beam toward a fixed spatial point, while the RF-EH device was moved across the area and its harvested DC power output was recorded.

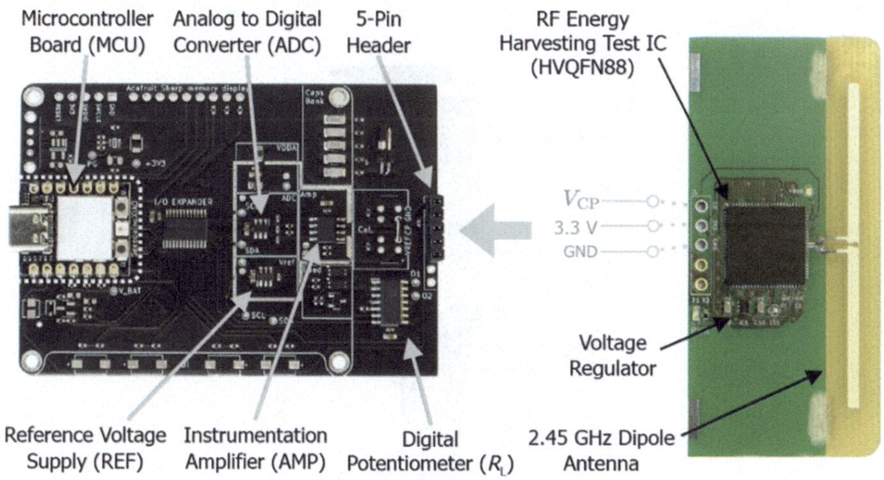

Fig. 13 RF energy harvesting test platform with the microPowerProfiler (left) and test IC module with a 2.45 GHz dipole antenna (right)

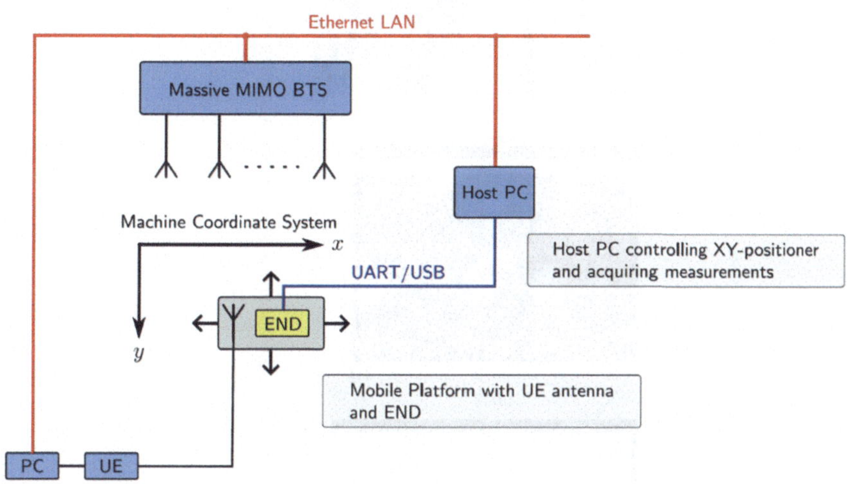

Fig. 14 Schematic overview of the wireless power transfer measurement setup

This setup provides valuable insights into practical RF-EH performance under controlled beamforming conditions and confirms the test IC's applicability for energy-neutral IoT systems operating within dense wireless environments.

Wideband Backscatter Channel Estimation

A Proof-of-Concept (PoC) platform for validating wideband backscatter channel estimation algorithms is shown in Fig. 16. The setup emulates a bistatic system, in which an Orthogonal Frequency-Division Multiplexing (OFDM) signal transmitted

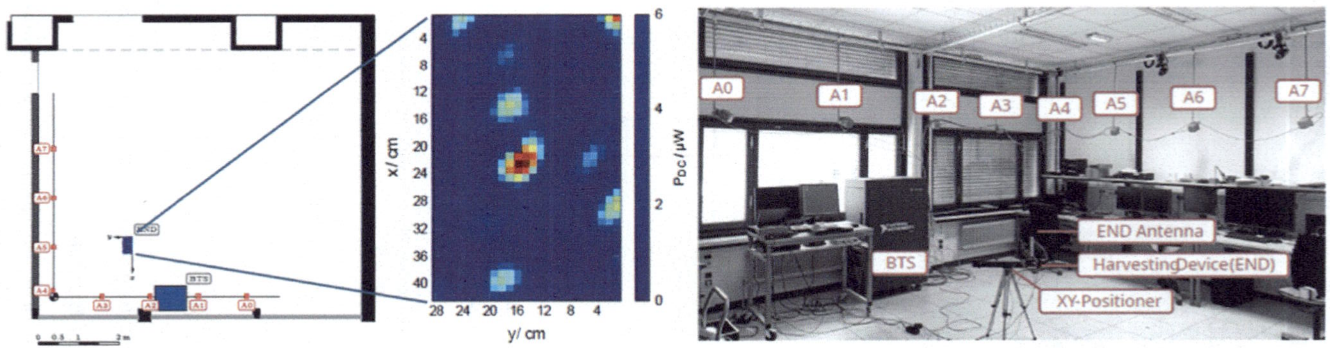

Fig. 15 Indoor RF-WPT beamforming experiment: layout and received power mapping from the BTS across a defined measurement area

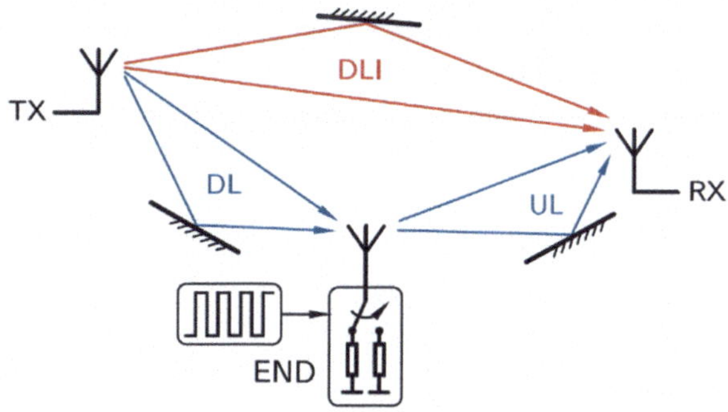

Fig. 16 System model: An OFDM signal transmitted from the TX impinges at the RX as the sum of the strong DLI with several multipath components and the weak backscatter-modulated signal from the END

by a source (TX) arrives at the receiver (RX) as a combination of strong Direct-Link Interference (DLI) and weak backscatter-modulated components from an END. The scenario reflects practical challenges in backscatter communications, such as channel estimation, interference suppression, and synchronization.

The algorithm proposed in [30] enables: (i) wideband Channel State Information (CSI) estimation in bistatic setups, (ii) mitigation of direct-link interference, and (iii) estimation of unknown backscatter modulation parameters for synchronization. The recovered CSI facilitates positioning of the END, optimization of WPT, and robust communication, even when ambient TX signals such as Wi-Fi or OFDM are used.

Figure 17 illustrates the wired test configuration using a Software-Defined Radio (SDR). The TX signal is split: one path feeds the receiver directly, while the second is routed to the backscatter END via a long coaxial cable to emulate propagation delay. A variable RF attenuator adjusts the Signal-to-Interference Ratio (SIR). The END is implemented using an ADG918 CMOS-based RF switch with external impedances modulated via a Pulse-Width Modulated (PWM) control signal from an arbitrary waveform generator.

While this setup does not include full energy harvesting capability, the modulation circuit requires only minimal energy beyond inherent parasitic storage. As such, the framework is representative of an UP-END implementation, in which the

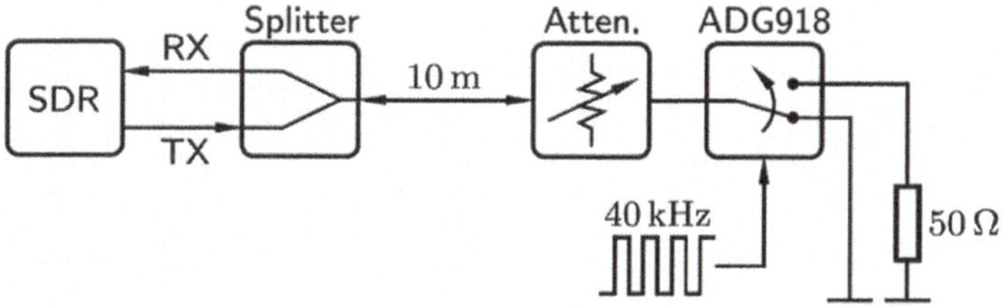

Fig. 17 Wired test setup for bistatic backscatter channel estimation using an SDR

device reflects and modulates incident signals without active power storage. The platform may be further extended to allow for wireless MIMO transmissions to explore advanced CSI estimation techniques and validate algorithms for ambient-powered backscatter networks.

2.3 Duty-Cycled Persistent END

Distinguishing Feature(s)

Presence of long-term rechargeable energy storage and persistent operation according to schedules/duty cycles.

2.3.1 Architectural and Operational Properties

DP-END devices include small, rechargeable energy storage components, such as micro-batteries or robust supercapacitors, which allow them to operate on scheduled duty cycles. This stored energy supports periodic wake-ups for performing moderate sensing tasks, logging data, and executing low-complexity actuation. When needed, these devices can preserve or restore system state using non-volatile memory, enabling robust operation across intermittent power cycles.

The extended energy budget also allows DP-ENDs to support lightweight data processing and communicate via backscatter or ultra-low-power active radio protocols. By accumulating enough energy between active periods, they can reliably measure environmental parameters, process or store data, and transmit information. These devices strike a practical balance between autonomy and energy efficiency, making them well-suited for applications such as industrial IoT systems, precision agriculture, and biomedical monitoring-domains where intermittent yet dependable functionality is essential. Compared to lower-tier END classes, DP-ENDs offer increased versatility and adaptability. A high-level block diagram of a representative DP-END is shown in Fig. 18.

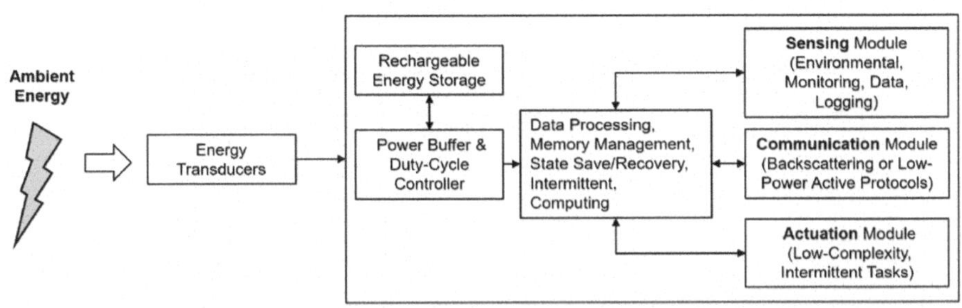

Fig. 18 High-level block diagram of DP-END

Energy: DP-END has long-term energy storage capability in the form of a sufficiently large supercapacitor or even a small rechargeable battery. Energy storage allows energy accumulation to handle scheduled duty cycles and maintain regular or event-driven wake cycles for moderate-complexity tasks. The most commonly used energy sources DP-END include small solar cells, thermoelectric gradients, dedicated RF, kinetic/vibration, piezoelectric, and electrodynamic.

Typical Functions: In addition to the typical functionalities of the DB-END (i.e., sensing and data processing), DP-END can also provide moderate on-device operations such as low-complexity actuation, data aggregation from multiple sensors, and low-complexity local decision-making. They may recover computation/data states using intermittent computation techniques.

Security: DP-END may support lightweight cryptographic and authentication measures, e.g., simple challenge-response protocols or key-based authentication. It is also possible to integrate PUF by storing PUF-based or material-based signatures with added capability to handle queries through basic logic. Finally, the NVM allows the device to retain credentials or encryption keys between cycles.

Applications: DP-END fits i) precision agriculture, by continuously monitoring soil moisture, temperature, and crop health with minimal power consumption; ii) industrial settings, by tracking machinery performance and environmental conditions, reducing maintenance costs and downtime; and iii) wearable biomedical applications, by reliably recording vital signs over extended periods. Moreover, their energy-efficient design and adaptive functionality make them excellent candidates for smart infrastructure and critical monitoring systems in various sectors.

DB-END and DP-END devices can support indoor localization by leveraging their energy-efficient operation for positioning tasks. Passive DB-END devices can be detected based on their impact on channel estimates at standard receivers. When illuminated by ambient RF signals, such as downlink transmissions from base stations or uplink signals from user equipment in mobile networks, these devices modulate their presence into the received signal [31–33]. This enables proximity-based localization by determining whether a DB-END is detectable within the receiver's range [34].

More accurate positioning becomes possible with DP-END devices, which can intermittently activate to measure received signal strength or phase differences. Using such measurements, these devices can estimate their relative position with greater precision [34]. Together, these methods enable robust indoor asset tracking and navigation while maintaining extremely low-power consumption. By exploiting existing ambient communication signals, the system remains scalable and cost-effective for deployment in smart buildings, industrial facilities, and other environments, where traditional localization methods may be impractical.

2.3.2 Prototypes and Case Studies

Low-Power Ambient Backscatter Device with MCU Timer Management.

A DP-END for Ambient Backscatter Communication (AmBC) is presented in [32] and shown in Fig. 19. The prototype integrates a Photovoltaic (PV) energy harvester, a low-power MCU, and a backscatter modulator based on an RF switch. Energy harvested from indoor lighting is stored in four 0.1 F supercapacitors, providing a total of 0.4 F of storage capacity.

The device implements Binary Frequency-Shift Keying (BFSK) modulation by toggling the RF load impedance between two switching frequencies. A timer peripheral in the MCU generates the baseband modulation signal with a 50% duty cycle, while another timer manages symbol timing. Both timers operate autonomously, allowing the MCU to remain in a low-power sleep state during most of the symbol duration. During active transmission, the device consumes approximately 22 mW with a 40 ms symbol duration. Under a 10-s transmission interval, the energy storage is sufficient to sustain autonomous operation for over 36 h without external illumination.

The backscatter modulator reflects ambient RF signals by switching the antenna load between open and short states. These impedance variations are superimposed on the incident RF signal, enabling a remote receiver to demodulate the reflected waveform. This architecture demonstrates how energy-aware timer control and compact energy storage can support sustainable, low-power wireless communication in indoor IoT scenarios.

Fig. 19 Low-power AmBC device based on MCU timer management

2.4 *Continuous Operation END*

Distinguishing Feature(s)

Presence of long-term rechargeable energy storage that allows (semi-)continuous operation.

2.4.1 Architectural and Operational Properties

CO-END represents the most capable category of END, featuring robust or continuous EH and sufficiently large rechargeable storage to enable near-continuous operation, including advanced computation/processing features. They typically support active communication protocols, making them highly suitable for smart city applications, predictive industrial maintenance, and autonomous environmental monitoring. CO-END is expected to function autonomously, self-adapt to changing conditions, and integrate more advanced analytics such as Machine Learning (ML) inference, all while remaining energy-neutral. They bridge the gap between energy-efficient IoT and fully autonomous wireless systems. A high-level block diagram for CO-END is provided in Fig. 20.

Energy: CO-END harvests energy from either a single high-power EH source or multiple sources, such as solar, thermal, and/or vibration, and store it in a long-term storage. Furthermore, these devices may also include smart power controllers to dynamically allocate energy among different types of tasks, e.g., computation, sensing, communication, and actuation.

Typical Functions: CO-END can perform moderate-to-high on-device operations such as sensing, data processing, real-time low-complexity actuation, data fusion, and even local ML inference. In addition, the moderate-to-high onboard logic for processing can be used to perform self-monitoring and adaptive scheduling.

Communication: CO-END may maintain active communication, including frequent and/or real-time data transmission, and implement adaptive communication protocols, e.g., based on channel conditions, frequency/time resource availability, etc.

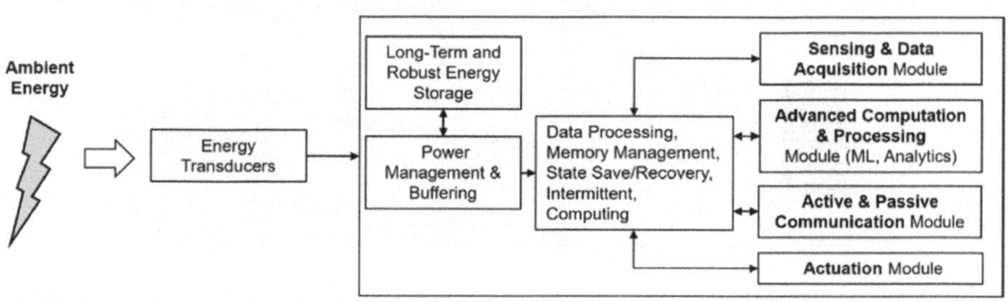

Fig. 20 High-level block diagram of a CO-END

Security: CO-END supports secure firmware updates, mutual authentication with gateways, and intrusion detection, e.g., through anomaly based monitoring. In addition to these network-based security procedures, CO-END can also run AI-based threat detection or anomaly detection locally, thus safeguarding the device and network in real time.

Applications: Generally, CO-END has their applications in smart city and infrastructure, and predictive industrial maintenance. Furthermore, properties such as hybrid EH and smart power controllers make them suitable for environmental monitoring, wildlife monitoring, high-end wearables, and healthcare devices.

2.4.2 Prototypes and Case Studies

Light-to-RF Backscatter Relay (Li2BC):

A continuously CO-END named Li2BC is presented in [35] and illustrated in Fig. 21. The device functions as a decode-and-forward relay, translating control information encoded in visible light into RF backscatter signals. By bridging the visible light and RF domains, Li2BC enables light-driven wireless control and sensing in scenarios where active RF transmission is impractical or energy-prohibitive.

Li2BC is powered by a PV cell, which simultaneously serves as an optical receiver. Information is embedded in the high-frequency components of the incident light and extracted using analog filtering. The filtered signal is amplified and passed to a low-power MCU, which demodulates and temporarily stores (caches) the received VLC data. This caching capability decouples the timing of light reception from RF transmission, allowing asynchronous and configurable backscatter operation.

The MCU also interfaces with onboard sensors, periodically acquiring measurement data. For transmission, Li2BC uses BFSK by configuring an internal timer to control an RF switch that modulates the antenna's reflection coefficient. The timer

Fig. 21 Light-controlled backscatter device with VLC decode and RF-forwarding capability (Li2BC)

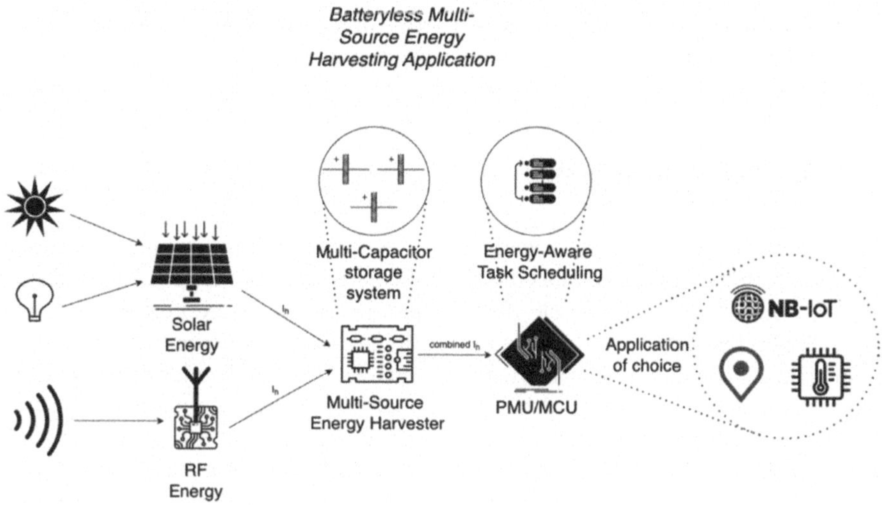

Fig. 22 System overview of the outdoor battery-less tracker

runs independently for each symbol, allowing the MCU core to remain in a low-power state between updates. The resulting backscattered signal can be received by standard RF receivers, enabling reliable and energy-efficient wireless data relay from visible light input to RF output.

Outdoor Battery-less Tracker with Multi-Source Energy Harvesting:

A duty-cycled buffered DP-END designed for outdoor asset tracking is shown in Fig. 22. The device monitors both the location and environmental conditions of an asset and operates entirely without batteries, relying on a hybrid energy harvesting and management architecture. Its main components include a supercapacitor, a multi-source energy harvester, a Power Management Unit (PMU), and application-specific modules [36].

The integrated energy harvester combines a solar panel with an RF energy harvesting front end and is designed to accommodate additional harvesting modalities. Harvested energy is stored in a supercapacitor, which buffers the system against fluctuations in ambient energy availability. To manage these dynamics, the PMU incorporates a low-power MCU that continuously monitors both the harvester input and the state of the energy storage. Based on available power, it dynamically allocates energy to system functions, enabling higher data transmission rates during periods of surplus and ensuring critical functionality, such as periodic heartbeat messages or alarm signals, even under constrained conditions.

The application layer includes a Global Positioning System (GPS), a temperature sensor, and a Narrowband IoT (NB-IoT) communication module. These components operate independently under the control of the PMU, which orchestrates their activation based on real-time energy availability. This adaptive control framework ensures reliable tracking and condition monitoring across diverse outdoor environments, even in the presence of intermittent or low-intensity energy sources.

3 Life Cycle Assessment

The ecological assessment presented in this section is preliminary and conceptual in nature. As this work is produced during the early phase of an ongoing research project, comprehensive life cycle data for the discussed device prototypes is not yet available. Instead, we outline the key ecological considerations relevant to each component class and introduce assessment frameworks that will guide future sustainability evaluations. This forward-looking perspective serves two purposes: first, to identify potential environmental hotspots early in the design process, and second, to inform the development of metrics and methodologies for evaluating ambient-powered IoT systems in real-world deployments. Future updates will expand on these foundations with quantitative results as more empirical data becomes available through project activities.

Each deployed IoT device and associated service consume resources, including raw materials and energy. Regardless of scale, every product or service has an inherent environmental cost. The electronics industry alone accounts for approximately 4% of global greenhouse gas emissions, placing it among the top eight sectors responsible for more than 50% of total emissions [37, 38]. Notably, 77% of these emissions originate from the supply chain. Compounding the issue, the electronics sector is heavily dependent on Critical Raw Materials (CRMs) for essential components such as batteries, energy harvesters, and integrated circuits. Furthermore, only 20% of today's IoT devices are formally recycled, contributing to electronic waste becoming the fastest-growing waste stream globally.

Understanding and quantifying these impacts is essential for the development of more sustainable IoT solutions. Life Cycle Assessment (LCA) is a widely recognized methodology for evaluating the environmental footprint of products and systems across their entire lifespan. The ISO 14040 and ISO 14044 standards [39, 40] define the internationally accepted framework for conducting such assessments. They provide a structured methodology for analyzing environmental impacts from raw material extraction through manufacturing, use, and end-of-life, and they specify principles, data quality requirements, and reporting procedures. By enabling the identification of environmental hotspots, trade-offs, and opportunities for improvement, these standards play a critical role in guiding sustainable design choices related to material selection, energy use, and system architecture.

While the assessments presented in this chapter do not yet implement these standards in full, they are informed by their underlying principles and structured to allow future alignment. This approach ensures consistency as the project matures, and more detailed data becomes available. It is particularly relevant for the future commercialization of ENDs, where adherence to standardized LCA methodologies will

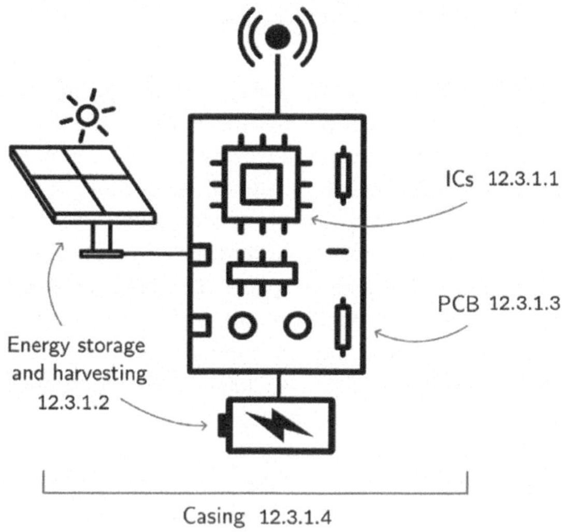

Fig. 23 Overview of ecological assessment key points

be important for sustainability certification, regulatory compliance, and industrial adoption.

The carbon footprint of a typical IoT device currently ranges between 1 kgCO2eq and 10 kgCO2eq [41], significantly lower than that of a modern smartphone, which has a Global Warming Potential (GWP) of approximately 70 kgCO2eq [42]. However, due to the projected massive scale of deployment, potentially tens of billions of devices [43], combined with increasing processing demands and the use of smaller semiconductor nodes [44], the environmental footprint of IoT as a sector is expected to grow dramatically.

Section 3.1 explores the environmental impact of different IoT device components, as illustrated in Fig. 23. Section 3.2 then outlines mitigation strategies aimed at reducing the overall environmental burden of future ambient-powered IoT systems.

3.1 Environmental Impact of IoT Device Components

The manufacturing phase of typical IoT devices accounts typically for more than 60% to 70% of the environmental impact of IoT device [41]. This section highlights this importance by providing an overview of the impact associated with individual IoT device components, highlighting the key contributors to this environmental impact.

3.1.1 IC Integrated Circuits

IC are the core of any IoT device, enabling computation, communication, and control. However, their environmental footprint is substantial [41]. The semiconductor industry relies on the extraction of rare minerals, e.g., tin, tungsten, tantalum, and gold, which must be mined to support production. Additionally, the manufacturing processes rely on various chemicals and materials, including acids, bases, and solvents, for deposition, patterning, selective removal, and cleaning.

Most environmental impacts of IC production stem from high electricity consumption during manufacturing and wafer usage at the raw material stage. Such manufacturing requires large amounts of clean water and generates hazardous waste. IC are therefore typically associated as the component with the highest environmental impact in a typical IoT device [41].

Advances in technology have improved efficiency per transistor but have led to an overall increase in the environmental footprint [44]. While the overall die area stays relatively constant, the environmental cost of IC increases when shifting to smaller technology nodes, therefore increasing the total environmental cost of IC fabrication. Optimizing environmental impacts per cm2 is therefore unlikely to significantly reduce the environmental footprint of IC production, particularly for advanced nodes, incorporating more processing power [45].

3.1.2 Energy Storage and Harvesting

Many IoT devices rely on batteries and supercapacitors for power. A major limitation of this approach is the finite battery capacity. Despite this, they also pose challenges due to material extraction, manufacturing emissions, and EOL disposal. Non-rechargeable batteries, commonly used in many IoT devices, are designed for single use, turning into inevitable waste. Once depleted, they are typically discarded, adding to landfill waste and slowly releasing harmful chemicals into the environment. Rechargeable battery types already bring benefits by their multiple use but also use scarce materials. NMC, NCA, LTO, and the older NIMH and NICD batteries use the most amount of scarce elements, i.e., one or two element(s) in the rising threat category and three or four elements in the future risk category. LFP, e.g., uses no elements from the rising threat category and only two elements from the future risk category [46].

An alternative or supplementary method of powering IoT devices is by using energy harvesting. The possibility to downscale or even eliminate polluting batteries could make these alternatives a "greener" solution to traditional power sources. However, it introduces its own environmental impact due to the additional hardware requirements, such as IC, materials, or antennas, as well as potential inefficiencies in energy transfer. The potential of EH is further discussed in Sect. 3.2.2.

3.1.3 PCB

PCBs are an essential component in IoT devices. Their impact is influenced by factors such as the choice of substrate material (FR4, PET, Polyimide, etc.), the number of layers, the surface finish (HASL, ENIG or OSP), and the specific fabrication methods involved. Three main categories of PCB types, each with their own characteristics, are depicted below:

Conventional PCB-Based Devices

Manufacturing and Materials: Conventional PCB-based devices are built using well-established PCB technologies [47]. Standard PCB substrates such as FR4 (a flame-retardant woven glass-reinforced epoxy laminate) or other rigid composite materials are most common. Automated SMT assembly lines are used to solder components onto the board with high throughput and excellent repeatability. The metallization (usually copper) and standard solder masks ensure robust electrical interconnections and mechanical durability.

Form Factor and Practicalities: Conventional PCB typically has a fixed, rectangular, or standard layout, which is dictated by design rules and available board sizes. They are optimized for high-density integration of components while maintaining structural rigidity. They offer excellent thermal management and mechanical strength, making them suitable for industrial or stationary applications, where shock, vibration, and temperature variations are expected. A major advantage is that these devices benefit from economies of scale, mature supply chains, and a well-understood design process. However, the inherent rigidity limits their use in wearable or conformal applications and may not be ideal for devices that must integrate onto curved or flexible surfaces.

Flexible and Thin-Film-Based Devices

Manufacturing and Materials: Flexible devices are fabricated using advanced printing and roll-to-roll processing techniques on plastic substrates such as PET or polyimide (e.g., Kapton) [22]. In place of conventional rigid substrates, these devices often use conductive inks or ultra-thin metallization layers. In some cases, organic semiconductors or printed dielectrics are used to further reduce weight and allow for conformability. Manufacturing processes are more specialized than traditional PCB production and may involve inkjet printing, screen printing, aerosol printing, or other additive manufacturing techniques [48].

Form Factor and Practicalities: Since the substrate is flexible, these devices can be made ultra-thin, conformal, and even foldable. They can easily be integrated into wearable electronics (e.g., smart textiles or medical patches) or onto non-planar surfaces such as curved enclosures [49]. Their form factor offers a low-profile design and can be tailored for applications, where weight and aesthetic integration are key. However, the use of flexible materials can sometimes mean trade-offs in terms of durability and long-term reliability. Flexible circuits might be more sensitive to environmental conditions (e.g., moisture, ultraviolet exposure) and may require additional protective coatings. Production volumes may be lower and costs higher (especially at smaller scales) compared with conventional PCB devices, although emerging high-volume printing techniques are steadily closing this gap.

Metamaterial-Based Devices

Manufacturing and Materials: Metamaterial-based devices use engineered materials whose structure is designed to yield properties not found in conventional materials. This might involve nanoscale or microscale patterning using advanced lithography, etching, or self-assembly techniques [50]. The materials can include composites of metals and dielectrics arranged in subwavelength patterns to achieve, e.g., negative refractive indices, enhanced electromagnetic absorption, or tunable resonances [51]. Fabrication often requires high-precision processes (e.g., electron-beam lithography or nanoimprint lithography), which can increase production complexity and cost.

Form Factor and Practicalities: Metamaterial-based designs allow for innovative form factors that are not limited to standard planar geometries. They can be integrated as ultra-thin layers, conformal coatings, or even as "smart skins" that combine sensing with adaptive electromagnetic functionality, e.g., reconfigurable antennas or cloaking layers. Their unique electromagnetic properties can improve the performance of wireless communication and EH in A-IoT devices, enabling, for instance, improved signal control or novel sensor modalities. On the practical side, while these devices can offer transformative performance enhancements, they often come with higher manufacturing complexity and may currently be limited to niche or high-value applications. They can also require tighter tolerances and more stringent quality control, which might limit scalability and increase production costs.

3.1.4 Casing

Materials like ABS, aluminum, and steel can contribute to this impact through raw material extraction, manufacturing processes, and end-of-life disposal. For IoT devices with large casings relative to their electronics, the casing can have a significant impact on the overall environmental footprint [41]. Traditional materials like ABS, aluminum, and steel can contribute to this impact through raw material extraction, manufacturing processes, and EOL disposal. LCA results show that the environmental impact increases significantly with material durability. For example, when comparing polycarbonate with aluminum and stainless steel, it shows that, on average, the environmental impact of aluminum is approximately 3.65 times higher than that of polycarbonate, while the impact of stainless steel is approximately 7 times greater [52]. Depending on the application, most electronic casings are made of petroleum-based plastics (e.g., polycarbonate, ABS, epoxy resin) [53], while there is growing interest in exploring more eco-friendly alternatives.

A promising strategy is the use of plant-derived natural materials (biomass). Their production, e.g., mycelium (the root network of a fungus), is less carbon-intensive than petroleum counterparts, and they are easy to grow, are home-compostable at their eventual EOL, and could even be bio-contributing, returning nutrients to the soil [54]. Furthermore, using plants that have already absorbed CO_2 during their growth, can reduce their environmental footprint.

Despite these advantages, bio-based materials often degrade faster, have lower mechanical strength and durability, and typically offer poor heat resistance. Additionally, the crystallization rate of bio-resins during molding is relatively slow, which affects their processability. As a result, lots of bioplastic casings still contain a significant share (often up to 50%) of petroleum-based components to ensure sufficient performance.

When considering biodegradable materials with a lower environmental impact, this trade-off must be weighed carefully against application requirements. Paradoxically, if these materials reduce product lifespan or performance, they could lead to a higher overall impact due to more frequent replacements. Ultimately, selecting casing materials requires a careful balance between reducing environmental impact and ensuring sufficient durability and longevity of the device.

3.2 *Reduction of the Environmental Impact of ENDs*

There are several strategies to significantly reduce the overall environmental impact of devices. While the previous section already briefly outlined some obvious mitigations, such as alternative material choices, improved power systems, and smart design choices like optimized casing to enhance EOL management, this section goes further by exploring these approaches in greater detail and introducing additional strategies. These include energy-efficient wireless communication, sustainable power solutions, eco-friendly manufacturing processes, design strategies that prioritize EOL considerations, and backcasting.

Various strategies have been emerging to address the environmental sustainability of IoT devices, with a focus on energy efficiency, miniaturization, sustainable materials, modularity, and EOL considerations.

3.2.1 Energy-Efficient Operation

One of the primary approaches to address sustainability in IoT today is increasing their energy-efficient operation [55]. Energy efficiency encompasses communication, sensing, processing, and actuation. It is often achieved through the use of ultra-low-power communication technologies, low-power sensors, energy-efficient MCU, optimized actuator duty cycles, and highly efficient DC/DC conversion. Reducing overall energy consumption across all domains allows for smaller batteries, conserving raw materials, and extending the time between charges or battery replacements.

Wireless communication, in particular, significantly contributes to the overall energy consumption of IoT networks, as this is often the most energy-intensive. As most IoT devices continuously transmit and receive data, optimizing communication protocols can improve energy efficiency.

One approach to achieving this is through the adoption of low-power communication protocols [56] such as BLE, Zigbee, and LoRaWAN, which enable efficient data exchange. Additionally, optimizing machine type communications MTC helps to eliminate redundant data transmissions, ensuring that devices only send necessary information rather than continuously transmitting unoptimized data streams. Another promising avenue is AI-driven network optimization, where intelligent algorithms dynamically adjust transmission power and data rates based on real-time conditions.

Wake-up receivers [57] provide an energy-efficient solution for an asynchronous wake-up of higher power and performance radio, without the need for frequent energy-intensive synchronization. Since these receivers remain active most of the time, minimizing their power consumption is essential to reducing overall energy consumption. Duty cycling the RF front end can further optimize energy efficiency by limiting their active time.

Backscatter communication [58] is a promising technology for battery-free IoT, enabling data transmission by passively reflecting incident RF signals to the receiver. Using this technique, devices can achieve microwatt-level power consumption by eliminating energy-intensive radio components such as RF synthesizers and ADC [18].

Beyond communication protocol and receiver optimizations, edge computing can reduce energy consumption at the data center level. By processing and filtering data locally, the overhead of transmitting large amounts of data to power-hungry data centers can be significantly reduced [56, 59].

3.2.2 Sustainable Powering

Beyond improving energy efficiency, EH presents a valuable opportunity to reduce the size of, or even eliminate, batteries in IoT devices by providing supplementary energy. Batteries contain hazardous materials and, if not properly managed at end-of-life, can leak toxic substances into the environment. The integration of EH technologies [56] enables smaller batteries or the complete removal of batteries in some cases, resulting in more self-sufficient IoT devices and potentially lowering their environmental footprint.

The next step toward sustainability involves designing IoT devices that operate without any external power sources, relying entirely on ambient energy harvested from sources such as solar and PV cells, kinetic and vibrational energy (via piezoelectric materials), ambient RF transmissions, or ambient heat. Each of these EH modalities have distinct strengths and limitations, and their suitability, whether used individually or in hybrid configurations, depends heavily on the deployment context. However, the environmental impact of EH technologies is not uniform and is largely influenced by the specific method employed.

Photovoltaic energy harvesting generally has a relatively low environmental impact during production and offers usable energy under daylight conditions. With a conversion efficiency of approximately 20%, it significantly outperforms RF-based energy harvesting, which typically achieves efficiencies below 1% [60].

Nonetheless, RF-EH has advantages in scenarios where solar energy is less viable, such as indoor or low-light environments. A promising approach to improving energy availability in constrained IoT systems is the integration of dedicated energy transmitters into the network architecture [60], which can provide a more stable and predictable energy supply. However, this strategy requires the deployment of additional infrastructure, increasing the overall environmental operational lifecycle of these components. Although the potential of RF-EH is frequently emphasized, many claims are not supported by comprehensive assessments of WPT's environmental impact. In particular, the low efficiency of wireless power transfer over meter-scale distances raises concerns [61].

For instance, a study has shown that supplying energy to a battery-less IoT sensor located 5 m from the transmitter over a 10-year period results in 5.7 times (for a single sensor) and 7.6 times (for eight sensors) higher environmental impact than an equivalent solution based on non-rechargeable batteries, even assuming optimally designed systems [61]. These findings suggest that, under certain conditions, the environmental cost of long-range WPT may outweigh the benefits of battery elimination.

A potentially more sustainable alternative involves equipping devices with ambient EH transducers designed with larger form factors and greater energy conversion capacity. Alternatively, RF-EH for END could be realized by harvesting energy from existing BS or UE transmissions. However, in many deployment scenarios, the received signal power, particularly at the cell edge, can be as low as -110 dBm, which is insufficient for practical energy harvesting. In backscatter communication systems, it is possible to decouple the energy source from the communication signal, using separate ambient energy source for EH while employing existing RF transmissions solely for backscatter modulation [31]. In uplink scenarios, where the backscatter-based END is located close to a transmitting UE, the signal emitted by the UE may be strong enough to support both energy harvesting and backscatter communication [33].

Despite current limitations, EH remains a promising pathway to enhance the energy autonomy of IoT devices while reducing battery dependence and associated environmental impacts.

3.2.3 Eco-friendly Manufacturing

Another common practice in modern IoT device design is miniaturization. By reducing the size of the device, PCB, and IC, lesser quantities of materials are needed, lowering the environmental footprint. However, this is not necessarily the case, as devices tend to incorporate more processing power and memory in smaller semiconductor nodes. While the overall die area stays relatively constant, the environmental cost of IC increases when moving to smaller technology nodes, therefore increasing the total environmental cost of IC fabrication [44].

Gallium Nitride (GaN) FET exemplifies how miniaturization can reduce emissions. Due to their superior electrical properties, GaN FETs match silicon performance at a much smaller size, enhancing sustainability in production, and improving efficiency during operation by reducing energy losses.

Looking ahead, biodegradable electronics [62] offer great potential in reducing waste in sensing applications. Conductive biopolymer, carbon-based or metal-based hybrid materials on biodegradable substrates, could be used replacing conventional rigid PCB. For example, mycelIoTronics [63] have great potential in electronic PCB and sensors, which can facilitate the EOL phase, as they break down more easily. It could potentially allow devices to be left out in nature and disappear by themselves.

However, even if the biodegradable alternative has a reduced impact during production, their shorter lifespan could lead to more frequent replacements and ultimately surpass the carbon footprint of a non-biodegradable approach.

Generally, when printing electronics, inks made from non-renewable resources are used. Using additive manufacturing techniques, such as aerosol and inkjet printing, the amount of ink used can be minimized compared to subtractive manufacturing processes [22, 64, 65]. Additional benefits for these non-contact printing techniques include offering the ability to produce small batches of custom designs.

Another approach to produce more environmentally friendly devices is to substitute conventional fossil-based substrates for more ecological alternatives. Cellulose has been considered an excellent alternative as a result of its abundant nature, non-toxicity, biodegradability, and renewability [66]. However, due to the high porosity, surface roughness, and absorptive disposition, cellulose can be challenging to use as a substrate for printed electronics [65, 67]. These challenges can be overcome by using cellulose on a nanoscale and thus improving its surface, optical, and mechanical properties, making it a suitable substitute for commonly used plastic substrates [68].

3.2.4 Design for End-of-Life

Typically 60–70% of the environmental impact of IoT devices comes from the production phase [41]. A modular approach to IoT can be beneficial by enabling component reuse and replacement. However, modularity is not universally beneficial, as additional materials, components, and potential energy losses associated with modular designs may outweigh their advantages in some cases. A case-by-case LCA should be conducted to determine the most sustainable approach.

A critical aspect is preparing the devices for recycling at their EOL. By incorporating the EOL aspects from the early stages of the design phase, manufacturers can ensure that components or modules can be reused or recycled efficiently. A promising, yet uncommon solution today for EOL recycling is the use of biodegradable materials, as discussed in Sect. 3.2.3.

To efficiently manage large-scale IoT deployments and extend device longevity, automation is essential. Future work focuses on Unmanned Aerial Vehicle (UAV) servicing of IoT nodes [46]. UAV could assist by either supplying additional energy to constrained nodes or replacing faulty components at the end of their life cycle [55]. Compared to traditional methods, such as using cars for servicing, small UAV offer a significantly reduced environmental footprint when covering the same distance. This approach is particularly suited and can mainly be applied in outdoor scenarios in remote or inaccessible locations where human intervention is difficult.

3.2.5 Backcasting

With the large-scale deployment of critical IoT applications, the risk of exceeding the environmental impact limits arises. It has been shown that backcasting [69] is a useful tool to keep IoT deployments within environmental limits. This approach starts from a future desired outcome and works backward to the present to figure out the steps needed to achieve this. Instead of predicting the future based on current trends (forecasting), you imagine the ideal future and determine what actions are required today to make that future a reality.

Backcasting is particularly effective in addressing rebound effects, which are especially relevant for IoT deployments, as these systems are highly susceptible to such effects [69]. We advocate from going from a typical LCA to prioritizing backcasting studies, particularly in high-impact applications, as backcasting studies are more correct and should become the methodological state of the art.

4 Conclusion

This chapter has presented a comprehensive case study on the sustainability of ambient-powered IoT systems, focusing on the architecture, classification, and ecological implications of ENDs. By eliminating or reducing reliance on batteries and enabling autonomous operation via energy harvesting or wireless power transfer, ENDs provide a promising pathway toward large-scale, maintenance-free, and environmentally responsible IoT deployments.

We have outlined four primary classes of ENDs, Ultra-Passive, Duty-Cycled Barebone, Duty-Cycled Persistent, and Continuous Operation, each offering distinct trade-offs between complexity, energy autonomy, and functional capability. The selected case studies demonstrate the practical feasibility of these devices across domains such as agriculture, logistics, environmental sensing, and structural monitoring.

A detailed ecological assessment indicates that while ambient-powered IoT systems can significantly reduce operational energy consumption and maintenance-related emissions, sustainability gains are contingent on holistic design strategies. These include minimizing the use of critical raw materials, optimizing circuit and casing designs for recyclability, and deploying energy-efficient communication and processing methods.

Moving forward, interdisciplinary collaboration is required to bridge engineering innovation with sustainability science. Research must address lifecycle optimization, environmentally benign materials, and system-level integration to fully realize the promise of ambient-powered IoT in achieving long-term sustainability goals. ENDs not only redefine what is technically feasible in IoT but also contribute to the broader objectives of the European Green Deal and global climate neutrality initiatives.

Acknowledgements This work has been supported by the AMBIENT-6G project, which has received funding from the Smart Networks and Services Joint Undertaking (SNS JU) under the European Union's Horizon Europe research and innovation program under Grant Agreement No 101192113.

Views and opinions expressed are however those of the authors only and do not necessarily reflect those of the European Union, or SNS JU. The European Union, or SNS JU cannot be held responsible for them.

Acknowledgment: The authors would like to sincerely thank Dr. Christophe Gaie and Dr Markus Mueck for their invaluable support and insightful contributions throughout the development of this book chapter. Their expertise, guidance, and constructive feedback greatly enhanced the quality and clarity of this chapter.

References

1. López OLA, Rosabal OM, Ruiz-Guirola DE, Raghuwanshi P, Mikhaylov K, Lovén L, Iyer S (2023) Energy-sustainable IoT connectivity: vision, technological enablers, challenges, and future directions. IEEE Open J Commun Soc 4:2609–2666. https://doi.org/10.1109/OJCOMS.2023.3323832
2. López O, Singh RK, Phan-Huy D-T, Katranaras E, Mazloum N, Ruttik K, Jäntti R, Khan H, Rosabal O, Alexias P, Raghuwanshi P, Ruiz-Guirola D, Singh B, Höglund A, Van DP, Azarbahram A, Famaey J (2025) Zero-energy devices for 6G: technical enablers at a glance. IEEE Internet Things Mag 8(3):14–22. https://doi.org/10.1109/IOTM.001.2400138
3. 3GPP (2023) TR 22.840 V19.0.0; Study on Ambient power-enabled Internet of Things; Release 19
4. Butt, M.M., Mangalvedhe, N.R., Pratas, N.K., Harrebek, J., Kimionis, J., Tayyab, M., Barbu, O.-E., Ratasuk, R., Vejlgaard, B.: Ambient IoT: A missing link in 3GPP IoT Devices Landscape (2023). https://arxiv.org/abs/2312.06569
5. Group, I...W.: IEEE 802.11 AMP SG Proposed CSD. IEEE 802.11–23/1212r3 (2023). https://mentor.ieee.org/802.11/dcn/23/11-23-1212-03-0amp-ieee-802-11-amp-sg-proposed-csd.docx
6. NGMN: 5G devices categorization (2020). https://ngmn.org/wp-content/ uploads/Publications/2020/200114–5G-Devices-Categorization v1.2.pdf

7. Balakrishnan, B., et al.: Hexa-X-II-Deliverable D5.2: Characteristics and classification of 6G device classes (2023). https://ngmn.org/wp-content/uploads/ Publications/2020/200114–5G-Devices-Categorization v1.2.pdf
8. 3GPP: TR 38.848 V18.0.0; Study on Ambient IoT (Internet of Things) in RAN; Release 18 (2023)
9. Marasinghe D, Tayyab M, Hailu S, Hsieh F, Mangalvedhe N, Butt MM, Ratasuk R, Hathiramani N (2024) Ambient Iot devices in future cellular networks: a system level study. In: 2024 IEEE international conference on communications workshops (ICC Workshops), pp 129–134. https://doi.org/10.1109/ICCWorkshops59551.2024.10615328
10. Alliance L (2015) LoRaWAN specification. https://resources.lora-alliance.org/technical-specifications/lorawan-specification-v1-0
11. Technologies S, What is a Pseudocapacitor? Capacitors explained. https://www.skeletontech.com/skeleton-blog/what-is-a-pseudocapacitor-capacitors-explained
12. Mulders JV, Callebaut G (2025) Designing RF-powered battery-less electronic shelf labels with COTS components. https://arxiv.org/abs/2503.13009
13. Subrahmannian A, Behera SK (2022) Chipless RFID: a unique technology for mankind. IEEE J Radio Freq Identif 6:151–163. https://doi.org/10.1109/JRFID.2022.3146902
14. Viswanathan B (2016) Energy sources: fundamentals of chemical conversion processes and applications. Newnes
15. Zheng K, Xu R, Mei J, Yang H, Lei L, Wang X (2024) Ambient IoT towards 6G: standardization, potentials, and challenges. https://arxiv.org/abs/2412.12519
16. Mahala S, Khosravinia K, Kiani A (2023) Unwanted degradation in pseudocapacitors: challenges and opportunities. J Energy Storage 67:107558. https://doi.org/10.1016/j.est.2023.107558
17. Qoitech: The hidden cost of IoT: why battery life must be a top priority. Accessed: 2025 May 02. https://www.qoitech.com/blog/the-hidden-cost-of-iot-why-battery-life-must-be-a-top-priority/
18. Talla V, Hessar M, Kellogg B, Najafi A, Smith JR, Gollakota S (2017) Lora backscatter: enabling the vision of ubiquitous connectivity. Proc ACM Interact Mob Wearable Ubiquitous Technol 1(3):1–24. https://doi.org/10.1145/3130970
19. Arduino (2024) Societal Benefits of the IoT. Accessed: 2025 May 02. https://www.arduino.cc/education/societal-benefits-of-the-iot
20. Baldé CP, Kuehr R, Yamamoto T, McDonald R, D'Angelo E, Althaf S, Bel G, Deubzer O, Fernandez-Cubillo E, Forti V, Gray V, Herat S, Honda S, Iattoni G, Khetriwal DS, Cortemiglia VL, Lobuntsova Y, Nnorom I, Pralat N, Wagner M (2024) Global E-waste monitor 2024. International Telecommunication Union (ITU) and United Nations Institute for Training and Research (UNITAR), Geneva/Bonn. Accessed: 20 May 2025. https://ewastemonitor.info/wp-content/uploads/2024/03/GEM
21. IMEC: 78 million batteries will be dumped every day by 2025. Accessed: 02 June 2025. https://www.imec-int.com/en/press/78-million-batteries-will-be-dumped-every-day-2025
22. Wiklund J, Karakoç A, Palko T, Yiğitler H, Ruttik K, Jäntti R, Paltakari J (2021) A review on printed electronics: fabrication methods, inks, substrates, applications and environmental impacts. J Manuf Mater 5(3). https://doi.org/10.3390/jmmp5030089
23. López OL, Han Z, Sabharwal A (2024) Leaky wave antenna-equipped RF chipless tags for orientation estimation. arXiv preprint arXiv:2409.00501. https://doi.org/10.48550/arXiv.2409.00501
24. Görtschacher LJ, Grosinger J (2019) UHF RFID sensor system using tag signal patterns: prototype system. IEEE Antennas Wirel Propag Lett 18(10):2209–2213. https://doi.org/10.1109/LAWP.2019.2940336
25. Xie B, Dowhuszko A, Koskinen K, Mela L, Lietzén J, Ruttik K, Jäntti R, Hämäläinen J (2024) Integration of visible light and backscatter communications for ambient internet of things. In: IEEE VTC2024-Spring, pp 1–6. https://doi.org/10.1109/VTC2024-Spring62846.2024.10682827

26. Xie B, Mela L, Lietzén J, Ruttik K, Dowhuszko A, Jäntti R (2024) Light-controlled thin film backscatter device using additive manufacturing. IEEE Sens Lett 8(5):1–4. https://doi.org/10.1109/LSENS.2024.3395450
27. Cox B, Buyle C, Delabie D, De Strycker L, Perre L (2022) Positioning energy-neutral devices: technological status and hybrid RF-acoustic experiments. Future Internet 14(5):156. https://doi.org/10.3390/fi14050156
28. Alamdar S, Tam S-W, Bagherkhani S, Flaviis FD, Chen Z, Poehl D, Towliat M, Cao R, Baxter J, Sun J, Ishimwe A, Amtmann F, Thueringer P (2025) A 16 × 16 array composed of 4 × 4 fully scalable reconfigurable intelligent surfaces (RIS) for 2.4 GHz Wi-Fi using RFID distributed control. IEEE Trans Microw Theory Tech 1–14. https://doi.org/10.1109/TMTT.2025.3547686
29. Poehl D, Muehlmann U, Amtmann F, Thueringer P, Grosinger J (2025) Design and analysis of a 2.45 GHz RF energy harvesting device. In: 2025 IEEE wireless power technology conference and expo (WPTCE), pp 1–5. https://doi.org/10.1109/WPTCE62521.2025.11062318
30. D'Angelo L, Deutschmann BJB, Witrisal K (2024) Combined wideband channel estimation and direct link interference mitigation in bistatic backscatter systems, pp 506–510. https://doi.org/10.1109/SPAWC60668.2024.10694033
31. Liao J, Wang X, Ruttik K, Jäntti R, Phan-Huy D-T (2023) In-band ambient FSK backscatter communications leveraging LTE cell-specific reference signals. IEEE J Radio Freq Identif.https://doi.org/10.1109/JRFID.2023.3280108
32. Liao J, Wang X, Koskinen K, Zhang T, Ruttik K, Jäntti R, Phan-Huy D-T (2024) Indoor backscattering communication by using commercial lte pilots. In: 2024 IEEE 99th vehicular technology conference (VTC2024-Spring), pp 1–5. https://doi.org/10.1109/VTC2024-Spring62846.2024.10683260
33. Liao J, Zhang T, Ruttik K, Jäntti R, Phan-Huy D-T (2025) Ambient backscatter communication in LTE uplink sounding reference signal. arXiv:2501.10952. https://doi.org/10.48550/arXiv.2501.10952
34. Yang S, Bénédic Y, Phan-Huy D-T, Gorce J-M, Villemaud G (2024) Indoor localization of smartphones thanks to zero-energy-devices beacons. In: 2024 18th European conference on antennas and propagation (EuCAP), pp 1–5. https://doi.org/10.23919/EuCAP60739.2024.10501308
35. Koskinen K, Xie B, Ruttik K, Jäntti R (2025) Li2BC: From visible light communication to ambient RF backscatter. In: 2025 16th German microwave conference (GeMiC), pp 346–349. https://doi.org/10.23919/GeMiC64734.2025.10979106
36. Puluckul PP, Weyn M (2023) Infiniteen: a multi-source energy harvesting system with load monitoring module for batteryless internet of things. In: 2023 IEEE 9th world forum on internet of things (WF-IoT), pp 1–8. https://doi.org/10.1109/WF-IoT58464.2023.10539555
37. Rahmani H, Shetty D, Wagih M, Ghasempour Y, Palazzi V, Carvalho NB, Correia R, Costanzo A, Vital D, Alimenti F, Kettle J, Masotti D, Mezzanotte P, Roselli L, Grosinger J (2023) Next-generation Iot devices: sustainable eco-friendly manufacturing, energy harvesting, and wireless connectivity. IEEE J Microw 3(1):237–255. https://doi.org/10.1109/jmw.2022.3228683
38. Al-Dhahir I (2023) Decarbonizing the electronics industry with sustainable manufacturing. Available from https://www.idtechex.com/en/research-article/decarbonizing-the-electronics-industry-with-sustainable-manufacturing/28329
39. Environmental Management—Life Cycle Assessment—Principles and Framework. International Organization for Standardization. Available from https://www.iso.org/standard/37456.html
40. Environmental Management—Life Cycle Assessment—Requirements and Guidelines. International Organization for Standardization. Available from https://www.iso.org/standard/38498.html
41. Pirson T, Bol D (2021) Assessing the embodied carbon footprint of Iot edge devices with a bottom-up life-cycle approach. J Clean Prod 322:128966. https://doi.org/10.1016/j.jclepro.2021.128966
42. Moreau N, Pirson T, Le Brun G, Delhaye T, Sandu G, Paris A, Bol D, Raskin J-P (2021) Could unsustainable electronics support sustainability? Sustainability 13(12). https://doi.org/10.3390/su13126541

43. White Paper: Economics of a Trillion Connected Devices-ARM Community Blog. https://community.arm.com/arm-community-blogs/b/internet-of-things-blog/posts/white-paper-the-route-to-a-trillion-devices Accessed 24 Jan 2022
44. Pirson T, Delhaye TP, Pip AG, Le Brun G, Raskin J-P, Bol D (2023) The environmental footprint of ic production: review, analysis, and lessons from historical trends. IEEE Trans Semicond Manuf 36(1):56–67. https://doi.org/10.1109/TSM.2022.3228311
45. Pirson T, Delhaye T, Pip A, Le Brun G, Raskin J-P, Bol D (2022) The environmental footprint of ic production: meta-analysis and historical trends. In: ESSDERC 2022-IEEE 52nd European solid-state device research conference (ESSDERC), pp 352–355. https://doi.org/10.1109/ESSDERC55479.2022.9947198
46. Van Mulders J, Cappelle J, Goossens S, De Strycker L, Perre L (2023) Uav-based servicing of Iot nodes: assessment of ecological impact. Sensors 23(4). https://doi.org/10.3390/s23042291
47. Canal Marques A, Cabrera J-M, de Fraga Malfatti C (2013) Printed circuit boards: a review on the perspective of sustainability. J Environ Manage 131:298–306. https://doi.org/10.1016/j.jenvman.2013.10.003
48. Buga CS, Viana JC (2022) The role of printed electronics and related technologies in the development of smart connected products. Flex Print Electron 7(4):043001. https://doi.org/10.1088/2058-8585/ac91de
49. Song C, Ding Y, Eid A, Hester JGD, He X, Bahr R, Georgiadis A, Goussetis G, Tentzeris MM (2022) Advances in wirelessly powered backscatter communications: from antenna/RF circuitry design to printed flexible electronics. Proc IEEE 110(1):171–192. https://doi.org/10.1109/JPROC.2021.3125285
50. Liu X, Hu J, Zhang H, Di B, Xu C, Song L (2024) Internet of metamaterial things: A new paradigm for passive wireless sensing systems. IEEE Commun Mag 62(4):136–142. https://doi.org/10.1109/MCOM.002.2300085
51. Hu J, Zhang H, Di B, Han Z, Poor HV, Song L (2022) Meta-material sensor based internet of things: design, optimization, and implementation. IEEE Trans Commun 70(8):5645–5662. https://doi.org/10.1109/TCOMM.2022.3187150
52. Clément L-PP-VP, Jacquemotte QES, Hilty LM (2020) Sources of variation in life cycle assessments of smartphones and tablet computers. Environ Impact Assess Rev 84:106416. https://doi.org/10.1016/j.eiar.2020.106416
53. Bioplastics for Electronic Equipment. Available from https://www.nec.com/en/global/rd/technologies/bioplastics/bioplastics2.html
54. Kim Y, Ruedy DR (2023) In: Dhiman S (ed) Mushroom packages: an ecovative approach in packaging industry. Springer, Cham, pp 199–223. https://doi.org/10.1007/978-3-031-28885-29
55. Almalki FA, Alsamhi SH, Sahal R, Hassan J, Hawbani A, Rajput NS, Saif A, Morgan J, Breslin J (2021) Green iot for eco-friendly and sustainable smart cities: future directions and opportunities. Mob. Netw. Appl. 28(1):178–202. https://doi.org/10.1007/s11036-021-01790-w
56. Alsharif MH, Jahid A, Kelechi AH, Kannadasan R (2023) Green Iot: a review and future research directions. Symmetry 15(3). https://doi.org/10.3390/sym15030757
57. Mercier PP, Calhoun BH, Wang P-HP, Dissanayake A, Zhang L, Hall DA, Bowers SM (2022) Low-power RF wake-up receivers: analysis, tradeoffs, and design. IEEE Open J Solid-State Circuits Soc 2:144–164. https://doi.org/10.1109/OJSSCS.2022.3215099
58. Jiang T, Zhang Y, Ma W, Peng M, Peng Y, Feng M, Liu G (2023) Backscatter communication meets practical battery-free internet of things: a survey and outlook. IEEE Commun Surveys Tutor 25(3):2021–2051. https://doi.org/10.1109/COMST.2023.3278239
59. Chidolue O, Ohenhen PE, Umoh AA, Ngozichukwu B, Fafure AV, Ibekwe KI (2024) Green data centers: sustainable practices for energy-efficient it infrastructure. Eng Sci Technol J 5:99–114. https://doi.org/10.51594/estj.v5i1.730
60. Rosabal OM, López OLA, Alves H, Latva-Aho M (2023) Sustainable RF wireless energy transfer for massive IoT: Enablers and challenges. IEEE Access 11:133979–133992. https://doi.org/10.1109/ACCESS.2023.3337214
61. Gonzalez M, Xu P, Dekimpe R, Schramme M, Stupia I, Pirson T, Bol D (2023) Technical and ecological limits of 2.45-ghz wireless power transfer for battery-less sensors. IEEE Internet Things J 10(17):15431–15442. https://doi.org/10.1109/JIOT.2023.3263976

62. Min J, Jung Y, Ahn J, Lee JG, Lee J, Ko SH (2023) Recent advances in biodegradable green electronic materials and sensor applications. Adv Mater 35(52):2211273. https://doi.org/10.1002/adma.202211273
63. Danninger D, Pruckner R, Holzinger L, Koeppe R, Kaltenbrunner M (2022) Myceliotronics: fungal mycelium skin for sustainable electronics. Sci Adv 8(45):7118. https://doi.org/10.1126/sciadv.add7118
64. Pammo A, Christophliemk H, Keskinen J, Björkqvist T, Siljander S, Mäntysalo M, Tuukkanen S (2019) Nanocellulose films as substrates for printed electronics. In: 2019 International conference on manipulation, automation and robotics at small scales (MARSS). IEEE, pp 1–6. https://doi.org/10.1109/MARSS.2019.8860931
65. Wiklund J, Miettinen A, Parkkonen J, Mela L, Karakoç A, Paltakari J (2024) A detailed investigation of acetylated cellulose nanofiber films as a substrate for printed electronics. Mater Des 245:113230. https://doi.org/10.1016/j.matdes.2024.113230s
66. Hoeng F, Denneulin A, Bras J (2016) Use of nanocellulose in printed electronics: a review. Nanoscale 8(27):13131–13154. https://doi.org/10.1039/C6NR03054H
67. Xie L, Mäntysalo M, Cabezas AL, Feng Y, Jonsson F, Zheng L-R (2012) Electrical performance and reliability evaluation of inkjet-printed ag interconnections on paper substrates. Mater Lett 88:68–72. https://doi.org/10.1016/j.matlet.2012.08.030
68. Välimäki MK, Sokka LI, Peltola HB, Ihme SS, Rokkonen TM, Kurkela TJ, Ollila JT, Korhonen AT, Hast JT (2020) Printed and hybrid integrated electronics using bio-based and recycled materials—increasing sustainability with greener materials and technologies. Int J Adv Manuf Technol 111:325–339. https://doi.org/10.1007/s00170-020-06029-8
69. Pirson T, Golard L, Bol D (2023) Evaluating the (ir)relevance of iot solutions with respect to environmental limits based on lca and backcasting studies. In: Ninth computing within limits. LIMITS. https://doi.org/10.21428/bf6fb269.6af396ff

Chapter 13
Digital Pathways to Sustainability: Insights Beyond Europe

Fabrício Ramos Neves, Polyana Batista Silva, Vasantha Shanmugam, Vimala Dharmasivam, Lina Bariah, Merouane Debbah, and John M. Roman

Abstract While Europe is placing significant political weight on sustainability objectives, other regions in the world also engage with similar ambitions. In this chapter, we outline the local actions of the United States, Brazil, India, and the UAE to understand how each has developed distinct approaches. Each context reflects a different combination of governance structures, institutional capacity, and sociopolitical priorities. We discuss how Brazil uses digital tools to protect ecosystems and reduce inequalities, how India focuses on large-scale administrative efficiency and inclusion, how the UAE aligns digitalization with its Net Zero 2050 agenda, and how the United States combines federal initiatives, state regulations, and industry-driven innovation in a complex and often politically divided landscape. By comparing these cases, we identify shared challenges and opportunities to advance digital sustainability beyond Europe.

F. R. Neves (✉) · P. Batista Silva
Guanambi, Brazil
e-mail: fabricioneves@alumni.usp.br

P. Batista Silva
e-mail: polybsilva@gmail.co

V. Shanmugam · V. Dharmasivam
Chennai, India
e-mail: vasanthashan06@gmail.com

V. Dharmasivam
e-mail: vimala.sivam1987@gmail.com

L. Bariah · M. Debbah
Abu Dhabi, United Arab Emirates
e-mail: lina.bariah@ku.ac.ae

M. Debbah
e-mail: merouane.debbah@ku.ac.ae

J. M. Roman
Rochester, USA
e-mail: jrbatghost1@gmail.com

C. Gaie and M. Mueck (eds.), *Digital Sustainability: Legislations, Standards, Solutions and Related Case Studies*, Advances in Sustainability Science and Technology,
https://doi.org/10.1007/978-981-92-0458-8_13

Keywords Sustainable digital transformation · Green ICT · Regulatory sandboxes · AI for sustainability · Climate tech · Low-carbon digital economy · Sustainable Development Goals (SDGs)

1 Introduction

Digital sustainability is not a mere technological frontier or a fashionable agenda item; rather, it is an arena where societies negotiate their priorities, capacities, and vulnerabilities in facing planetary crises. Examining the experiences of Brazil, India, the UAE, and the USA reveals digital sustainability as an assemblage of material infrastructures, policy designs, socio-technical imaginaries, and contested values. In these regions, sustainability through digitization is not simply a matter of efficiency or optimization. It reflects political economies of development, historical trajectories of state formation, and divergent ecological urgency. Thus, in this chapter, we unpack the multilayered strategies that these countries employ to integrate digital technologies with sustainability objectives, tracing not only innovations but also silences, contradictions, and emergent possibilities within each context.

2 Global Landscape of Digital Sustainability

Globally, digital sustainability efforts are framed by initiatives such as the UN Sustainable Development Goals (SDGs), the OECD's digital economy guidelines, and ITU standards for green ICT. These frameworks promote digital inclusion, decarbonization, and innovation as interconnected objectives. Yet, their implementation varies widely.

For example, many countries focus on using digital tools to support environmental goals, like AI-based monitoring of deforestation or smart grids for energy efficiency. At the same time, there is growing recognition that digital systems themselves have environmental costs: Data centers consume large amounts of electricity, device production relies on mining, and e-waste management remains inadequate in most regions.

Across regions, some common themes emerge. First, digital infrastructure is both an enabler of sustainability and a source of environmental impact. Second, regulatory approaches differ, but there is a trend toward creating sandboxes and flexible frameworks that allow innovation while managing risks. Third, inequalities in digital access and skills remain a major barrier. Without addressing them, the potential of digital technologies for sustainability remains unevenly distributed.

When we look at Brazil, India, the UAE, and the USA, these global dynamics take specific forms. All four countries use digital technologies to advance sustainability, but they differ in their focus. Brazil emphasizes ecological protection alongside digital innovation. India priorities digital inclusion and service delivery. The UAE integrates digital strategies into its national decarbonization agenda. In the USA, the Environmental Protection Agency uses digital platforms for real-time monitoring of air and water quality, but policy continuity is vulnerable to political cycles. These differences reflect each country's political economy, institutional capacity, and development path.

3 Digital Sustainability in Brazil

3.1 Introduction: The Multidimensionality of Digital Sustainability

Far from being just a buzzword, digital sustainability is a multidimensional concept that includes both the sustainability of digital technologies—how green, efficient, and resilient are our networks, data centers, and devices—and the sustainability enabled by digital transformation—how can digital tools help us tackle environmental, social, and economic challenges [1, 2]. For instance, deploying AI is not enough in itself—we must ask whether the energy and resources it consumes are justified by the sustainability outcomes it delivers, such as better climate forecasting or real-time pollution monitoring [1, 3].

This dual perspective—making digital greener and making sustainability more digital—is at the heart of Brazil's approach. Brazilian policymakers, scholars, and practitioners are wrestling with both sides: reducing the carbon footprint of digital infrastructure and using digital innovation as a lever to reach the Sustainable Development Goals (SDGs), all while navigating stark regional inequalities and the ecological imperative to protect biomes like the Amazon [1, 4].

3.2 Historical and Political Context: Brazil's Policy Turn

Brazil's digital journey, as in much of the world, started with a focus on digital inclusion and e-government—putting computers in schools and public offices, and creating access points for citizens [4]. This democratization of digital access was critical. By 2020, Brazil had reached over 80% internet penetration, but with stark divides—urban access rates often doubled those in rural and Amazonian regions [5].

By the mid-2010s, the narrative began to change. The E-Digital strategy implemented in 2018 and updated for 2022–2026 [6, 7] positioned digital transformation as a lever not only for productivity and competitiveness, but explicitly for sustainable development. The Brazilian Artificial Intelligence Strategy (EBIA) added layers of ethics, inclusion, and risk management, mirroring the broader global debates on responsible AI [8].

Legislation moved in parallel: the Brazilian Civil Rights Framework for the Internet (Marco Civil da Internet) in 2014 established rights and principles for internet use, while the General Law for the Protection of Personal Data (LGPD) in 2018 placed Brazil at the forefront of Latin America for data protection. These frameworks, often inspired by the European model, gave Brazil a strong platform to pursue digital sustainability in both governance and the private sector [4].

3.3 National and Multilevel Initiatives: Policy Meets Practice

E-Digital and the Institutional Ecosystem The E-Digital strategy's five pillars, ranging from digital inclusion, open government, sustainable infrastructure, innovation, and SDG alignment, translate into concrete initiatives at all levels of government [4, 9]. Over R$ 1 billion (about $200 million USD) was invested in expanding broadband to the most remote areas between 2018 and 2021 [10], and the number of public digital services offered through the federal "Gov.br" portal grew from 1700 in 2019 to over 4,200 by 2023, reducing both bureaucratic friction and the carbon footprint of physical service delivery [4].

- **Regulatory Sandboxes for Sustainable Innovation** Brazil has embraced regulatory sandboxes as adaptable and forward-thinking frameworks to foster digital sustainability innovations [11]. Originally emerging in financial markets to balance innovation with regulatory stability, sandboxes in Brazil now span diverse sectors, creating controlled environments for experimentation while reducing compliance burdens [11]. For example, the 2021 Startup Legal Framework enabled agencies to create sandboxes, leading to initiatives by the Central Bank to incubate fintech products and by the Securities Commission to explore tokenization of ESG-linked assets. These initiatives demonstrate how Brazilian regulators are leveraging sandboxes to test blockchain applications that can advance sustainability, such as carbon credit tokenization.

  Importantly, we argued before [11] that regulatory sandboxes can go beyond digital transformation to address complex public challenges through co-production with citizens. In this spirit, SUSEP (the insurance regulator) launched its third sandbox edition in 2024 prioritizing climate risk insurance and ESG-linked products, directly aligning with Brazil's Ecological Transformation Plan. At the municipal level, Sandbox.Rio fosters urban sustainability by piloting projects like electric scooter sharing and vertical farming containers. These examples illustrate how sandboxes can integrate citizen participation and agile regulation to generate innovations that not only meet market needs but also promote environmental and social benefits, advancing digital sustainability agendas in Brazil.
- **Artificial Intelligence for Environmental Monitoring** Brazil's AI strategy emphasizes practical sustainability applications. For example, AI-driven satellite analysis supports the real-time monitoring of illegal deforestation in the Amazon,

with platforms like MapBiomas and DETER providing open access data to public agencies and NGOs [3, 8]. The result: In 2022 alone, over 2,500 alerts of illegal clearing were generated, triggering rapid-response interventions and feeding into global supply chain transparency initiatives.

AI is also transforming agriculture. In the Cerrado biome, the EMBRAPA research network uses big data and AI to recommend the optimal use of fertilizers, reducing nitrogen runoff and improving yields for soy and corn farmers. In 2022, pilot projects saw up to 25% reduction in fertilizer use with no yield loss, directly impacting SDG 2 (zero hunger) and SDG 12 (responsible production) [12].

3.4 Case Studies: Local Innovation, National Impact

Smart Green Governance in Action Few Brazilian cities illustrate digital sustainability's promise better than São José dos Campos, a municipality far 97 km from São Paulo, capital. Here, remote sensing and high-resolution GIS are combined with citizen science and open data platforms to monitor over 20,000 urban trees. Residents scan QR codes attached to trees, accessing species, health, and maintenance history, while city officials use this live inventory to target climate adaptation efforts, plan green corridors, and engage citizens in reforestation campaigns [4, 13].

IoT for Sustainable Aquaculture Another striking example occurs in Santa Catarina, where artisanal fish farmers partnered with universities to deploy low-cost IoT sensors for monitoring water quality, oxygen, and feeding times [14]. As detailed by Ramanathan et al. [14], these sensors led to a 15% reduction in feed waste and increased fish yields, while enabling small producers to comply with stricter environmental regulations and reduce their ecological impact. However, the project also found that uptake depended heavily on training and ongoing technical support, echoing broader findings that digital inclusion is as much about skills and ecosystem as about technology [4, 14].

Indigenous Digital Innovation: The digital bow Brazil's experience with digital sustainability is also shaped by the voices and agency of Indigenous peoples. A compelling example is the "Digital Bow" initiative, launched during the pandemic, which empowered Indigenous communities to harness smartphones, solar-powered laptops, and online platforms to share ancestral knowledge, promote eco-tourism, and advocate for land rights. This initiative not only diversified income sources but also reinforced cultural resilience and strengthened environmental stewardship—demonstrating that digital sustainability must be grounded in local realities and driven by social inclusion [15].

Green Technology in Agriculture Brazil's agricultural sector, responsible for a third of national exports, demonstrates digital sustainability at scale, albeit unevenly. Large agribusinesses deploy AI, satellite imaging, and drones for precision farming, optimizing inputs and reducing waste, while public research bodies like EMBRAPA have developed open data platforms (e.g., AgroAPI) to make climate, soil, and crop data accessible to all producers [12]. Still, as Martha Júnior and Lopes [12] argue, smallholder adoption lags without targeted support, echoing global concerns about the "digital divide" within agriculture.

3.5 Challenges: Inequality, Institutional Fragmentation, and Capacity

Despite these advances, Brazil's digital sustainability pathway is marked by three stubborn challenges:

1. **Digital Inequality:** Rural connectivity rates remain 30–40 percentage points below those in metropolitan regions, with the North and Northeast lagging most [4, 5]. Indigenous and riverine communities face even greater barriers, often relying on solar-powered radio for connectivity.
2. **Fragmented Governance:** Policy is often siloed between ministries (e.g., science, environment, agriculture) and levels of government, hampering integrated approaches. Projects are too often pilots, rather than scaled, sustainable systems [3, 4].
3. **Capacity Gaps:** Many state and municipal agencies struggle to hire and retain skilled digital talent, limiting their ability to design, implement, and monitor sophisticated digital sustainability programs.

3.6 Socioeconomic Impacts: Toward just and Inclusive Digitalization

Brazil's experience makes clear that digital sustainability can boost productivity, transparency, and resilience, but only if structural inequities are addressed. Evidence shows that digital public services can save citizens millions of hours annually [5], and digital agriculture increases yields and reduces input costs. Yet, the distribution of these benefits is not automatic. Ongoing investment in digital literacy, targeted support for small producers and marginalized groups, and robust social policies are vital [4, 5].

3.7 Final Remarks: Insights from Brazil for a Global Agenda

What does Brazil's digital sustainability story offer to the world? First, it demonstrates that digital sustainability is fundamentally multidimensional: Success requires making digital systems greener and ensuring those systems accelerate social and environmental progress [2, 3]. Second, Brazil's experience highlights the need for context-sensitive, inclusive approaches: Policies that work in São Paulo may not fit the Amazon, and solutions must reflect local realities and knowledge. Third, Brazil's journey reveals the importance of combining national frameworks with space for local innovation. Federal strategies (like E-Digital and EBIA) provide direction, but progress depends on active, empowered local actors—municipalities, Indigenous leaders, farmers, and civil society.

Finally, the Brazilian case reminds us that digital sustainability is a process, not a destination. New technologies, challenges, and social needs will always exist. The lesson for the world is that digital pathways to sustainability must remain open, adaptive, and deeply attuned to questions of justice, participation, and long-term resilience.

4 Digital Sustainability in India

4.1 Introduction: The Multidimensionality of Digital Sustainability

The best practices in e-governance are evaluated in terms of the broad range of servicing, e-filing, and e-residency. Stakeholders' engagement and active feedback are certified through transparency, trust, and inclusive engagement. The effectiveness of state of the art is demonstrated through one-stop-shop services. The existence of continuous monitoring is assessed through the presence of comprehensive digital infrastructure and 24/7 services. Generally, in any country, the sustainability of e-governance is measured in terms of following three pillars, namely:

1. Environmental sustainability;
2. Social equity;
3. Operational efficiency.

The section focuses on how the dimensions of e-governance sustainability enablers, namely best practices, stakeholder engagement and feedback, the state of the art, and continuous monitoring, enable the Indian government to attain social equity, environmental sustainability, and efficiency. The research primarily evaluates how the key drivers, namely best practices, stakeholder engagement and feedback, state of the art, and continuous monitoring, support the sustainability of e-governance in India and contribute to social equity, environmental sustainability, and operational efficiency of India. Subsequently, the challenges associated in achieving these outcomes are also explained. Best practices include broad range of e-governance services, such as e-tax and e-residency. The achievements gained out of these best practices are citizen engagement, transparency, and social inclusion. Stakeholders' engagement and feedback performance are assessed through transparency, trust, and inclusive engagement. The benefits gained from these are citizen satisfaction, trust, user-friendly, and convenience. The state-of-the-art technology is judged with user experience with one-stop-shop services and resulting benefits such as data security, user centric, data in real time, and personalized services. The continuous monitoring support is measured through the modernization of comprehensive digital infrastructure and user-centric 24/7 services. The benefits generated on account of these are continuous feedback, regular auditing, and continuous improvement of infrastructure.

India is committed to digitizing its government services in order to increase overall transparency, competency, and citizen participation. Even with impressive advancements, these e-government databases still have a long way to go before they can be sustained. This case study examines these issues and provides insights based on future explanations tailored to the Indian context. It is difficult to establish an unwavering technological structure in India due to its vast and diverse topography. Rural and rustic areas frequently lack reliable power sources and internet connectivity. Assimilation of numerous state and federal government agencies is thought-provoking, given the variety of prevalent technical backgrounds and criteria. The increasing frequency of cyberattacks forces robust cybersecurity procedures. Protecting sensitive information across a wide range of disciplines and diplomacies is essential.

4.2 Enablers

Best Practices The swift expansion of digitalization has prompted numerous governments worldwide to integrate technology into their administrative procedures. The use of internet and advanced technology by the regime to recommend and smooth the progress of administration services, sequential substitute, communication transactions, and integration of numerous standalone systems and services is known as electronic governance, or e-governance. To put it another way, it is the application of technology to carry out official tasks and accomplish governance goals. With e-governance, citizens and businesses can access government services quickly, simply, and transparently. Examples of e-governance include the Digital India initiative, the Prime Minister of India portal, the National Portal of India, Aadhaar, online tax filing and payment, digital land management systems, and more.

Stakeholder Management and Feedback Shareholders have the right to acquire information about the general operations, performance, and financial health of the company. This transparency aids in their capacity to assess the company's management and render informed decisions. Access to relevant information regarding the company's strategies, risks, and other key information that could affect an investor's decision to invest is guaranteed. Viewing the registers gives shareholders the chance to look for any irregularities, discrepancies, or potential fraud. This deters dishonest behavior and promotes adherence to good governance practices inside the organization. As watchdogs, shareholders can play a critical role in preventing bad management and ensuring that the company abides by the law and moral standards.

State-of-the-art technology Since 1999, STQC has offered quality assurance and conformity assessment services in the IT and e-government area in order to provide its esteemed clients with cutting-edge technology-based quality assurance services and to comply with MeitY's mandate to concentrate on this area. After then, governance has advanced along with technology. There are numerous e-government programs underway right now, both at the federal and state levels. The Department of Electronics and Information Technology and the Department of Administrative Reforms and Public Grievances developed the National e-Governance Plan (NeGP) in 2006 with the goal of ensuring that all government services are affordable and accessible to the general public, while also meeting the needs of the latter by ensuring that these services are reliable, efficient, and transparent.

Continuous Monitoring All tiers of government must be transformed by e-government, but as local governments are the closest to the public and serve as many people's first point of contact with the government, they should receive the most attention. Citizens and local authorities typically have a close connection since their interests are closely aligned on matters like public services, local development, education, and other matters.

The enablers introduced above are summarized in Table 1.

The introduction of e-governance efforts in India has resulted in several benefits, including increased administrative efficiency, transparency, and public empowerment. In order to expedite the countrywide adoption of e-governance at the state,

Table 1 Pillars of sustainability and their enablers

e-Governance sustainability enablers	Benefits from digital transformation	Pillars of sustainability	Examples of services offered
Innovative best practices	Reduced corruption, transparency and accountability, convenience	Social equity	Cybersecurity, Digi Locker, State data centers, Aadhaar, mygov.in
Stakeholder management and feedback	Digital communication, speedy administration, seamless interaction	Environmental sustainability	Inclusive participation, transparency, government support, UMANG, Paygov, computerization of land records
State-of-the-art technology	Digital literacy, digital infrastructure, digital delivery of services	Operational efficiency	National e-governance Plan, Digital India, e-Pramaan, G-I Cloud
Continuous monitoring	Efficient public services, payment of utility bills, handling of grievances		E-governance Development Index, Interoperability Framework for e-governance

local, and federal levels, the government has embraced a pragmatic approach predicated on a shared objective and structure. This approach has the potential to save a substantial amount of money because it has a common core and support infrastructure, standardized procedures that facilitate interoperability, and a smooth public perception of government operations.

Administrative reforms in local governments based on e-government can have the greatest possible impact on the public. Information technology has not benefited everyone equally. It has been observed that the wealthier segments of society typically gain from e-government as well. Thus, directing e-government reforms requires a concentrated effort.

4.3 Review of Literature

Sustainable Development In 2025, Neha Saini and Ravita Kharb carried out a paper titled Enhancing Sustainable Development via Digital Transformation: Perspectives from Digital India [16]. After a comprehensive literature analysis, ten important facilitators were found and examined in this study utilizing the total interpretive structural model (TISM). This new method has been used to examine the hierarchical relationship between the enablers that have been found. To assess the driving and reliance power among enablers, the *Matrice d'impacts croisés multiplication appliquée à un classement* (MICMAC) analytical technique is employed. The results indicate that the most powerful strategic enabler is the government's supportive policy, which is at the bottom of the model. Initiatives for digital empowerment must be supported by the government to guarantee that everyone has access to the data they need to make informed decisions.

In order to generalize the findings and concentrate on the driving force behind India's digital transformation in order to meet sustainable development objectives, this study reviews the literature. According to this study, hierarchical models based on TISM and MICMAC assist practitioners, managers, and other stakeholders in concentrating more on strategic elements to improve the performance aspects of digital transformation. Collaboration between government, industry, and academia fosters innovation and the adoption of digital approaches, which advances a holistic approach to environmental sustainability. Furthermore, it provides fresh perspectives on the enablers, including how the identified enablers interact with one another based on the motivation and power of dependency to accomplish the objectives of the transformation of digital India to meet sustainable development goals.

Digital Transformation A 2019 study by Aiswariya Vijayan [17] titled A Roadmap to Sustainability demonstrates that nations who have adopted technology for the good of their people have seen a significant shift in every area, which spurs GDP development and raises both the national and per capita incomes. The government has a responsibility to raise citizens' quality of living by launching initiatives that spur economic expansion. "Technology," a key component of the "Digital India

Program," was the driving force behind the digitization. One nation that has seen the implementation of development programs by its government to boost economic growth and provide jobs for the younger generation is India.

The primary goal was to make all services available to all citizens electronically or through web portals in order to facilitate seamless and transparent transactions. The government is increasing its investment in technology in an effort to eradicate corruption and black money from public life. Although India has begun the digital transition, it will take some time to fully experience its effects. India will accomplish the United Nations Sustainable Goals Agenda by 2030 thanks to this program. This study aids in comprehending the economic impact of Digital India, determining the extent to which its objectives have been met, and creating a model for achieving the Sustainable Development Goals using the foundations of Digital India.

Environmental Sustainability Dutta and Pandya [18] proved that e-governance positively impacts on environmental sustainability through declining paper consumption, advancing data management, and improving transparency. He highlighted that PARIVESH is an online platform designed for environmental clearances and has minimized processing time by 50% and enhanced transparency. According to Sharma et al. [19], In India, the initiatives of e-governance are progressing well; as a result, its aims reduce carbon emissions by 20–30%. The e-Green Watch platform is designed to monitor and improve transparency in forest conservation effort; it provides assisted online supervising of environmental compliance. It contributes to 30% reduction of the non-compliance. Digital India initiatives empowered the citizens to engage interactions with the Government online to facilitate reduction of paper.

Social Equity Indorsing, intensifying, and upholding e-government edges necessitate fragmentary patronage. Nevertheless, pecuniary constrictions and incongruous environs unswervingly constrict fiscal monies. It can be problematic to validate the enduring welfares of e-government exertions to sustenance maiden stashes. A scarcity of skillful mavens equipped for dealing and retaining e-government collaboration subsists. This is only deceptively alongside in bucolic areas where coherent proficiency is restricted. Inhabitants and personnel of the government can be diffident to cuddle fresh digital encroachments and would have a preference to twig with the status quo. It might be problematic to traverse the multifarious web of societal and governmental desires. It is indispensable to bond policies in order to hasten unified digital governance. Governmental upheavals that transpire recurrently might source metamorphoses in plan primaries, undermining the durability and viability of e-government initiatives. Afar socioeconomic groups, there are substantial gaps in digital technology access between borough and rural communities. It is hazardous to bond this embargo to comprehensive e-governance. In order to foster public confidence in e-government services, it is imperative to address persistent concerns over privacy, data security, and usability of services.

4.4 Environmental Sustainability

E-government in India has evolved gradually, starting with the computerization of government agencies and continuing with initiatives that cover the finer points of administration, such as transparency, service orientation, and citizen centricity. The lessons learned from previous e-government programs have had a significant impact on the nation's evolving e-government approach. It has been properly recognized that in order to accelerate the adoption of e-government across the various departments of government at the national, state, and municipal levels, a program approach that is guided by a common vision and strategy must be adopted. This approach might possibly save a substantial sum of money, promote interoperability through standards, and provide citizens with a seamless interaction with government by integrating core and support infrastructure.

4.5 Operational Efficiency

To promote e-government holistically, a number of projects and policy initiatives have been started to develop core and supporting infrastructure. The government has approved NeGP's objective, method, strategy, key components, implementation procedure, and management structure [20]. The introduction of e-governance efforts in India has resulted in several benefits, including increased administrative efficiency, transparency, and public empowerment. In order to expedite the country-wide adoption of e-governance at the state, local, and federal levels, the government has embraced a pragmatic approach predicated on a shared objective and structure. By sharing core and support infrastructure, this strategy can result in significant cost savings. It also promotes interoperability through standardized practices and giving the public a smooth perspective of how the government is run.

4.6 Legal and Policy Framework for Electronic Governance in India

In India, large number of e-governance schemes have been implemented continuously both at Union and State level driven by technological advancement. In India, some prominent examples of e-governance achievements are Digi Locker, National Portal of India, e-filing of Income tax, Aadhaar, computerization of land records, e-Seva portal, etc. In India, IT Act of 2000 was established to address the legal challenges posed by digital technologies in governance and was amended substantially in the 2008 to enhance its scope. This legislation establishes the basic framework for electronic records, digital signatures, and prevention of cybercrime.

Legal Framework for E-governance in India The primary anchor for the legal policy in India is Information Technology Act, 2000. This act encompasses the basic authentication for e-governance. This act is most significant in the transformation of paper-based records to digital platforms [21, 22]. The main aspects of this amendment comprises of e-records, digital signatures, and internet banking. All the above-mentioned dimensions play an important role in e-governance.

a. **Information Technology Act, 2000**

This Act is the basis of the country's e-government policy. It ascertains the legitimacy of digital documents and signatures. This will ensure proper e-service delivery. This amendment deals with the complexities regarding security, data privacy, and accountability. These factors are important to withstand confidentiality and data protection in the online connections [21].

b. **Digital India Act, 2023**

This Digital India Act of 2023 was positioned to reinstate the Information Technology Act, 2000. It is set to resolve the difficulties in this timeline. It establishes an advanced policy and framework for e-governance. This Act comprises the factors, namely Internet problem solving, accountability, and the amendments in the e-services. It will also bring the legal structuring and framework into focus with high resolution [23]. This enables India to be the pioneer in economy.

c. **Cybersecurity Laws**

An essential part of the legal framework for e-governance is cybersecurity. The National Cyber security Policy, 2013 and the creation of the Indian Computer Emergency Response Team (CERT-In) are two of the steps India has taken to improve cybersecurity. These steps are necessary to safeguard online transactions and the digital infrastructure [23, 24].

Policy Framework for E-Governance in India A number of projects and programs that support digital government and enhance service delivery define India's e-governance policy framework. These regulations are intended to guarantee that all citizens benefit from e-government and to foster an atmosphere that encourages the use of technology in governance.

a. **National E-Governance Plan (NeGP)**

One of the most important policy initiatives in the field of e-governance is the National E-Governance Plan (NeGP). NeGP was established in 2006 with the goal of using information and communication technology (ICT) to enhance the provision of public services. By combining several projects into a single framework, the plan aims to establish a common vision and objective for e-governance throughout the nation [25, 26].

b. **Digital India Initiative**

The Government of India's flagship program, Digital India, was introduced in 2015 with the goal of making the nation a digitally empowered society. Digital infrastructure, digital literacy, and digital governance are the three main areas of concentration for the effort. It encompasses a variety of initiatives, including the creation of Common Service Centers (CSCs), the Unified Payments Interface (UPI), and the National Digital Health Mission [23, 25, 26].

c. **National Policy on Open Standards for E-Governance**

To guarantee the consistent and dependable application of e-governance systems, the National Policy on Open Standards for e-government was created. In order to improve interoperability and prevent vendor lock-in, the policy encourages the use of open standards. Additionally, it seeks to level the playing field for all parties involved and enhance the technology options accessible to the government [27].

d. **Digital Literacy and Capacity Building**

Building capacity and promoting digital literacy are essential elements of the e-governance policy framework. The Pradhan Mantri Gramin Digital Saksharta Abhiyan (PMGDISHA) and the National Digital Literacy Mission (NDLM) are two of the government's initiatives to encourage digital literacy. These programs seek to provide citizens with the knowledge and abilities necessary to engage in the digital economy and efficiently access digital services [26, 28].

This study highlights the key enablers of e-governance sustainability, as exemplified by India. India prioritizes social equity by leveraging technology to empower citizens and improve government efficiency. India, in particular, has created a robust digital infrastructure that enables seamless access to public services, while India has focused on empowering citizens through information and improving government–citizen interactions.

Policy initiatives and their objectives are summarized in Table 2.

4.7 Emerging Trends and Future Directions

India's e-governance legal and policy framework is always changing to meet new opportunities and problems. Among the major patterns and potential paths are as follows:

Machine Learning and Artificial Intelligence The efficiency and efficacy of service delivery are anticipated to increase with the incorporation of machine learning (ML) and artificial intelligence (AI) into e-governance. Predicting citizen requirements, personalizing services, and improving decision-making are all possible with AI and ML. Nevertheless, the application of these technologies also presents moral and legal issues that require attention [23, 24].

Blockchain Technology Blockchain technology offers a transparent and safe method of conducting transactions and keeping data, which has the potential to

Table 2 Policy initiatives and their objectives

Policy/Initiative	Objective	Citation
National E-Governance plan	To improve service delivery through ICT and integrate e-governance initiatives	Pathak and Barnwal [25], Muttoo [26]
Digital India Initiative	To create a digitally empowered society and promote digital governance	Tiwari et al. [23], Pathak and Barnwal [25], Muttoo, 2019 [26]
National Policy on Open Standards	To ensure interoperability and avoid vendor lock-in in e-governance solutions	Macdonald [27], MacDonald (n.d.)
Digital Personal Data Protection Act	To protect personal data and ensure responsible data processing	Tiwari et al. [23]
Digital India Act, 2023	To replace the IT Act of 2000 and address digital age challenges	Tiwari et al. [23]
National Digital Health Mission	To provide digital health services and improve healthcare delivery	Tiwari et al. [23]
Unified Payments Interface	To facilitate digital payments and promote financial inclusion	Tiwari et al. [23]
Pradhan Mantri Gramin Digital Saksharta Abhiyan	To promote digital literacy in rural areas	Prasad [28]
National Cybersecurity Policy	To strengthen cybersecurity and protect digital infrastructure	Tiwari et al. [23], Ebert [24]

completely transform e-governance. Blockchain is being investigated by the government for a number of uses, such as supply chain management, identification verification, and land records [23, 24].

Digital Accessibility and Inclusion To guarantee that the advantages of e-governance reach all residents, including those in rural and marginalized communities, digital inclusion and accessibility are essential. By increasing broadband connectivity, encouraging digital literacy, and making sure that digital services are usable by people with disabilities, the government is attempting to close the digital divide [28, 29].

International Standards and Cooperation India is actively collaborating with international organizations and other nations to exchange best practices and create global standards for digital governance as e-government becomes more widespread. In order to solve cross-border issues and guarantee that India's e-governance system conforms to international standards, this collaboration is crucial [23, 24].

4.8 Challenges and Opportunities

India's e-governance legal and legislative framework is not without its difficulties. Among the main difficulties are as follows:

The Digital Gap In India, the digital divide continues to be a major obstacle to e-governance. Even while internet availability has increased significantly, many people still do not have access to digital technologies, especially those living in rural and underprivileged areas. To guarantee that the advantages of e-governance are shared fairly, this gap must be closed [28, 29].

Risks to Cybersecurity India's e-governance is becoming increasingly concerned about cybersecurity concerns. The risk of cyberattacks and data breaches rises as the government depends more and more on digital platforms to provide services. To reduce these hazards, citizens' knowledge must be increased, and cybersecurity measures must be strengthened. In the digital age, data protection and privacy are crucial concerns. The collection, storage, and responsible use of citizen data must be guaranteed by the government. To solve these issues, the Digital Personal Data Protection Act and other relevant legislation would be essential [28, 29].

4.9 Execution and Management

Initiatives pertaining to e-governance must be implemented and governed effectively. To combat corruption and guarantee accountability, the government must make sure that policies are carried out effectively and that there is sufficient oversight [25, 29].

Table 3 emphasizes the crucial role of digital identification and cybersecurity in fostering the growth of digital public services. Prioritizing user experience is essential, ensuring simple, accessible, and user-friendly services that encourage feedback and improvement. Investing in cutting-edge digital technologies is vital to deliver state-of-the-art public services that inspire trust and widespread adoption. Lastly, continuous monitoring of user behavior is necessary to adapt to evolving needs, introduce innovative digital services, streamline administrative processes, and safeguard against emerging threats. The above table also encompasses the various enablers of e-governance sustainability and their responsibilities. The different enablers are innovative best practices, stakeholder management, state-of-the-art technology, and continuous monitoring.

4.10 Conclusion

The government's dedication to using technology to improve governance has formed India's dynamic and changing legal and policy framework for e-governance. A variety of laws, regulations, and programs that seek to establish a strong digital ecosystem

Table 3 Pillars of sustainability and their roles

Pillars of sustainability (triple bottom line)	Role of the pillars of sustainability
Social equity	Providing information to empower people, increasing government efficiency, and enhancing communication with business and industry
Environmental sustainability	Simplicity, efficiency, citizen centricity, sustainability, and cost-effectiveness
Operational efficiency	Competency, reliability, scalability, and a holistic approach

back the framework. Even though there are obstacles to overcome, e-governance in India appears to have a bright future thanks to new technologies and international collaboration that present fresh chances for development and creativity.

The case studies of India offer valuable lessons in designing and implementing sustainable e-government services, highlighting the need to align technological innovation with strong governance principles and meaningful citizen engagement. What stands out in both countries' experiences is the understanding that digital transformation of public services requires more than just new technology—it calls for a thoughtful approach to policy, collaboration with key stakeholders, and the flexibility to adapt to changing needs.

The unified framework presented in this article, which combines theoretical ideas with real-world examples, provides a clear pathway for other nations looking to advance their own digital government efforts. By focusing on operational efficiency, social equity, and environmental sustainability, governments can create digital services that are not only more effective but also accessible, transparent, and fair for all citizens.

5 Digital Sustainability in the UAE

5.1 Legislative and Regulatory Enablers

The UAE has actively established a range of legislative and regulatory frameworks to enable sustainable digital transformation. These policies are aimed to create an environment where technological innovations are developed in alignment with the international and UAE national sustainability goals. They include high-level strategies (establishing long-term vision and targets), specific policy frameworks (providing guidelines for sustainable tech adoption), and novel regulatory mechanisms (like sandboxes to pilot new ideas). These initiatives are designed to strike a balance between economic growth, driven by digitization, and environmental sustainability prospects. Key enablers in the UAE include the following:

UAE Net Zero 2050 Strategic Initiative. One key move in the UAE is the *Net Zero by 2050 Strategic Initiative* [30], which commits to achieve net-zero carbon emissions by 2050. This initiative, which is the first of its kind in the MENA region, provides a comprehensive national climate strategy that guides all sectors, including the digital and technology domains. This initiative is designed as a tool to stimulate growth, with expectations to create 200,000 new jobs in clean tech (e.g., solar, battery, and hydrogen industries) and to boost the UAE's GDP by around 3%, while also enhancing green export opportunities.

To achieve the net zero, the strategy proposes more than 25 initiatives across six key sectors, namely energy, industry, transport, buildings, waste, and agriculture. More importantly, the strategy identifies several key enablers in order to achieve the needed goals, including dedicated climate finance mechanisms, the development of new low-carbon technologies through advanced R&D efforts, and a dynamic plan for empowering human capital with the needed green skills. By highlighting innovation and capacity building as enablers, the UAE acknowledges that achieving net zero will require continuous technological advancement. This is clearly emphasized in this strategy through explicitly recognizing digital technologies as a key enabler for achieving its climate goals. From smart grids and AI-driven energy forecasting to IoT-based emission monitoring and blockchain-enabled carbon trading, digital tools are envisioned as critical for decarbonizing key sectors.

The UAE's transition to a low-carbon economy will require real-time data analytics for emissions tracking, digital twins for infrastructure planning, and scalable digital platforms in sustainability programs. Moreover, the shift to e-government and paperless services contributes to reducing the carbon footprint of public services and daily life. Hence, digitization is not a marginal component but a pillar and driver of the UAE's roadmap to net zero, enabling smarter resource management, faster policy execution, and more efficient sustainability governance.

General Framework for Sustainable Digital Transformation. The UAE has introduced the General Framework for Adopting Sustainable Digital Transformation in 2024 [31], with the aim to ensure the integration of sustainability goals and principles into the digital transformation agenda of the UAE. This general framework acts a guiding policy for all government sectors in the UAE, in order to ensure that services and operations digitization is performed in an environmentally sustainable manner. This has been further emphasized during the UAE's Year of Sustainability, in which the UAE leadership has directed that digital initiatives align with climate and sustainability goals.

The core objectives of this framework are twofold. First, to integrate sustainability principles into all activities of digital transformation within government, thereby enhancing the country's sustainable digital readiness. Second, by achieving the first goal, the UAE aims to set a global benchmark in digital sustainability while supporting its commitments to combat climate change and build a green future enhanced by technology. The UAE's general sustainable digital transformation framework lays down a comprehensive approach for aligning environmental goals into digitization progress across the government entities. This comprises eight pillars which serve as guiding actions levers:

1. Paperless government, promoting e-services to reduce paper consumption and associated waste;
2. Cashless government, encouraging digital payments to reduce physical visits to government locations and subsequently associated emissions;
3. Green data centers and cloud computing, which is focused on optimizing energy use, minimizing e-waste, and leveraging renewable energy sources;
4. Sustainable digital procurement, which integrates environmental goals into the consideration of digital suppliers in procurement and contracts;
5. Green websites and software, through advancing resource-efficient design and supporting open-source software and APIs;
6. Sustainable management of digital devices, incorporating circular economy principles such as the reuse and recycling of digital equipment;
7. Sustainable digital collaboration, which reduces carbon footprints through remote work and virtual meetings;
8. Sustainable digital skills, aimed at equipping the human resources with the knowledge to efficiently implement and manage digital systems.

Following these pillars, the outcomes are envisioned to introduce significant enhancement to the UAE's global position in sustainable tech, reducing the environmental footprint of government operations, and ensuring that going digital also means going green.

This framework has already led to projects like merging government data centers for efficiency, considering only digital documentation in federal procedures, and auditing websites for energy consumption. It is worth highlighting as well that the General Framework was one of the flagship initiatives launched during 2024, at the same year in which the UAE government approved more than 87 sustainability-related policies (including this framework, a new climate law, a biofuels policy, etc.) as part of its plan to host COP28.

ICT and Digital Economy Sandboxes. Another important enabler in the UAE's policy toolkit is the use of regulatory sandboxes and innovative regulatory laboratories to enable experimentation in the digital era while maintaining alignment with national standards. Regulatory sandboxes [32] are controlled environments where new technologies, business models, or services can be tested under relaxed regulatory requirements and careful supervision. The UAE has adopted this concept to accelerate digital innovation in sectors like telecommunications, fintech, and autonomous mobility, which allowed companies to pilot new solutions without immediately being restricted by full regulatory compliance, under identified safeguards requirements.

In the ICT sector, the Telecommunications and Digital Government Regulatory Authority (TDRA) has established an ICT Regulatory Sandbox program. This allows the participants to pilot and test new ICT innovations in a safe and controlled manner; meanwhile, it enabled TDRA to monitor and gather conclusions on whether existing regulations need to be updated. To serve its needs, the TDRA sandbox is designed to be adaptive enough to accommodate applications pertinent to Internet of Things

(IoT) solutions, blockchain-based services, AI, AR/VR, metaverse platforms, and digital twins, among other key emerging technologies, all of which have potential sustainability benefits (including optimizing resource use, enabling remote services, etc.). At the same time, the flexibility feature allows temporary exemptions or tailored rules from some regulations, while ensuring particular safety and customer protection requirements are met. Such flexibility is critical for sustainable digital transformation because it allows environmentally beneficial innovations to be tested quickly and brought to market faster than traditional regulatory processes. Target participants range from licensed telecom operators testing a new service to startups working on innovative digital products that normally require regulatory approval.

In addition to the ICT sandbox, the UAE has also established the Regulations Lab (RegLab) at the federal level [33], which is a government initiative to accelerate legislative innovation for emerging technologies. The RegLab relies on issuing temporary licenses or minimal viable legislation (MVL) for pilot projects in emerging tech domains, effectively granting early regulatory approval with strict conditions. This has led to several important trials that support sustainability and digital advancement, including the trial of the first electric cargo plane, as well as self-driving vehicles on UAE roads. Once the pilot period concludes and results are evaluated, the UAE Cabinet will design and formalize new laws or regulations to fully enable the technology. This flexible model of legislate by learning ensures that regulations keep pace with innovation, a critical factor when trying to embrace new digital tools for sustainability.

Expanding beyond governmental entities, sandbox approaches has been extended to ensure a comprehensive coverage of the broader digital economy of the UAE, and this ensures that key sectors including financial sector and free zones are part of these initiatives. For example, the Central Bank of the UAE and financial centers like the Abu Dhabi Global Market (ADGM) and Dubai International Financial Centre (DIFC) have all launched fintech sandboxes or innovation hubs. These allow testing of fintech solutions (digital payments, green fintech, crypto assets, etc.) under regulatory supervision. It has been highlighted that by using sandboxes, policymakers can co-create regulations with industry in real time, addressing legal questions and technical standards collaboratively.

5.2 National Programs and Flagship Initiatives

Building on the strong legislative and regulatory foundations, the UAE has launched a range of national programs and flagship initiatives that transfer a policy into an action. These projects reflect the UAE's ambition to become a global leader in both digital transformation and environmental sustainability.

One of the most prominent examples is Masdar City, which is located in Abu Dhabi. It is designed specifically to represent sustainable urban development, through integrating renewable energy, smart grid technology, and AI-driven energy optimization. It serves as a living laboratory for clean tech innovation, hosting research institutions, startups, and corporate offices dedicated for green technology. Another example is Sharjah Sustainable City, a residential project that uses solar power, grey-water recycling, and smart building systems to minimize its environmental footprint. These initiatives showcase how digital tools such as IoT sensors and data platforms are enabling real-time energy management, waste reduction, and sustainable mobility within urban environments.

The UAE has also implemented several national-level smart government initiatives that support sustainability goals. For example, the Dubai Paperless Strategy has reduced over one billion pieces of paper annually through end-to-end digitization of government services. On the same spirit, the UAE Blockchain Strategy and the Artificial Intelligence Strategy 2031 aim to enhance efficiency, transparency, and sustainability across key sectors. The Expo 2020 Dubai (Programme for People and Planet) further highlighted the UAE's commitment to digital sustainability. The event featured technology-driven exhibits and forums focused on climate action and digital inclusion.

The success of digital sustainability in the UAE is supported by a strong collaborative ecosystem that comprises government bodies, academia, and the private sector. Key entities such as the Ministry of Climate Change and Environment (MOCCAE), TDRA, and the Ministry of AI play strategic roles in shaping the national policy and enabling innovation. Local authorities like Abu Dhabi and Dubai Digital Authorities take the leadership in the implementation aspect, while academic institutions, including Khalifa University (particularly the 6G Research Center) and MBZUAI, contribute through advance research and talent development in AI, energy, and climate. The 6G Research Center plays an important role in exploring how next-generation networks can drive sustainability, focusing on energy-efficient communication systems, AI-powered network optimization, low-power edge computing, and digital twins for smart cities and industries. These innovations are key to enabling green digital infrastructure at scale, supporting the UAE's broader goals of sustainability and technological leadership. Public–private partnerships with global technology firms such as Huawei, Nokia, and Nvidia further accelerate the deployment of sustainable digital solutions across sectors.

5.3 Challenges and Future Directions

While the UAE has made substantial progress, several challenges still exist in scaling and maintaining digital sustainability. One concern is the energy footprint of digital infrastructure, especially as demand for cloud computing, AI, and data centers continues to grow. Managing this growth sustainably will require stricter energy efficiency standards, incentives for renewable-powered data centers, and innovation in low-carbon computing. There is also a need to further develop green digital skills within the workforce to ensure that public and private sectors can design, deploy, and manage sustainable technology systems. Moreover, interoperability and data sharing across government entities remain a key challenge for coordinated digital sustainability efforts. Strengthening governance mechanisms, such as unified digital identity systems and open data platforms, will be critical in next stages. Finally, future policy development should address ethical, privacy, and cybersecurity concerns associated with emerging technologies. As the UAE is adopting advanced AI, autonomous systems, and the metaverse, ensuring that these tools support sustainability without compromising rights or reliability will be vital.

Looking ahead, the UAE has the opportunity to continue refining its policy frameworks, scaling successful pilot initiatives, and strengthening deeper international collaboration. Its proactive approach to digital sustainability serves as a valuable model for aligning technology, economy, and environment in the coming years.

6 Digital Sustainability in the USA

Digital sustainability in the USA is guided by a combination of federal laws, state-level regulations, and industry-led initiatives. This multilayered approach has evolved over decades as governments and companies recognized the need to address the sustainability of digital products and to harness digital technologies for environmental and social goals. At the federal level, Congress enacts legislation that shapes national priorities and mandates, while federal agencies are tasked with enforcement and oversight. States also exercise significant autonomy and can pass their own laws that are binding for businesses and operations within their territories. Complementing these legal frameworks, industries and private organizations have increasingly developed practices aligned with Environmental, Social, and Governance (ESG) principles, recognizing the growing importance of sustainable operations in global markets.

6.1 Federal Laws and Agencies Focused on Digital Sustainability

Several major U.S. environmental laws, such as the **Clean Air Act (USA EPA** [34]), the **Clean Water Act (USA EPA, 1972** [35]), the **Energy Policy Act (USA EPA** [36]), and the **Food Quality Protection Act (USA EPA** [37]), have created opportunities for integrating digital technologies into sustainability enforcement and monitoring. The **Environmental Protection Agency (EPA) (USA EPA** [38]), which oversees compliance with these laws, has embraced digital transformation through its **Digital Strategy (USA EPA** [39]). This initiative promotes the use of data platforms and metadata sharing to support more effective environmental protection.

For example, the EPA has digitized reporting processes to monitor air and water quality in real time, allowing faster interventions when pollution thresholds are exceeded (EPA 2023).

Similarly, the **U.S. Department of Energy (DOE)** has established a **Sustainability Program Office (US Department of Energy** [40]), which leverages data to optimize energy efficiency across federal facilities. This office supports data sharing and analysis to track greenhouse gas emissions and to develop strategies for decarbonization.

One of the most significant pieces of legislation in recent years was the **Inflation Reduction Act (IRA)** of 2022 [41], which allocated incentives for clean energy production, renewable technologies, and energy efficiency measures. The IRA applied to a wide range of stakeholders, from government agencies to private companies and individual consumers. Businesses benefitted from tax credits for investing in solar and wind energy as well as electric vehicles (EVs), while households received rebates for energy-efficient appliances. Additionally, the **Internal Revenue Service (IRS)** [42] began digitizing administrative processes to better manage these incentives and reduce reliance on paper-based systems.

However, the policy landscape shifted dramatically in 2025 with the **One Big Beautiful Bill (OBBB) act** [43]. This bill substantially reduced the sustainability incentives established by previous administrations. The World Resources Institute (WRI) reported that the OBBB would eliminate $6.8 billion in public funding and at least $9.5 billion in anticipated private investments, leading to the loss of nearly 4900 jobs in cleaner industrial projects [44].

6.2 Federal Executive Orders

In addition to formal legislation, U.S. presidents issue executive orders (EOs) to set short-term federal priorities. During his tenure (2021–2025), President Biden signed multiple EOs aimed at improving energy efficiency and advancing digital sustainability, including measures for the responsible development of digital assets. These EOs encouraged agencies to adopt sustainability-focused procurement policies and infrastructure planning. However, most of these executive orders were revoked by the subsequent administration, highlighting the political volatility of U.S. sustainability initiatives and the challenges of establishing long-term policy continuity.

6.3 State Laws and Practices

States play a vital role in filling regulatory gaps left by federal law, particularly in areas such as **electronic waste (e-waste)** management. Although there are no nationwide regulations for e-waste disposal, states like Oregon and California have implemented stringent local policies. Oregon's **Department of Environmental Quality (DEQ)** requires manufacturers to meet annual recycling targets based on the weight of their digital waste. California, one of the largest technology markets in the USA, has banned the disposal of e-waste in landfills and mandates certified recycling practices. A comprehensive list of these state-level initiatives is maintained by the **Electronics Recycling Coordination Clearinghouse website** [45].

Not all states, however, prioritize ESG objectives. Some have passed legislation restricting ESG-related investments. For instance, Alaska introduced **House Bill 6 (HB 6)** (Pleiades Strategy, 2025) in 2025, which seeks to prevent state pension funds from being used to advance ESG agendas. These measures reflect a broader political divide in the USA, where some regions resist or challenge ESG-driven investment strategies.

6.4 Industry Practices and Examples

Beyond government action, U.S. industries have developed their own digital sustainability strategies. These efforts span energy management, carbon emission reductions, and water conservation. A major focus is the rapid growth of **data centers**, which serve as the backbone of digital infrastructure. The adoption of artificial intelligence (AI) and cloud services has led to an unprecedented expansion of data centers, creating debate about their environmental footprint. Data centers already account for approximately 4% of total U.S. electricity consumption, and projections suggest this could rise to as much as 23% by 2033 [46, 47]. These facilities rely heavily on energy-intensive cooling systems and often operate in regions with carbon-intensive power grids, exacerbating their environmental impact.

Companies are increasingly adopting innovative practices to address these challenges. For example, **Google** uses AI-powered tools to optimize data center cooling, and its **Flood Hub** [48] system now provides up to seven days of advance warning for potential flooding in over 100 countries [49, 50]. Similarly, semiconductor manufacturer **Intel Corporation** [51] has committed to using 99% renewable energy in its production processes and partners with healthcare and industrial customers to improve energy efficiency through Intel technologies.

The automotive sector also illustrates industry-led sustainability efforts. **Ford Motor Company** aims to achieve carbon neutrality by 2050, largely through the development of electric vehicles (EVs) [52]. **Chevrolet**, another major automaker, has set a target for carbon neutrality by 2030. Both companies publish annual sustainability reports detailing their progress, a practice that promotes accountability and transparency.

In the data center sector, companies such as **EdgeConneX** and **Iron Mountain** are recognized for integrating renewable energy sources and adopting waste-reduction measures. EdgeConneX, for instance, has committed to achieving carbon neutrality by 2030 and earned top certifications for zero waste to landfill [53]. These examples highlight how industry actors are not only adopting cleaner energy sources but are also innovating in areas such as circular economy practices and advanced cooling systems to reduce their environmental footprint.

6.5 Challenges and Future Directions

Digital sustainability in the USA faces both opportunities and significant hurdles. Technological advancements such as AI and renewable energy innovations present opportunities to reduce environmental impact. However, the regulatory environment remains highly uncertain. The repeal or scaling back of sustainability initiatives, such as those introduced under the OBBB, demonstrates how political shifts can quickly undermine long-term planning and investments.

Despite these challenges, many U.S. companies with global operations continue to pursue ambitious digital sustainability goals to remain competitive and meet international standards. Looking ahead, the success of digital sustainability will depend on the ability to establish consistent policies and to balance economic growth with environmental responsibility. Efforts must also ensure equitable access to the benefits of digital sustainability, so that communities most affected by environmental risks are not left behind.

7 Lessons Learned Across Regions

Comparison of Brazil, India, the UAE, and the USA reveals both convergence and divergence in the way digital sustainability is understood and operationalized. All three countries see digitalization as a way to improve efficiency and achieve broader development goals, but the directions they take reflect their distinct contexts and priorities.

In Brazil, digital sustainability combines two interlinked efforts: reducing the environmental footprint of digital infrastructures while deploying digital tools for ecological monitoring and social inclusion. For example, AI-based real-time deforestation alerts in the Amazon illustrate practical environmental applications, linking data analytics to enforcement and transparency. However, challenges remain in turning these alerts into effective policy responses, given issues of fragmented governance and limited capacity at subnational levels.

India focuses its digital sustainability approach on enhancing administrative efficiency and expanding public service delivery through programs like Digital India and Aadhaar. These initiatives have improved access to services and strengthened transparency in governance, creating a digital backbone that supports welfare and financial inclusion goals. Yet, they continue to face structural challenges such as the persistent urban–rural digital divide, disparities in digital literacy, and concerns about data privacy and security. The country's scale amplifies both achievements and gaps, highlighting the need for tailored local implementation within national frameworks.

The UAE integrates digital sustainability directly into its Net Zero 2050 strategy, positioning digital technologies as drivers of decarbonization across energy, mobility, and governance sectors. Its approach combines regulatory coherence, such as through the Sustainable Digital Transformation Framework, with sandboxes that accelerate innovation by allowing safe testing of emerging technologies. This creates an environment where policy and technology evolve together. Nonetheless, questions remain about the environmental costs of rapidly expanding data centers, AI applications, and digital infrastructures, especially given the country's high per capita energy consumption.

The USA represents a hybrid model. Federal policies can mobilize large investments, as seen in the Inflation Reduction Act, but they are highly vulnerable to political reversals, such as the OBBB. States like California and Oregon often lead with strong e-waste and renewable energy policies, while others actively resist ESG-driven strategies. Meanwhile, US companies innovate to reduce the environmental footprint of their operations, using AI to optimize energy use in data centers and pursuing carbon neutrality targets, as seen in Google, Intel, Ford, and Chevrolet.

Despite these different approaches, there are shared challenges in all three countries. Digital inequality persists as a structural barrier in both Brazil and India and also affects rural and marginalized communities in the USA, restricting the reach and impact of sustainability-focused digital policies. In contrast, the UAE's challenges relate more to managing the environmental impacts of its digital expansion and ensuring sufficient green skills within the workforce to support its ambitious goals. Governance structures play a critical role in shaping these outcomes. The fragmentation of governance limits coordinated action in Brazil and the USA, while the centralized approach of the UAE enables rapid execution, but may reduce local adaptability. India occupies an intermediate position, with robust national programs but uneven implementation capacities across states and municipalities. Overall, these experiences show that digital sustainability is not a universal formula to be replicated. Rather, it is a set of practices deeply shaped by each country's political economy, institutional capacity, and socio-environmental realities. Success in advancing digital sustainability requires more than technological deployment; it depends on coherent governance, continuous capacity building, and aligning digital strategies with social inclusion and environmental justice objectives. As these countries demonstrate, the pathways to digital sustainability are multiple, context-dependent, and necessarily adaptive to changing ecological, technological, and social conditions.

8 Conclusion

This chapter set out to explore how digital sustainability is approached in contexts beyond Europe, focusing on Brazil, India, USA, and the UAE. As discussed in the introduction, digital sustainability is not just about deploying new technologies; it is about how these technologies are embedded in each country's social, political, and environmental realities.

The comparison of Brazil, India, the UAE, and the USA shows that while similar digital tools are used across these countries—such as AI, IoT, and regulatory sandboxes—their priorities and outcomes diverge in meaningful ways. Brazil combines digital innovation with a strong focus on ecological protection and social inclusion, using digital tools to monitor deforestation, improve agriculture, and amplify Indigenous voices. Yet, its fragmented governance and persistent inequalities limit the reach of these initiatives. India's approach centers on expanding digital public services and administrative efficiency through programs like Digital India and Aadhaar. These have transformed governance and inclusion, but challenges remain in bridging the rural–urban digital divide and safeguarding privacy. The UAE integrates digital sustainability into its national decarbonization strategy, supported by strong policies, investment in green technologies, and regulatory sandboxes that accelerate innovation. However, concerns persist about the environmental footprint of its rapidly growing digital infrastructures. The USA, however, shows both the strengths and vulnerabilities of decentralized governance. Federal laws and executive orders can generate strong momentum, but they are often fragile, as seen in the transition from the Inflation Reduction Act to the OBBB. State-level leadership and industry innovation fill gaps but lead to a patchwork of approaches.

Looking across these regions, some common lessons emerge. First, technology alone does not guarantee sustainability. What matters is how digital strategies are designed, governed, and aligned with local priorities. Second, inequalities in digital access and skills remain structural barriers. Third, policy coherence and institutional capacity are critical enablers of effective digital sustainability pathways.

Reflecting back to the abstract and the global focus of this volume, these insights underline that while Europe's experiences shape much of the international digital sustainability discourse, countries beyond Europe are experimenting with diverse models that combine technological, social, and ecological goals in context-sensitive ways. Their experiences show that digital sustainability is not a universal pathway but a set of negotiated practices, shaped by political economy, governance capacity, and ecological urgency.

As digital sustainability becomes a central pillar of development agendas worldwide, future research and policy must go beyond techno-centric or efficiency focused paradigms. Instead, it must ask: Whose sustainability is prioritized? Who benefits from digital transitions? And how can digital infrastructures, governance frameworks, and knowledge systems be shaped to create more just, inclusive, and ecologically resilient futures?

In this sense, the lessons from Brazil, India, the UAE, and the USA offer valuable insights beyond Europe. They remind us that achieving sustainability through digitalization requires continuous adaptation, collaboration across sectors and scales, and a commitment to align technological advances with social justice and environmental stewardship in each unique context.

Acknowledgements The authors would like to sincerely thank Dr. Christophe Gaie and Dr Markus Mueck for their invaluable support and insightful contributions throughout the development of this book chapter. Their expertise, guidance, and constructive feedback greatly enhanced the quality and clarity of this chapter.

References

1. Meinhold J (2024) Digital sustainability and eco-environmental sustainability: a review of concepts and approaches. Sustain Dev 32:201–220
2. Santarius T, Dencik L, Diez T, Ferreboeuf H, Jankowski P, Hankey S, Hilbeck A, Hilty LM, Höjer M, Kleine D, Lange S, Staab P (2023) Digitalization and sustainability: a call for a digital green deal. Environ Sci Policy 147:11–14https://doi.org/10.1016/j.envsci.2023.04.020
3. Rijal S, Erwin MK, Aman A, Judijanto L (2025) Digitalization and the sustainable development goals (SDGs): a bibliometric analysis of the role of technology. SDI J 2(1). https://doi.org/10.58812/sdi.v2i01.1788
4. Avila TJT, Lanza BBB, Valotto DDS (2023) Digital transformation, technology and innovation in Brazilian states: the pathways proposed by governors elect 2023–2026. In: Proceedings of the 16th international conference on theory and practice of electronic governance. ACM, pp 332–339. https://doi.org/10.1145/3614321.3614366
5. OECD (2020) Going digital in Brazil. OECD reviews of digital transformation. OECD Publishing.https://doi.org/10.1787/e9bf7f8a-en
6. See Decree No. 9.319 of 21 March 2018, at http://www.planalto.gov.br/ccivil_03/_ato2015-2018/2018/decreto/D9319.htm; E-Digital was updated on 17 November 2022 by the Ministry of Science, Technology, and Innovation (MCTI). See https://www.gov.br/mcti/pt-br/acompanhe-o-mcti/transformacaodigital
7. Estratégia Brasileira para a Transformação Digital (E-Digital) 2022–2026
8. Pigola A, da Costa PR, Carvalho LC, Silva LFD, Kniess CT, Maccari EA (2021) Artificial intelligence-driven digital technologies to the implementation of the sustainable development goals: a perspective from Brazil and Portugal. Sustainability 13(24):13669. https://doi.org/10.3390/su132413669
9. Costa Vazquez K (2022) Uma agenda estratégica e de futuro para o desenvolvimento sustentável do Brasil. CEBRI-Revista: Braz J Int Aff (4):89–103. https://cebri-revista.emnuvens.com.br/revista/article/view/80. Environmental Protection Agency (EPA) (2023) EPA digital strategy for sustainability. https://www.epa.gov/developers/digital-strategy
10. Brazilian Ministry of Science, Technology and Innovations (MCTI) (2022)
11. Silva PB, Neves FR (2024) Thinking inside the sandbox: beyond public services digitalization with co-production. In: Gaie C, Mehta M (eds) Transforming public services—combining data and algorithms to fulfil citizen's expectations. In: Intelligent systems reference library, vol 252. Springer. https://doi.org/10.1007/978-3-031-55575-63
12. Martha Júnior GB, Lopes MA (2023) Charting new sustainable agricul tural innovation pathways in Brazil. Scientia Agricola 80:e20230067. https://doi.org/10.1590/1678-992X-2023-0067
13. Neves FR, Aquino ACB (2025) Bringing the green into the screen: remote sensing and smart green governance in urban forest management. In: Kerrache CA, Sahraoui Y, Calafate CT, Vegni AM (eds) Mobile crowdsensing and remote sensing in smart cities. Internet of Things. Springer. https://doi.org/10.1007/978-3-031-72732-04
14. Ramanathan R, Duan Y, Valverde J, Van Ransbeeck S, Ajmal T, Valverde S (2023) Using IoT sensor technologies to reduce waste and improve sustainability in artisanal fish farming in Southern Brazil. Sustainability 15(3):2078. https://doi.org/10.3390/su15032078
15. Freitag AM, Camargo-Borges C (2024) The digital bow: exploring the potential of digital presence for Indigenous community-based tourism in Brazil. Tour Plan Dev 21(6):755–772. https://doi.org/10.1080/21568316.2024.2307974

16. Saini N, Kharb R (2025) Empowering sustainable development through digital transformation: insights from digital India. J Asia Bus Stud
17. Vijayan A (2019) Digital India—a roadmap to sustainability. Int J Innov Technol Explor Eng 8(5):571–576
18. Dutta S, Pandya M (2024) From gridlock to growth: how leadership enables India's PRAGATI ecosystem to power progress. Retrieved from: https://ora.ox.ac.uk/objects/uuid:34268453-91fb-4dd3-b052-90fbb52fe247/files/s1j92g9820
19. Sharma G, Kalra S (2019) Advanced multi-factor user authentication scheme for E-governance applications in smart cities. Int J Comput Appl 41(4):312–327
20. Mathur D, Gupta P, Sridevi A (2009) E-governance approach in India: the National E-Governance Plan (NeGP). Transform Government 3(1):3–50
21. Gupta S, Mishra ON, Kumar S (2021) Citizen empowerment and adoption of E-governance services: the role of online citizen skills, awareness, and engagement. Int J Electron Governance 13(4):386–407. https://doi.org/10.1504/IJEG.2021.121237
22. Bhushan A (2019) Understanding of information technology act for E-Governance in India. IIBM's J Manage Res 1–6
23. Tiwari AA et al (2024) An overarching conceptual framework for ICT enabled responsive governance. Inf Syst Front 26(3):1161–1182
24. Ebert H (2020) Hacked IT superpower: how India secures its cyberspace as a rising digital democracy. India Rev 19(4):376–413. https://doi.org/10.1080/14736489.2020.1797317
25. Pathak P, Barnwal SK (2013) E-governance in India—a critical appraisal. J Adv Manage Sci 1(2)
26. Muttoo SK et al (2019) Regional analysis of e-governance in India. In: E-Governance in India: the progress status, pp 37–78
27. MacDonald UU, Nnaji I, Edokobi TD (2022) Assessing the effects of E-Governance and its application by the Independent National Electoral Commission during the 2019 general election in Nigeria. NIU J Soc Sci 8(2):53–62
28. Prasad K (2012) E-governance policy for modernizing government through digital democracy in India. J Inf Policy 2:183–203
29. Roy SK, Chowdhury S (2024) Citizen's attitude toward the use of open data in e-governance. In: Perspectives and practices of public administration in South Asia: post-pandemic recovery and sustainable development agenda. Springer, pp 79–94
30. UAE Gov. (2024) The UAE's net zero 2050 strategy, https://u.ae/en/about-the-uae/strategies-initiatives-and-awards/strategies-plans-and-visions/environment-and-energy/the-uae-net-zero-2050-strategy
31. UAE Cabinet (2024) The general framework for adopting sustainable digital transformation. https://www.uaelegislation.gov.ae/en/policy/details/the-general-framework-for-adopting-sustainable-digital-transformation
32. TDRA (2024) ICT regulatory sandbox. https://tdra.gov.ae/en/pages/ict-regulatory-sandbox
33. UAE Government (2024) Regulations lab. https://reglab.gov.ae/sec2
34. USA EPA (1970) Summary of the clean air act. https://www.epa.gov/laws-regulations/summary-clean-air-act.
35. USA EPA (1972) Summary of the clean water act. https://www.epa.gov/laws-regulations/summary-clean-water-act
36. USA EPA (2005) Summary of the energy policy act. https://www.epa.gov/laws-regulations/summary-energy-policy-act
37. USA EPA (1996) Summary of the food quality protection act. https://www.epa.gov/laws-regulations/summary-food-quality-protection-act
38. USA EPA (2025) Summaries of environmental laws and executive orders. https://www.epa.gov/laws-regulations/laws-and-executive-ordersmajorlaws
39. USA EPA (2025) Digital strategy. https://www.epa.gov/developers/digital-strategy
40. US Department of Energy (2025) Sustainability performance office. https://www.energy.gov/managemenperformance-office

41. US Congress (2022) H.R.5376—inflation reduction act of 2022. https://www.congress.gov/bill/117th-congress/house-bill/5376/text
42. IRS (2022) Inflation reduction act of 2022. https://www.irs.gov/inflation-reduction-act-of-2022
43. US Congress (2025) H.R.1—one big beautiful bill act. https://www.congress.gov/bill/119th-congress/house-bill/1/text
44. Byrum Z, Carlsen W (2025) The US budget bill's impact on American industry, jobs and innovation. World Resources Institute. https://www.wri.org/update/reconciliation-budget-bill-industry-divestment
45. ERCC (2025) Map of states with legislation. https://www.ecycleclearinghouse.org/contentpage.aspx?pag Pleiades Strategy (2024) Live Anti-ESG state action tracker. https://www.pleiadesstrategy.com/pleiades-anti-esg-bill-tracker-state-legislation-attacks-on-responsible-investing
46. Guidi G, Dominici F, Gilmour J, Butler K, Bell E, Delaney S, Bargagli-Stoffi FJ (2024) Environmental burden of United States data centers in the artificial intelligence era. https://arxiv.org/pdf/2411.09786
47. Accenture (2025) Powering the future of US data centers. How electricity providers can overcome bottlenecks and supercharge growth. https://www.accenture.com/content/dam/accenindustry/document/Accenture-Powering-the-Future-of-US-Data-Centers-TL.pdfzoom=40
48. Google (2025) Flood forcasting. https://sites.research.google/gr/floodforecasting/
49. Nearing G, Cohen D, Dube V et al (2024) Global prediction of extreme floods in ungauged watersheds. Nature 627:559–563. https://doi.org/10.1038/s41586-024-07145-1
50. Nevo S, Morin E, Gerzi Rosenthal A, Metzger A, Barshai C, Weitzner D, Voloshin D, Kratzert F, Elidan G, Dror G, Matias Y (2022) Flood forecasting with machine learning models in an operational framework. Hydrol Earth Syst Sci 26(15):4013–4032.https://doi.org/10.5194/hess-26-4013-2022
51. Intel (2023) Sustainable computing. https://www.intel.com/content/www/us/en/environment/sustainabil
52. Ford Motor Company (2025) 2025 integrated sustainability and financial report. https://corporate.ford.com/content/dam/corporate/us/en-us/documents/reports/2025-integrated-sustainability-and-financial-report.pdf
53. Jackson A (2025) Top 10: sustainable data centres. Data Centre Magazine. https://datacentremagazine.com/top10/top-10-sustainable-data-centres

The manufacturer's authorised representative in the EU is Springer Nature Customer Service Centre GmbH, Europaplatz 3, 69115 Heidelberg, Germany. If you have any concerns regarding our products, please contact ProductSafety@springernature.com

Printed and bound by CPI Group (UK) Ltd, Croydon, CR0 4YY
21/09/2026
02220015-0012